P. S. Mellor
M. Baylis
C. Hamblin
C. H. Calisher
P. P. C. Mertens (eds.)

African Horse Sickness

SpringerWienNewYork

Dr. Philip S. Mellor
Dr. Matthew Baylis
Dr. Christopher Hamblin
Dr. Peter P. C. Mertens
Institute for Animal Health, Pirbright Laboratory,
Pirbright, Woking, Surrey, U.K.

Prof. Dr. Charles H. Calisher
Arthropod-borne and Infectious Diseases Laboratory, Colorado State University,
Fort Collins, Colorado, U.S.A.

Softcover reprint of the hardcover 1st edition 1998

Graphic design: Ecke Bonk

Printed on acid-free and chlorine-free bleached paper

SPIN: 10633570

With 86 partly coloured Figures

CIP data applied for

ISBN-13:978-3-211-83132-8 e-ISBN-13:978-3-7091-6823-3
DOI: 10.1007/978-3-7091-6823-3

Preface

African horse sickness virus (AHSV) is a dsRNA virus which causes a non-contagious, infectious arthropod-borne disease of equines and occasionally dogs. Nine distinct, internationally recognised serotypes of the virus have so far been identified.

In naive populations of horses the clinical signs of African horse sickness (AHS) can be particularly severe and mortality rates commonly exceed 80 or 90%. AHSV is therefore rightly considered as the cause of one of the most lethal of horse diseases and accordingly has been allocated OIE List "A" status. Disease in other solipeds (mules, donkeys and zebra) is progressively less severe with clinical signs being minimal or totally absent in zebra and most donkeys.

The virus is transmitted in the field almost entirely by certain species of *Culicoides* (Diptera: Ceratopogonidae) biting midge and in particular by *C. imicola sl* a complex of several sibling species. Competent individuals within this complex transmit the virus only by bite, there is no evidence that the virus can be transmitted vertically (transovarially). It has been suggested that other genera of haematophagous insects may also be involved in AHSV transmission to a lesser extent. This has not yet been proven and the absence of field isolations from blood-free specimens of other insects suggests that any contributions to AHSV transmission by insect genera other than *Culicoides* is likely to be very small.

AHSV is enzootic in tropical and sub-tropical regions of sub-Saharan Africa but is capable of sudden extensions beyond these areas. In such circumstances the virus can cause severe losses both directly in terms of the disease itself and also indirectly due to disruption to international trade and equine sporting events. However in the past the virus apparently has been unable to sustain itself outside sub-Saharan Africa for extended periods of time (ie. more than 2 years). This characteristic was attributed to an absence of long-term vertebrate reservoir hosts from these areas and also to a low prevalence or seasonal incidence of efficient vector insects.

This "traditional view" of the situation changed in July 1987 when AHSV appeared in Spain after an absence from Europe of over 20 years. The virus was identified by the Spanish veterinary authorities as AHSV serotype 4. During subsequent years the same virus serotype was recorded in Spain during 1988, 89 and 90; in Portugal during 1989; and in Morocco during 1989, 90 and 91, before it was finally eliminated from the whole geographical region by the cumulative zoosanitary and vaccination measures taken by the veterinary authorities of the 3 affected countries.

These outbreaks were unusual for a number of reasons. First because AHSV 4 had never previously been identified outside southern Africa; all previous epizootics being due to AHSV 9. Second, the outbreaks were the furthest north that an AHSV had ever been recorded in the field. Third, the succession of annual bouts of disease spread over 5 years (1987–91) implied that the virus was able to persist for extended periods of time in the western Mediterranean region whereas previously this had manifestly not been the case.

Part of the response of the European Commission (EC) to the outbreaks was to fund two interrelated programmes of work based upon AHSV diagnostics, molecular biology, vector studies and epidemiology. These two programmes entitled "AHSV in Europe" (Contract No. 8001-CT91-0211) and "The AHSV Epizootic in Morocco" (Contract No. TS3-CT92-0151) involved laboratories and scientific organizations in France, Morocco, Portugal, Spain and the UK. These programmes were intended to advance our knowledge of the virus serogroup itself on a broad scientific front. However, implicit in their overall remit was the requirement to provide more, basic information, that would also be of real practical value in reducing the likelihood of further incursions of AHS or similar vector-borne diseases into Europe or North Africa and to minimise the costs from any incursions that might occur.

Furthermore, as part of its long established policy of encouraging the dissemination of scientific discoveries made as a result of its funded work, the EC also requested that following completion of the two projects an international workshop on AHS should be held to discuss their findings. It was also the wish of the EC that the fruits of the work and the subsequent discussions should be published so that they might be more readily available to the scientific community and to legislative authorities. The meeting, sponsored jointly by the EC, the Office International des Epizootics, the Government of Morocco and DAKO Ltd, took place in Rabat, Morocco during April 1997. Now through the generosity and enthusiasm of many people and organizations, this volume has been published as a proceedings of that meeting.

P. S. Mellor
M. Baylis
C. Hamblin
C. H. Calisher
P. P. C. Mertens

Contents

Epidemiology

Entomology

Molecular biology

Vaccines and diagnosis

Epidemiology

Taxonomy of African horse sickness viruses

C. H. Calisher[1] and **P. P. C. Mertens**[2]

[1] Arthropod-borne and Infectious Diseases Laboratory, Foothills Campus, Colorado State University, Fort Collins, Colorado, U.S.A.
[2] The Institute for Animal Health, Pirbright Laboratory, Working, Surrey, U.K.

Summary. Nine distinct genera are currently recognised within the virus family *Reoviridae*, which include a total of 63 virus groups or species (species = virus group = electropherotype or serogroup), comprising 214 virus serotypes or subtypes, as well as 20 provisional types or subtypes, most of which (149 + 9 tentative) are assigned to the genus *Orbivirus* [5, 9, 16]. The 19 species of orbiviruses (serogroups), were established principally on antigenic (serologic) grounds but many of these placements have been supported by molecular analyses. This introductory paper defines the taxonomy and classification of these viruses and establishes guidelines for use in other papers to be presented at this symposium and elsewhere.

Introduction

The International Committee for the Taxonomy of Viruses (ICTV) has provided fundamental descriptions of virus families, including the family *Reoviridae* [5, 9, 16]. Within this family are nine genera, comprising viruses isolated from a wide variety of terrestrial and nonterrestrial vertebrates, terrestrial and nonterrestrial invertebrates, and plants – truly a remarkable assortment of viruses found in a remarkable variety of econiches (Table 1). Of the 63 virus species or groups in the family *Reoviridae*, the largest number have been assigned to the genus *Orbivirus* (the orbiviruses). The 149 serotypes of orbiviruses are placed in one or other of 19 "serogroups" or species, established originally on antigenic (serologic), or molecular grounds. Nearly all orbiviruses have been isolated from arthropods and many have been isolated from vertebrates. Many of the orbiviruses cause serious diseases of livestock and a few cause febrile illnesses in humans. Recently, Swanepoel et al. [20] reported the involvement of at least two African horse sickness viruses in cases of serious human illness in laboratory workers, believed to have been caused by inhalation of freeze dried virus "dust".

Nine virus serotypes comprise the African horse sickness virus (AHSV) group (Table 3), which also represents a single virus species [6]. Any AHSV group virus

Table 1. Viruses of the family *Reoviridae*[a] by genus

Genus	No. of member species (serotypes) [tentative or unassigned isolates]
Orthoreovirus	4 (6) [+2 unnassigned]
Orbivirus	19 (149) [9 tentative]
Cypovirus	14 (14) {> 230}[b]
Aquareovirus	6 (17) [+5 tentative]
Rotavirus	6 (6)
Coltivirus	4 (10) [+4 tentative]
Fijivirus	5 (7)
Phytoreovirus	3 (3)
Oryzavirus	2 (2)
Total	63 (214 + 20 tentative or unassigned)

[a] Based on data from Holmes et al. [5] as modified Mertens et al. [9]

[b] Large numbers of cypovirus isolates {> 230} have been recovered from a range of different insect host species. Many of these are related viruses and can be grouped into 14 distinct electropherotypes, which are also serologically distinct (14 species) [9]

can cause African horse sickness (AHS – the disease) which, in acute and typically severe cases, is characterised by progressive respiratory disease leading to death [2] but which Laegreid [7] reports may also be manifest as cardiac or simple febrile illnesses. Laegreid explained "that the immune status of horses influences the form of disease they develop ...". McIntosh [8] stated that "differentiation of (disease) forms appears to be related to variation in susceptibility as a result of immunity from previous infection, rather than to any viral property". Erasmus [1] reiterated this view, suggesting that the cardiac and febrile forms of AHS are usually observed in immune animals that have been infected with a heterologous virus type. However, the existence of attenuated vaccine strains of each AHSV serotype, demonstrates that "viral properties" can also dramatically affect the severity of the disease caused by each of the AHSV serotypes, even in serologically naive animals. Clearly, the specific serotype of AHSV, as well as other important factors, including attenuation state, immune status of the host, as well as the target host species must be known in order to understand individual cases of the disease in equines, the epizoology of viruses in the AHSV group, and cross-protectivity conferred by vaccines.

Although oedema of the supraorbital fossae is characteristic of AHS, pathognomonic signs are not specific by serotype. Therefore, the specific aetiologic agent cannot be determined from the characteristics of the clinical disease that it causes. The virus must be isolated and identified or, if the animal survives long enough, seroconversion to a particular AHSV must be determined before a specific AHSV serotype can be confirmed as causing that particular disease. Thus any of the nine AHSV serotypes may be referred to as "an African horse sickness virus" but it would be incorrect to call any one of them "African horse sickness virus" because use of that expression does not indicate which of the nine serotypes is being discussed. African horse sickness virus is a collective term, referring to the entire serogroup or species, containing all nine serotypes. For example: if a virus isolate is reported as "a member of the family *Reoviridae*" we would know a great deal about that virus. If a virus isolate is reported as "a member of the genus *Orbivirus*" we would know a great deal more about that virus and be able to conclude that it is one of the 19 species and represents one of the 149 recognised serotypes, one of the 9 ungrouped virus isolates, or is an isolate of a new serotype. If a virus isolate is reported as "a member of the AHSV serogroup/species" we would know that it is one of the nine recognised serotypes in that group, or that it is an isolate of a tenth serotype.

The definition of species (orbivirus serogroup) is the lowest level of classification that has any formal taxonomic significance. However, if you isolate a virus and report it as "AHSV 7" we would also know the neutralisation characteristics of the virus and understand some of its serological relationships to other virus isolates/strains within the species AHSV. Had the different isolates of AHSV and other orbiviruses been given names rather than numbers to indicate their serotype, no one would have confused them one for another but such a nomenclature would give no indication of serological differences or similarities. Although the identification of virus by serotypes does record important serological differences and relationships, the simple use of numbers to identify viruses tends to suggest that all isolates of the same serotype are essentially identical and all isolates of different serotypes are very different. However, "serotype" is only determined by the outer capsid proteins and the genome segments from which they are translated (in bluetongue virus (BTV) and AHSV these are genome segments 2 and 5 or 6, which encode proteins VP2 and VP5, respectively).

During coinfection of individual cells by distinct but related orbiviruses (members of a single serogroup/species), genome segment "reassortment" can occur with high frequency and has been demonstrated in cell culture as well as in insect vectors and mammalian hosts [12, 14, 17–19]. This exchange of genome segments and genetic information between parental strains from different serotypes can generate progeny virus strains, which themselves belong to different serotypes but are identical in eight out of ten genome segments (those encoding core and nonstructural proteins [12]. Other viruses from the same coinfection and reassortment process may be different in eight out of ten genome segments but still contain identical outer coat proteins and therefore would be identified as the same serotype. Specifically in order to avoid such problems in taxonomy, caused by the exchange of genome segments, the lowest formally recognised taxon (virus species) is defined to include all of these different progeny and parental virus strains and therefore all of the different serotypes in a single virus species (as suggested by Gorman [3]).

Within the family *Reoviridae,* the primary determinant for inclusion of distinct virus isolates within a single virus species is their ability to exchange genetic information by genome segment reassortment during co-infection, thereby generating viable progeny virus strains [9]. However, direct evidence concerning genome segment reassortment between different virus isolates, in most of the genera within the *Reoviridae*, is limited or non-existent. Evidence of similarity and of the genetic compatibility required for reassortment, can therefore be provided by other methods ("polythetic definition" of virus species). The parameters used to define members of the different species (serogroups) of orbiviruses are: 1) Their ability to exchange genetic material by genome segment reassortment, thereby producing viable progeny virus strains; 2) High levels of serological cross reaction, for example by ELISA, or by more traditional assays such as complement fixation (CF), or agar gel immunodiffusion (AGID) assays (using either polyclonal sera, or monoclonal antibodies against VP7(T13) or other "conserved" antigens). Distinct but related species may show low levels of serological cross reaction; 3) High levels of RNA sequence homology in "conserved" genome segments (e.g. $<24\%$ sequence variation in genome segment 3, encoding the major subcore structural protein, VP3(T2) [10]); 4) Cross hybridization of "conserved" genome segments under high stringency conditions ($>90\%$ homology); 5) PCR, using primers to conserved segments such as segment 3 or 7, which can be coupled with cross hybridisation analysis (northern or dot blots); 6) Identification of the virus serotype with a virus type already classified within a specific species of *Orbivirus*; 7) Analysis of "electropherotype" by agarose gel electrophoresis but not by PAGE. (Some similarities can exist between closely related species); 8) Identification of similar "conserved terminal regions" of the genome segments. (Some closely related species may also have identical terminal sequences on at least some segments).

In order to comply with these "species parameters", the Kemerovo serogroup of orbiviruses, which previously contained four distinct "complexes" of virus isolates, has been reclassified as three distinct species (Chenuda virus, Great Island virus and Wad Medani virus) (Table 2). Viruses previously classified within in the Kemerovo and Great Island complexes, which can "reassort" genome segments [14], are now reclassified as a single species (the Great Island virus species) [9].

Cross hybridisation and sequencing studies of the orbiviruses (particularly BTV) have demonstrated that significant levels of sequence variation do exist in the eight "conserved" segments of the genome (encoding core and nonstructural proteins) some of which correlate with the geographic origins of individual isolates [11]. These differences are not reflected in strain identification by serotype. Gould [4] has proposed the term "topotype" to identify orbiviruses having common geographic origins and which can be identified by representative sequence analysis of genome segment 3 (coding for the conserved inner core "T2" protein, VP3) (reviewed by Mertens, [10]). It is therefore a useful practice to identify precisely the geographic origin of any orbivirus isolate, possibly introducing an appropriate abbreviation, for example for BTV1 from South Africa (BTV1 SA), or from Australia (BTV1 AUS) [4, 11]. Although the geographic distribution of AHSV is more limited than that of BTV, significant variations exist in proteins other than those which form the outer capsid (for example the small nonstructural protein NS3 (see Martin et al.,this publication). Variations in strain

Table 2. Viruses of the genus *Orbivirus*[a] by serogroup (species), serotypes, hosts and principal vector [5, 9]

Virus serogroup (species)	Serotypes	Host species	Principal vector
1. African horse sickness virus species (AHSV)	AHSV 1 to 9	Equids, dogs, elephants, camels, cattle, sheep, goats, predatory carnivores and in special circumstances humans	*Culicoides*
2. Bluetongue virus species (BTV)	BTV-1 to 24	Cattle, sheep, goats, camels, elephants, (domestic and wild ruminants), predatory carnivores	*Culicoides*
3. Changuinola virus species (CGLV)	Twelve "named" serotypes	Humans, rodents, sloths	Phlebotomine flies, mosquitoes
4. Chenuda virus species[c] (CNUV)	Seven "named" serotypes	Seabirds	Ticks: *Ornithodoros*; *Argas*
5. Chobar Gorge virus species[b] (CGV)	Two "named" serotypes	Bats	Ticks: *Ornithodoros*
6. Corriparta virus species (CORV)	Three "named" serotypes	Humans, rodents	Culicine mosquitoes
7. Epizootic haemorrhagic disease of deer virus species (EHDV)	EHDV 1 to 8	Cattle, sheep, deer, camels, lamas, wild ruminants, marsupials	*Culicoides*
8. Equine encephalosis virus species (EEV)	EEV 1 to 7	Equids	*Culicoides*
9. Eubenangee virus species (EUBV)	Four "named" serotypes	Unknown (isolated from insect vectors)	*Culicoides* and Anopheline, Culicine mosquitoes
10. Ieri virus species[d] (IERIV)	Three "named" serotypes	Birds	*Culex* mosquitoes
11. Great Island virus species[c, e] (GIV)	Thirty six "named" serotypes	Seabirds, rodents, humans	Ticks: *Argas*; *Ornithodoros*; *Ixodes*
12. Lebombo virus species (LEBV)	Single serotype (LEBV-1)	Humans, rodents	Culicine mosquitoes
13. Orungo virus species (ORUV)	ORUV-1 to 4	Humans, camels, cattle, goats, sheep, monkeys	Culicine mosquitoes
14. Palyam virus species (PALV)	Eleven "named" serotypes	Cattle, sheep	*Culicoides* and Culicine mosquitoes
15. Umatilla virus species (UMAV)	Four "named" serotypes	Birds	Culicine mosquitoes

(continued)

Table 2 *(continued)*

Virus serogroup (species)	Serotypes	Host species	Principal vector
16. Wad Medani virus species[c] (WMV)	Two "named" serotypes	Domestic animals	Ticks: *Boophilus*; *Rhipicephalus*; *Hyalomma*; *Argas*
17. Wallal virus species (WALV)	Two "named" serotypes	Marsupials	*Culicoides*
18. Warrego virus species (WARV)	Two "named" serotypes	Marsupials	*Culicoides* and Anopheline, Culicine mosquitoes
19. Wongorr virus species[c] (WGRV)	Eight "named" or "numbered" serotypes	Cattle, macropods	*Culicoides* and mosquitoes
Tentative species and unassigned viruses within the genus			
Andasibe virus (ANDV)			mosquitoes
Ife virus (IFEV)			unknown
Itupiranga virus (ITUV)			mosquitoes
Japanaut virus (JAPV)			mosquitoes
Kammavanpettai virus (KMPV)			unknown
Lake Clarendon virus (LCV)			ticks
Matucare virus (MATV)			ticks
Ndelle virus (NDEV)			unknown
Tembe virus (TMEV)			mosquitoes
Total		158	

[a] Using data from Holmes et al. [5] with recent modifications [9]

[b] Chobar Gorge and Fomede are related but distinct as analysed by complement fixation tests. These viruses form the Chobar Gorge virus serogroup (species) [21]

[c] The 36 virus isolates, currently grouped within the Great Island virus species, as well as the 8 virus isolates in the Wogorr virus species [15] have not been extensively compared in serum neutralisation assays. However, preliminary serological studies of the GIV species [13] indicate that they may represent different serotypes

[d] Ieri, Arkonam and Gomoka are closely related viruses and form the Ieri virus serogroup (species) [21]

[e] The newly recognised Great Island virus species [9] contains virus isolates previously classified in the Kemerovo virus and Great Island virus complexes of the Kemerovo serogroup [14]

pathogenicity also exist (for example those caused by attenuation during multiple passage in mouse brain), which will inevitably be reflected in some changes in genomic RNA sequence. It is therefore useful to the reader, and may be an aid to interpretation of results or in planning future studies, if the passage history and attenuation (if any) of a virus isolate are also specifically stated.

Table 3. Recognized serotypes of African horse sickness virus listed by prototype strain designation/location, source, year of original isolation and geographic distribution of all isolates

Subtype	Strain/location	Source	Year	Geographic distribution
AHSV 1	A-501 / S.A.[a]	horse	1948	Southern Africa
AHSV 2	OD / S.A.	horse	1937	Southern Africa
AHSV 3	L / S.A.	horse	1940	Southern Africa
AHSV 4	47/58[b] / S.A.	horse	1958	Southern Africa, Iberia, North Africa
AHSV 5	VH / S.A.	horse	1940	Southern Africa
AHSV 6	114 / S.A.	horse	1940	Southern Africa
AHSV 7	Karen / Kenya	horse	1952	Southern Africa
AHSV 8	18/60 / S.A.	horse	1960	Southern Africa
AHSV 9	7/60 / Pakistan	donkey	1959	Southern Africa, Pakistan, Spain, India, North Africa, West Africa

[a] S.A. Republic of South Africa
[b] Used by Howell [6] to replace the poorly antigenic (original prototype of McIntosh) strain Vry (VRY)

We consider as proper the following terms for use in describing the disease aetiology and the viruses:

1. African horse sickness (AHS): the name of the disease, as in "The horse died of AHS".
2. African horse sickness virus (AHSV) group (serogroup and species): the nine serotypes of this virus, as in "The isolate belongs to the AHSV group".
3. An AHSV: an unspecified member of the AHSV group, as in "We isolated an AHS virus from this horse".
4. AHSV 1, AHSV 2, AHSV 3, etc.: names of the serotypes, 1, 2 and 3 in the AHSV group, as in "We isolated AHSV 8 from this horse".
5. *Orbivirus*: the name of the genus in which the AHSV serogroup of viruses has been placed, as in "AHSV is a member of the genus *Orbivirus*".
6. An orbivirus: any (or a specific) member of the genus, as in "AHSV is an orbivirus", or "AHSV 5 is an orbivirus".

In order to make this book clear and readable, the Editors will use these terms and identify specific viruses as correctly as possible and as available data permit. We have strongly urged authors of the papers presented here, as well as in other journals, to use the correct terminology when referring to or identifying orbiviruses, other members of the family *Reoviridae* and all other viruses that have been numbered rather than named.

Suggestions for changes, additions or corrections to the current taxonomy of the *Reoviridae* along with references to appropriate published data, can be made either to the current Chairman (1997–2000: Peter P. C. Mertens, E-mail: <peter.

mertens@bbsrc.ac.uk>) or to appropriate members of the ICTV Reoviridae Study Group. They will be evaluated for submission to ICTV. News and information concerning the Study Group and taxonomy of the family *Reoviridae* is currently available on the Internet: <http://www.iah.bbsrc.ac.uk/virus/Reoviridae/>

Acknowledgements

The authors thank Dr. Robert Swanepoel, National Institute for Virology, University of Witwatersrand, Sandringham, Transvaal, South Africa, for good humoredly providing insight and certain references, and Nick Knowles for help in preparation of the "web" pages.

References

1. Erasmus B (1973) The pathogenesis of African horse sickness. In: Bryans JF, Gerber H (eds) Equine-infectious diseases III. Proceedings of the 3rd International Conference on Equine Infectious Diseases, Paris, 1972. Karger, Basel, pp 1–11
2. Fenner F, Bachmann PA, Gibbs EPJ, Murphy FA, Studdert MJ, White DO (1987) Veterinary virology. Academic Press, Orlando, pp 587–590
3. Gorman BM (1983) On the evolution of *Orbiviruses.* Intervirology 20: 169–180
4. Gould AR (1987) The complete nucleotide sequence of bluetongue virus serotype 1 RNA 3 and a comparison with other geographic serotypes from Australia, South Africa and the United States of America and with other orbivirus isolates. Virus Res 7: 169–183
5. Holmes IH, Boccardo G, Estes MK, Furuichi MK, Hoshino Y, Joklik WK, McCrae M, Mertens PPC, Milne RG, Samal KSK, Shiata E, Winton JR, Uyeda I (1995) Family, *Reoviridae.* In: Murphy FA, Fauquet CM, Bishop DHL, Ghabrial SA, Jarvis AW, Martelli GP, Mayo MA, Summers MD (eds) Virus Taxonomy. Classification and Nomenclature of Viruses. Sixth Report of the International Committee on Taxonomy of Viruses. Springer, Wien New York, pp 208–237 (Arch Virol [Suppl] 10)
6. Howell PG (1962) The isolation and identification of further antigenic types of African horse sickness virus. Onderstepoort J Vet Res 29: 139–149
7. Laegreid WW (1996) African horse sickness. In: Horzinek MC (ed) Virus infections of vertebrates, vol 6: Studdert MJ (ed) Virus infections of equines. Elsevier, New York, pp 101–123
8. McIntosh BM (1958) Immunological types of horse sickness and their significance in immunization. Onderstepoort J Vet Res 27: 465–538
9. Mertens PPC et al (in preparation) Family Reoviridae. In: Virus Taxonomy: Classification and Nomenclature of Viruses. Seventh Report of the International Committee on Taxonomy of Viruses
10. Mertens PPC (1994) Orbiviruses and coltiviruses – general features. In: Webster RG, Granoff A (eds) Encyclopaedia of virology. Academic Press, London, pp 941–956
11. Mertens PPC, Pedley S, Cowley J, Burroughs JN (1987) A comparison of six different bluetongue virus isolates by cross hybridisation of the dsRNA genome segments. Virology 161: 438–447
12. Mertens PPC, Pedley S, Cowley J, Burroughs JN, Corteyn AH, Jeggo MH, Jennings AM, Gorman BM (1989) Analysis of the roles of bluetongue virus outer capsid proteins VP2 and VP5 in determination of virus serotype. Virology 170: 561–565
13. Moss SR, Ayres CM, Nuttall PA (1987) The Great Island subgroup of tick borne orbiviruses represents a single gene pool. J Gen Virol 69: 2721–2727

14. Nuttall PA, Moss SR (1989) Genetic reassortment indicates a new grouping for tick borne orbiviruses. Virology 171: 156–161
15. Parkes H, Gould AR (1996) Characterization of Wongorr virus, an Australian orbivirus. Virus Res 44: 111–122
16. Pringle CP (1996) Virus taxonomy 1996 – A Bulletin from the Xth International Congress of Virology in Jerusalem. Arch Virol 41: 2251–2256
17. Samal SK, El-Hussein A, Holbrook FR, Beaty BJ, Ramig RF (1987) Mixed infection of *Culicoides variipennis* with bluetongue virus serotypes 10 and 17: evidence for high frequency reassortment in the vector. J Gen Virol 68: 2319–2329
18. Samal SK, Livingston CW Jr, McConnell S, Ramig RF (1987) Analysis of mixed infection of sheep with bluetongue virus serotypes 10 and 17: evidence for genetic reassortment in the vertebrate host. J Virol 61: 1086–1091
19. Stott JL, Oberst RD, Channell MB, Osburn BI (1987) Genome segment reassortment between two serotypes of bluetongue virus in a natural host. J Virol 61: 2670–2674
20. Swanepoel R, Erasmus BJ, Williams R, Taylor MB (1992) Encephalitis and chorioretinitis associated with neurotropic African horse sickness virus infection in laboratory workers. Part III. Virological and serological investigations. S Afr Med J 81: 458–461
21. Zeller HG, Karabatsos N, Calisher CH, Digoutte JP, Cropp CB, Murphy FA, Shope RE (1989) Electron microscopic and antigenic studies of uncharacterized viruses III. Evidence suggesting the placement of viruses in the family *Reoviridae.* Arch Virol 109: 253–261

Authors' address: Dr. P. P. C. Mertens, Institute for Animal Health, Pirbright Laboratory, Ash Road, Pirbright, Working, Surrey GU24 ONF, U.K.

Epidemiology of African horse sickness and the role of the zebra in South Africa

B. J. H. Barnard

Onderstepoort Veterinary Institute, Onderstepoort, South Africa

Summary. Zebra are the only equine species native to South Africa. These animals roamed over much of the country in the 17th century when horses and donkeys were first imported. The first cases of African horse sickness (AHS) then occurred in the horses of hunters who entered zebra territory. AHS continued to occur on a country-wide basis until the beginning of the 20th century, though the number of outbreaks decreased as the populations of zebra collapsed through overhunting. For most of the 20th century almost all free-living zebra have been confined to the north-eastern parts of South Africa which are now the only areas in the country where AHS is endemic; though when climatic conditions are favourable, temporarily, it spreads beyond these areas. The minimum size of a zebra population necessary to maintain a focus of AHS virus is unknown but the small, isolated populations that have inhabited the majority of South Africa for most of the 20th century are apparently insufficient to maintain the virus in the long term. In this context, the restocking of many parts of the country with zebra should be viewed with caution since conditions may be generated that will facilitate the re-establishment of permanent foci of AHS virus.

Introduction

African horse sickness (AHS) virus is endemic in eastern and central Africa and occurs regularly in many parts of sub-Saharan Africa including parts of South Africa [11]. Occasionally the virus also spreads into North Africa, countries around the Mediterranean Sea and Asia [15]. The disease of AHS, which is caused by this virus, is an ancient one and was probably first recognized as early as 1569 by a certain Father Monclaro who described losses occurring at that time among horses in East Africa [21].

The major susceptible hosts in southern Africa

Horses

Horses are not native to southern Africa and were first introduced into the area in 1657 [10].

Initially these animals played a crucial role in transportation, in military operations and as draught power in agriculture. Their numbers therefore increased rapidly due both to importation and to an active breeding programme and by 1854 there was an estimated population of 175,000 horses in South Africa. The population of horses in the country probably peaked at about 300,000 animals at the time of the Anglo-Boer war (1900–1902) but subsequent to that period their numbers declined due to increasing mechanisation. At the present time (1997) there are approximately 260,000 horses in the country which are used mainly in racing and leisure activities.

Donkeys

Donkeys were probably introduced into South Africa some time after horses but rapidly spread throughout the country, being used widely in transport and as draught animals. However in recent times the population has fallen, though in localised areas throughout South Africa small populations can still be found.

South African donkeys are considered to be less susceptible to AHS than are donkeys from Mediterranean countries and they generally develop only a very mild form of the disease [5]. Furthermore viraemia following experimental infection is of low titre ($< 1.5 \times 10^2 \log_{10}TCID_{50}/ml$) and infection is often only detectable through serology (Venter G, Barnard BJH – unpubl. obs.). These observations coupled with their small numbers suggest that in South Africa donkeys are unlikely to play a significant role in the epidemiology of AHS.

Zebra

The modern zebra is regarded as being the descendant of animals that crossed into north-eastern Africa during the late Miocene Epoch some 5–15 million years ago [4].

Equus burchelli mauritianius from North Africa is believed to be the direct ancestor of the present day *E. burchelli* subspecies and the now extinct quagga (*E. quagga*) *is* regarded as being very closely related to *E. burchelli* [9, 20]. The origins of the Cape mountain zebra (*E. zebra*) and Hartmann's mountain zebra (*E. zebra hartmannae*) are obscure.

Today the genus is represented by 3 living species: *E. grevyi* from East Africa, *E. zebra* (Hartmann's zebra and Cape mountain zebra) and *E. burchelli* from southern Africa [19]. Hartmann's zebra and the Cape mountain zebra occur in small numbers in a few restricted areas of South Africa, Namibia (north of the Orange River) and SW Angola [16]. *Equus burchelli* is the most numerous species and is distributed widely in Somalia, Tanzania, Zambia, Mozambique, southern Angola, Katanga Province of Zaire and it also occurs in South Africa though its present day distribution in this country is much reduced compared to former times [16].

According to the records of the Dutch East India Company [10] hunting reduced the populations of the Cape mountain zebra and the quagga to such an eztent that the quagga became extinct by the end of the 19th Century and the Cape montain zebra was pushed to the verge of extinction [17]. Then in 1937 a reserve was established for the protection of the remaining Cape mountain zebra [16] though their total population had only risen to 474 by 1985 [18].

The numbers of Burchell's zebra were also greatly reduced in South Africa and at one time they had completely disappeared from the open plains in the interior of the country. At the present time large naturally occurring populations are still found only in the north-eastern parts of the country, particularly the Kruger National Park with about 33,000 zebra [19] and certain parks in NE KwaZulu-Natal which in 1996 held a further 8,000 (Rowe-Rowe, Natal Parks Board – pers. comm.). However within the last 25 years Burchell's zebra has made something of a come back and it has been reintroduced into private game parks on a country-wide basis. The total number present in 1996 in South African parks probably lay between 60,000 and 100,000 animals (Ebedes, Onderstepoort Veterinary Institute – pers. comm.) and furthe increase is likely.

African horse sickness in South Africa

AHS virus was apparently already present in southern Africa when the first horses were introduced into the region in the 17th century [10]. However the disease caused by the virus was not recognized until the first major outbreak of AHS in 1719, some 52 years after the introduction of horses, when approximately 1,700 animals died from the disease [10]. During the next 217 years at least 10 major and several lesser outbreaks of AHS were recorded [5]. In the largest outbreak ever in South Africa (1854–1855) which was particularly severe in the southern parts of the country, 70,000 horses succumbed to the disease. However during the last 100 years the frequency, severity and extent of AHS outbreaks has declined, particularly in southern South Africa. This decline has coincided with a decrease in the horse population, a greatly reduced zebra population and the introduction of AHS vaccines.

Vaccines

Vaccination has played a important role in the diminishing occurrence of AHS in South Africa by reducing the number of susceptible horses. The first AHS vaccines became available in 1936 and during the following 22 years only 3 outbreaks, all associated with above average rainfall, were recorded [5]. In 1958 improved AHS vaccines were introduced [8] and subsequent to that time only 3 further, minor outbreaks have been experienced. At the present time approximately 130,000 doses of vaccine are used annually in South Africa for the vaccination of horses (Broekman, Onderstepoort Veterinary Institute, unpubl. data).

Zebra as reservoirs

According to early records of the Dutch East India Company losses of horses in southern Africa, from an unknown disease that we now know resembles AHS, were first encountered during the late 16th century [10]. These deaths originally occurred among the horses of hunters who had travelled inland in search of game but did not apparently occur among the horses of farmers who had remained in close proximity to the area of the first settlement on the Cape Peninsula. The pattern of deaths suggested that the infection was acquired both in mountainous areas some 80 km or more inland and also on the plains to the north of the mountain ranges where mountain zebra and quagga respectively, were hunted. These initial isolated cases culminated in the first major outbreak of AHS in 1719 and in further outbreaks in subsequent years. AHS continued to occur, on an almost country-wide basis, throughout the 19th century and for the first few decades of the 20th century, though the number of outbreaks declined throughout the period, particularly in the south of the country. This decrease coincided with the country-wide disappearance of large populations of zebra.

For most of the 20th century virtually all zebra have been confined to game parks in the north-eastern parts of the country. On farms adjacent to these parks cases of AHS may be encountered in unprotected horses throughout the year. This endemic region for AHS virus is characterised by climatic conditions that facilitate the development of high populations of vector *Culicoides* and by an absence of frosts during the winters so that adult vectors are present throughout the year [14].

From the endemic area AHS virus sometimes spreads southwards into the central parts of South Africa, as far as the 27th parallel but rarely further (Gerdes, Onderstepoort Veterinary Institute – unpubl. data). The occurrence of AHS cases in the central parts of the country is terminated typically by the first frosts of winter [22]. The reappearance of AHS cases in successive years however does not continue from where the disease left off in the previous autumn and is often caused by different serotypes of the virus [5]. These factors suggest that the new outbreaks are caused by fresh introductions of virus and not by a reactivation of overwintering virus [5]. The seasonal and somewhat irregular southward spread of AHS virus in South Africa is attributed to unusually large increases in abundance and distribution of vector *Culicoides,* particularly in years with a higher than average rainfall [5].

In areas endemic for AHS virus the only potential vertebrate reservoirs for the virus are game of which species zebra is the most likely candidate. This likelihood was substantiated initially by the demonstration of AHS virus specific antibodies in the serum of free-living zebra [6]. More recently the susceptibility of zebra to infection with AHS virus has been proven unequivocally and these animals have been shown to be viraemic for periods as long as 6 weeks and to maintain infectious virus in the spleen for up to 7 weeks [3, 7]. It has also been shown that a zebra population is able to maintain an infection with AHS virus probably via the classical method of continuous circulation of virus within the herd, provided that the population is large enough and that an efficient vector is continually present in sufficient numbers to mediate transmission [1]. The practical implications of these findings were demonstrated recently when AHS virus serotype 4 was introduced into Spain in 1987 via the importation of infected zebra [13].

Zebra are probably the only large game animal in southern Africa to be involved in the epidemiology of AHS. Of the 16 most numerous game species in the region, including black and white rhinocerus which belong to the same family, only zebra have been shown to contain neutralising antibodies to AHS virus [2].

The importance of zebra in the epidemiology of AHS virus in South Africa can be strongly inferred by a consideration of the geographical distribution of those animals which contain specific neutralising antibodies to the virus, in relation to the incidence of AHS in unprotected horses. The prevalence of AHS virus antibodies in zebra and of AHS in unprotected horses is very similar. In the north-east of the country where AHS virus is endemic and where most cases of AHS occur in horses, seroconversion in zebra occurs in every month of the year and virtually all adult animals have antibodies to all 9 AHS virus serotypes [1]. In the central part of South Africa, where AHS occurs seasonally in horses, the prevalence of specific antibodies in local zebra is low [2]. In the southern part of the country where cases of AHS in horses are rare and can usually be traced back to their recent introduction from further north, zebra are completely AHS virus antibody free [2]. Nevertheless in considering these facts it should also be borne in mind that vector *Culicoides* are essential for the transmission of AHS virus and though these vectors may be present throughout the year in the more tropical (north eastern) parts of South Africa elsewhere in the country they have a seasonal incidence [14]. The minimum size of a zebra population necessary to maintain a permanent focus of infection with AHS virus is at present unknown. However clearly, sufficient numbers are required to ensure that the virus always has a regular and plentiful supply of new, susceptible foals to infect. In addition the local climatic conditions must be such as to permit the presence of large numbers of active, adult vectors throughout the year [15].

The sporadic and seasonal occurrence of AHS in horses in the central part of South Africa suggests that the small isolated zebra populations which at present inhabit this part of the country (usually consisting of less than 100 individuals per herd) are of insufficient size to maintain the virus in the long term. In addition it is also likely that vector abundance in this area, particularly in the winter, is lower than further north [12, 14] and is insufficient at present zebra population levels to support continuous circulation of the virus throughout the year. In the light of this information the ongoing restocking of many parts of South Africa with Burchell's zebra, should be viewed with a degree of caution, since at some ill-defined level of population density, conditions may well be generated that will facilitate the establishment of permanent foci of AHS virus, as was the case in the 19th century.

References

1. Barnard BJH (1993) Circulation of African horse sickness virus in zebra (*Equus burchelli*) in the Kruger National Park, South Africa, as measured by the prevalence of type specific antibodies. Onderstepoort J Vet Res 60: 111–117
2. Barnard BJH (1997) Antibodies against some viruses of domestic animals in southern African wild animals. Onderstepoort J Vet Res 64: 95–110
3. Barnard BJH, Bengis R, Keet D, Dekker EH (1994) Duration of viraemia in zebra (*Equus burchelli*). Onderstepoort J Vet Res 61: 391–393
4. Churcher CS, Richardson ML (1978) *Equidae*. In: Maglio VJ, Cook HBS (eds) Evolution of African mammals. Harvard University Press, Cambridge/Mass., pp 379–422
5. Coetzer JAW, Erasmus BJ (1994) African horse sickness. In: Coetzer JAW, Thomson GR, Tustin RC (eds) Infectious diseases of livestock. Oxford University Press, Cape Town, pp 460–475
6. Davies FG, Lund LJ (1974) The application of fluorescent antibody technique to the virus of African horse sickness. Vet Rec 17: 128–130
7. Erasmus BJ, Young E, Pieterse LM, Boshoff ST (1978) The susceptibility of zebra and elephants to African horse sickness virus. In: Bryans JT, Gerber H (eds) Equine infectious diseases. Proceedings of the 4th international conference on equine infectious diseases, Lyon, France, 24–27th September 1978. Veterinary Publications Inc, Princeton, NJ, pp 409–413
8. Erasmus BJ (1965) Attenuation of viscerotropic horse sickness virus in tissue culture. Bull Off Int Epiz 30
9. George M, Ryder GA (1986) Mitochondrial DNA evolution in the genus *Equus*. Mol Biol Evol 3: 535–546
10. Henning MW (1956) Animal diseases in South Africa, 3rd ed. Central News Agency, Pretoria
11. Howell PG (1963) Emerging diseases of animals. 11, African horse sickness. FAO Agric Stud 61: 71–108
12. Jupp PG, McIntosch BM, Nevill EM (1986) A survey of the mosquito and *Culicoides* of two localities in the Karoo region of South Africa with some observations on bionomics. Onderstepoort J Vet Res 47: 1–6
13. Lubroth J (1988) African horse sickness and the epizootics in Spain. Equine Prac 10: 26–33
14. Meiswinkel R, Nevill EM, Venter GJ (1994) Vectors, *Culicoides* spp. In: Coetzer JAW, Thomson GR, Tustin RC (eds) Infectious diseases of livestock. Oxford University Press, Cape Town, pp 68–89
15. Mellor PS (1993) African horse sickness: transmission and epidemiology. Vet Res 24: 199–212
16. Penzhorn BL (1975) Behaviour and population ecology of the Cape mountain zebra (*Equus zebra* Linn. 1758) in the Mountain Zebra National Park. DSc Thesis, University of Pretoria, South Africa
17. Sidney J (1965) The past and present distribution of some African ungulates. Trans Zoo Soc Lond 30: 1–397
18. Skinner JD, Smithers RHM (1990) The mammals of the southern African subregion. University of Pretoria, Pretoria
19. Smithers RHM (1986) South African red data book: Terrestrial Mammals. South African National Programmes Report Number 125: 1–216. Pretoria, CSIR
20. Smuts GL (1974) Growth, reproduction and population characteristics of Burchell's zebra (*Equus burchelli antiquorum* H Smith 1841) in the Kruger National Park. DSc Thesis, University of Pretoria

21. Theal EM (1899) Records of South Eastern Africa collected in various libraries and archive departments in Europe. Gov of Cape Colony vol 3, pp 224
22. Theiler A (1921) African horse sickness (pestis equorum). Science Bulletin No. 19 Department of Agriculture, Union of South Africa
23. Viljoen PC (1978) Ecological aerial survey in the Kruger National Park. Scientific Report 4–92, Kruger National Park, 1992

Authors' address: Dr. B. J. H. Barnard, Onderstepoort Veterinary Institute, Private Bag X05, Onderstepoort 0110, South Africa.

Transmission and distribution of African horse sickness virus serotypes in South African zebra

C. C. Lord[1,*], **M. E. J. Woolhouse**[1,**], and **B. J. H. Barnard**[2]

[1] Department of Zoology, University of Oxford, Oxford, U.K.
[2] Onderstepoort Veterinary Institute, Onderstepoort, Republic of South Africa

Summary. The prevalences of African horse sickness (AHS) virus serotypes in zebra foals from the Kruger National Park, South Africa were examined for possible associations between serotypes. Serotypes known to cross-react were combined for analysis. The distributions of serotypes between zebra were not always independent; in 7–8 month old zebra positive pairwise associations were observed between 3 serotypes. This could be generated by biological interactions between serotypes or heterogeneity in host-vector transmission. The data were also used to estimate the basic reproduction number, R_0. For AHS virus overall, estimates of R_0 ranged from 31–68. This underlines the need for a better understanding of serotype transmission and interactions in AHS.

Introduction

The importance of multiple serotypes of the virus in the epidemiology of African horse sickness (AHS) depends upon many factors, including the number and identity of the serotypes involved in a particular outbreak, the immune status of the local hosts, and variations in susceptibility to different serotypes in hosts and vectors. In some situations, such as the 1987–90 incursion of AHS virus (*Reoviridae: Orbivirus*) into Spain, only one serotype was involved. In other situations, such as the current distribution in sub-Saharan Africa, multiple serotypes coexist. This must be considered when interpreting epidemiological data and in the design of control programmes. Similar variation exists in other viruses (e.g. bluetongue [14], foot and mouth [4], dengue [7]), and has also been found in protozoan diseases such as malaria [5] and trypanosomiasis [17]. Recent studies [6, 8] have begun to examine the epidemiological consequences of serotype variation. It is clear that there will be consequences for the basic reproduction number, R_0 [6].

Current addresses: * Florida Medical Entomology Laboratory, University of Florida, U.S.A.
** Centre for Tropical Veterinary Medicine, University of Edinburgh, Midlothian, U.K.

This number is a measure of the strength of transmission of a parasite or pathogen, and is defined as the average number of secondary cases generated by one primary case in a susceptible population. It has been suggested by Gupta et al. [6] that conventional methods of calculating R_0 may overestimate the true value if the disease is caused by a composite of several independently transmitted serotypes. As R_0 is used in the design of control campaigns, it is crucial that we critically examine the potential impact of multiple serotypes upon it.

One aspect of multiple serotypes which will affect R_0 is the degree to which the serotypes are transmitted independently. Here, we define independent transmission as transmission without biological interactions between the serotypes. Thus, if infection with one serotype affects potential for infection with another serotype, transmission would not be independent. One potential method to test for independent transmission is by looking at the distribution of serotypes among hosts. However, the distribution of serotypes among hosts will also be affected by other types of heterogeneity in transmission rates, such as variation in vector biting rates. The detection of non-independent distributions of serotypes is analogous to the problem of determining independence between species distributions in community ecology, and methods developed to test for associations between species in communities can be used to study serotype distributions among hosts.

The effect of interactions between serotypes on R_0 is complex and not yet clearly understood [6, 8]. If independently transmitted serotypes are considered as a group, R_0 is likely to be overestimated [6]. When serotypes interact, however, the impact on R_0 will depend on the precise nature and strength of the interaction.

Here, we present further analysis of previously published data [2, 8], to investigate the independence of serotype transmission and distribution and the effect of multiple serotypes on calculations of R_0 for AHS.

Materials and methods

Serum collection

Serum was collected from free-living zebra foals in the Kruger National Park, Republic of South Africa, and analysed as described by Barnard [2]. Briefly, between August 1991 and May 1992, foals were captured, age estimated based on tooth eruption and wear, and serum collected. Animals were assigned to monthly age classes except for 7 and 8 months, which could not be distinguished and so were combined into one class. The serogroup specific agar gel immunodiffusion test (GP) was used to assay for antibodies to AHS virus, and microneutralisation tests were used to determine serotype-specific antibody titres (NA). Any titre greater than zero was considered positive. The cross-reacting pairs of serotypes (1 and 2, 3 and 7, 5 and 8, and 6 and 9 [3]) were combined. If an animal was positive for either of the two members of a crossreacting pair, it was considered positive for the combination (e.g. if positive for either serotype 1 or 2, it was positive for the combination 1–2).

It has been estimated [2] that zebra foals lose maternal antibodies to AHS virus at 4–6 months of age. In the data analysed here, foals aged 5 months had a bimodal distribution of serotypes, most having positive titres to 3 or fewer serotypes and 4 foals having positive titres to 8 or more serotypes. To prevent maternal antibodies from confounding the analysis, all foals under 5 months of age and the 4 foals aged 5 months which tested positive to 8 or more serotypes were excluded from the analysis. Final sample sizes for each age class are given in Table 1.

Associations

Schluter's variance test [9, 13] can be used to test for the presence of covariance between serotypes. This test compares the observed variance in the number of serotypes per zebra with that expected under the hypothesis of no covariance, and can be interpreted as comparing the observed distribution of the number of serotypes per zebra with that expected from the sum of independent binomial distributions.

Fisher's exact test (2-tailed; FREQ procedure, SAS [12]) was used to test for pairwise associations between serotypes. As pairwise associations would not be expected if the distributions of the serotypes were independent, this test was used only on those age classes where there was covariance between the serotypes. The significance level was adjusted for multiple tests using the Bonferroni correction [11].

Basic reproduction number

R_0 provides an estimate of the transmission potential of a disease and, in an endemic situation, can be calculated using age-prevalence data [1, 8]. This analysis has been previously published; the details can be found in Lord et al. [8]. The method assumes that zebra acquire infection with serotype *i* with an age-independent constant force of infection, λ_i. Maximum likelihood methods [8, 15], were used to estimate λ_i and the age at which maternal antibodies become undetectable. Here, the data for all 9 serotypes were used rather than the combined serotypes, as cross reactions would not bias the results. A common λ was fitted using the data for all 9 serotypes, and tested by likelihood ratios to the fit obtained with individual values of λ_i for each serotype. A significant lack of fit in this test indicates that the 9 λ_is are different. Confidence intervals were also calculated for pairwise comparisons.

R_0 for each serotype was then calculated from λ_i and the average lifespan of the zebra, assuming that the serotypes are independently transmitted. In addition, the calculation method assumes that immunity to AHS virus serotypes is life long [3], that virus prevalence is in equilibrium and that the host population is static (i.e. is not growing or decreasing and has a stable age structure). Mortality rates and population densities have been studied in a zebra population slightly south of those sampled here [10]. In that study, the population was reasonably stable and the lifespan assuming all deaths were due to lion predation was estimated to be 72 months. To account for other causes of death, a lifespan of 60 months was

Table 1. Results of Schluter's *W* tests for independent distributions of serotypes within each age class

Age class	Sample size[a]	Schluter's W[b]
5	11	16.1
6	36	65.2[c]
7–8	30	74.3[c]
9	19	19.0

[a] Number of zebra in age class
[b] Schluter's W distributed as $\chi^2_{\alpha/2,\,z}$ where z is sample size
[c] $P < 0.005$

used. R_0 was calculated for each serotype from the individual NA tests and for AHS virus overall from the GP and summed NA tests (i.e. an animal was considered positive for AHS virus if positive in any of the 9 NA tests).

Results

The overall pattern of prevalence for the combined serotypes is shown in Fig. 1. By 9–12 months old, the prevalence of all serotypes approached 100%.

Associations

The distributions of the number of combined serotypes per zebra in each age class are shown in Fig. 2. There was significant covariance in both the 6 and 7–8 month old age classes (Table 1), or in other words, the number of serotypes per zebra was

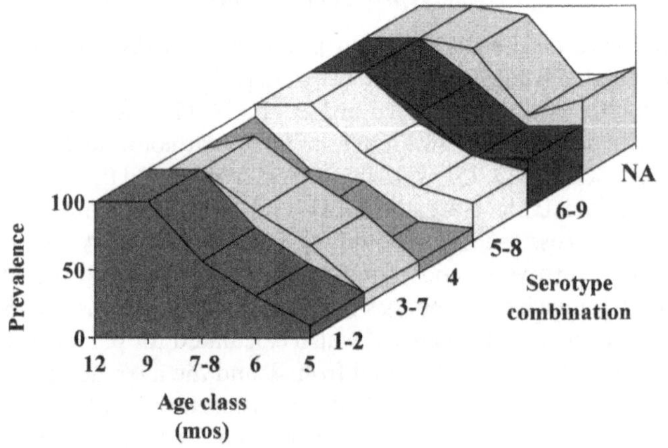

Fig. 1. Overall prevalence of serotype combinations of AHS virus by age. *NA* Pevalence of neutralising antibodies to any serotype. Serotypes combined based on known cross reactions (see text)

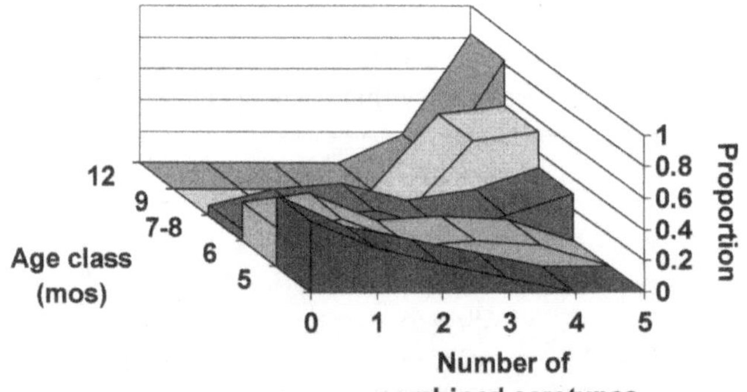

Fig. 2. The distribution of the number of combined serotypes per zebra by age. Serotypes combined based on known cross reactions (see text)

different from that expected under the assumption of independent serotype transmission. In 6 month old zebra there were, however, no significant pairwise associations. There were three significant pairwise associations in 7–8 month old animals: serotypes 4 and 5–8, 4 and 6–9, and 5–8 and 6–9 (Table 2).

Basic reproduction number

The basic reproduction numbers from these data have been previously published [8]; we recapitulate the results here for convenience. Estimates of λ_i for each serotype were significantly different, and λ_1 and λ_4 were significantly smaller than several other serotypes ($\lambda_1 < \lambda_5, \lambda_6, \lambda_7, \lambda_8, \lambda_9$; $\lambda_4 < \lambda_5, \lambda_6, \lambda_7, \lambda_9$). R_0 for individual serotypes varied from 10.5 (serotype 1) to 23.4 (serotype 6, Fig. 3). As R_0 is a

Table 2. *P* values for pairwise associations in 6 (above diagonal) and 7–8 (below diagonal) month-old zebra

	Combined serotype				
Combined serotype	1–2	3–7	4	5–8	6–9
1–2		0.483	1	0.00968	0.00636
3–7	0.0166		0.637	0.734	1
4	0.275	0.132		0.357	0.328
5–8	0.128	0.683	0.00015[a]		0.0732
6–9	0.433	0.385	0.00131[a]	0.000166[a]	

[a] Significantly different from expected under the hypothesis of no association at the $\alpha = 0.05$ level ($P < 0.0025$, with the Bonferroni correction [11] for multiple tests)

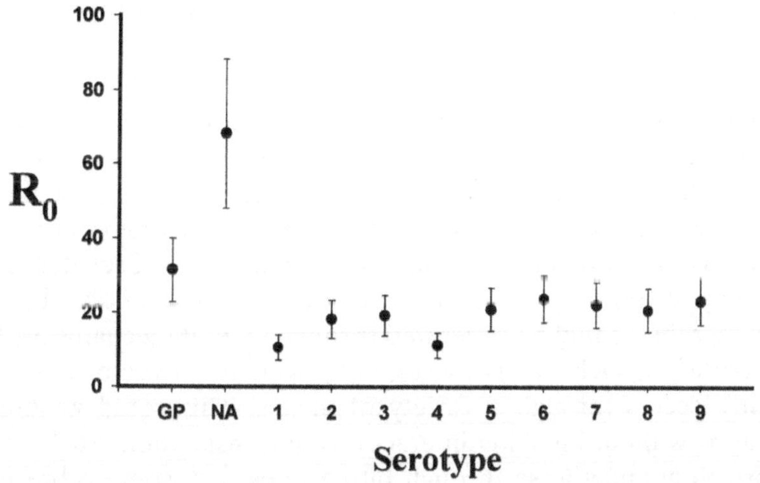

Fig. 3. R_0 (mean ± 95% c.i.) for AHS virus overall and for each of the 9 serotypes individually. *GP* Based on prevalence of group-specific antibodies to AHS virus. *NA* Based on prevalence of neutralising antibodies to any serotype as in Fig. 1. *S1–S9* Based on prevalence of neutralising antibodies to each serotype alone

multiple of λ_i, the same significance relationships apply; estimates with overlapping confidence intervals in Fig. 3 are not significantly different. For AHS virus overall, the estimate of R_0 is lower ($R_0 = 31$) using the GP data than when the combined NA data are used ($R_0 = 68$); the confidence intervals do not overlap (Fig. 3).

Discussion

The serotypes of AHS virus were not distributed independently in this zebra population but the cause cannot be determined from these data. Several mechanisms could generate the patterns observed here, such as facilitation, cross-reactive immunity or heterogeneity in host-vector transmission rates. However, not all of these mechanisms require non-independent transmission, i.e. biological interactions between serotypes.

Facilitation can act in two ways, either sequentially or simultaneously, and would lead to positive associations between specific serotypes. If infection with one serotype increases susceptibility to another serotype, sequential facilitation is present. In cohort data, the prevalence of one of the associated serotypes alone should be higher than the combination of serotypes while the prevalence of single infections with the other serotype (the one which requires the facilitation) should be the lowest. No such patterns are apparent in these data, as there were relatively few single infections with any of the 3 serotypes involved in pairwise associations (serotypes 4, 8 and 9). If a transmission event is more likely to result in infection when more than one serotype is present, simultaneous facilitation is acting. This would result in the pattern observed in the 7–8 month old zebra, where some serotypes are present primarily together and rarely or never alone. Further information on the course of infection in animals infected with more than one serotype is necessary to determine if facilitation is responsible for the associations observed.

Negative associations between serotypes might be expected, considering the known cross-immunity between AHS virus serotypes. However, if serotype identification is compromised by cross-immunity (as for AHS virus), then artefactual positive associations may result. Tests with cross-reacting serotypes combined are more conservative, and will reduce this problem. Known patterns of cross-immunity must be considered when interpreting the epidemiology of serotypes.

Heterogeneity in transmission rates, leading to non-random distributions of serotypes among zebra, could be generated through several mechanisms. With a vector-borne pathogen such as AHS virus, heterogeneity in vector-host contact rates frequently occurs through a variety of means. This could generate the observed patterns without invoking interactions in transmission. Such variation in host vector-contact may arise through biting rates (e.g. some zebra may be bitten more often than others) or spatial or temporal variations in the distribution of hosts or vectors. In addition, differences in susceptibility to different serotypes in either hosts or vectors could also generate non-independence in serotype distributions.

The stability of the associations would provide further information as to the mechanism generating the observed patterns. If the same patterns are observed repeatedly, in other populations or in repeated samples of the same population, facilitation, cross-immunity or genetic differences in susceptibility are plausible mechanisms. However, if non-independent distributions are consistently observed but the specific pairwise associations vary, other types of heterogeneity are more likely explanations.

The variation in R_0 by serotype indicates that there are likely to be differences in infectivity of different serotypes for vector midges or zebra, or both. The method used here to calculate R_0 for the individual serotypes assumes independent transmission. However, the relative differences are still informative, even if there are some interactions between the serotypes. These values of R_0 are among the highest reported for veterinary diseases [16]. The large population and year-round addition of new susceptible animals to the population apparently allow the virus to persist with high levels of transmission.

The variation observed in estimates of R_0 for AHS virus emphasizes the need for further analyses of this type and more detailed study of the interactions between serotypes. Until more detailed information is available, varied estimates of R_0 from epidemiological data can be considered as limits, rather than precise estimates. An upper limit on the estimate of R_0 which does not depend on assumptions of independent transmission can be derived from the prevalence of exposure to any serotype of AHS virus (NA in Figs. 1 and 3). If, however, the assumption of independent transmission is made, a reliable lower limit on the estimate is the highest individual R_0. Any control measures designed using this estimate would exceed the levels required for the other serotypes. Such limits may be used in the design of control programmes until further data are available on the independence of transmission and the consequences for estimating R_0.

For AHS virus, the highest individual value of R_0 was 23, whereas if all serotypes are considered together the upper estimate of R_0 was 68. This is a substantial difference and would have a significant effect on the formulation of a control programme. It is important, therefore, to establish the extent to which AHS virus serotypes are independently transmitted.

Acknowledgements

We thank Dan Haydon and Sunetra Gupta for discussions of strain variation. CCL was supported by a BBSRC Link Grant and MEJW by the Royal Society.

References

1. Anderson RM, May RM (1991) Infectious diseases of humans: dynamics and control. Oxford University Press, Oxford
2. Barnard BJH (1993) Circulation of African horsesickness virus in zebra (*Equus burchelli*) in the Kruger National Park, South Africa, as measured by the prevalence of type specific antibodies. Onderstepoort J Vet Res 60: 111–117

3. Coetzer JAW, Erasmus BJ (1994) African horse sickness. In: Coetzer JAW, Thomson GR, Tustin RC (eds) Infectious diseases of livestock with special reference to southern Africa. Oxford University Press, Oxford, pp 460–475
4. Domingo E, Mateu MG, Martinez MA, Dopazo J, Moya A, Sobrino F (1990) Genetic variability and antigenic diversity of foot-and-mouth disease virus. In: Kurstak E, Marusyk RG, Murphy FA, van Regenmotel MHV (eds) Applied virology research, vol 2. Plenum Press, New York, pp 233–266
5. Forsyth KP, Philip G, Smith T, Kum E, Southwell B, Brown GV (1989) Diversity of antigens expressed on the surface of erythrocytes infected with mature *Plasmodium falciparum* parasites in Papua New Guinea. Am J Trop Med Hyg 41: 259–265
6. Gupta S, Trenholm K, Anderson RM, Karen KP (1994) Antigenic diversity and the transmission dynamics of *Plasmodium falciparum*. Science 263: 961–963
7. Kettle DS (1990) Medical and veterinary entomology. CAB International, Wallingford
8. Lord CC, Woolhouse MEJ, Barnard BJH (1997) Transmission and distribution of virus serotypes: African horse sickness in zebra. Epidemiol Infect 118: 43–50
9. McCulloch CE (1985) Variance tests for species associations. Ecology 65: 1676–1681
10. Mills MGL, Biggs HC, Whyte IJ (1995) The relationship between rainfall, lion predation and population trends in African herbivores. Wildlife Res 22: 75–88
11. Rice WR (1989) Analyzing tables of statistical tests. Evolution 43: 223–225
12. SAS (1988) SAS User's guide. SAS Institute, Cary, NC
13. Schluter D (1984) A variance test for detecting species associations, with some example applications. Ecology 65: 998–1005
14. Verwoerd DW, Erasmus BJ (1994) Bluetongue. In: Coetzer JAW, Thomson GR, Tustin RC (eds) Infectious diseases of livestock with special reference to southern Africa. Oxford University Press, Oxford, pp 443–459
15. Williams BG, Dye C (1994) Maximum likelihood for parasitologists. Parasitol Today 10: 489–493
16. Woolhouse MEJ, Bundy DAP (1997) Epidemiological aspects of vaccination programs. In: Pastorett PP (ed) Veterinary vaccinology. Elsevier, London, pp 565–573
17. Woolhouse MEJ, McNamara JJ, Hargrove JW, Bealby KA (1996) Distribution and abundance of trypanosome (subgenus *Nannomonas*) infections of the tsetse fly *Glossina pallidipes* in southern Africa. Mol Ecol 5: 11–18

Authors' address: Dr. C. C. Lord, Florida Medical Entomology Laboratory, University of Florida, 200 9th St. SE, Vero Beach, FL 32962, U.S.A.

Serological and virological responses in mules and donkeys following inoculation with African horse sickness virus serotype 4

H. El Hasnaoui[1], **M. El Harrak**[2], **S. Zientara**[3], **M. Laviada**[4], and **C. Hamblin**[5]

[1] Laboratoire Régional d'Analyses et de Recherches Vétérinaires, Casablanca, Morocco
[2] Laboratoire Biopharma, Rabat-Akkari, Morocco
[3] Centre National d'Etudes Veterinaires et Alimentaires, Laboratoire Central de Recherches Vétérinaires, Maisons-Alfort, France
[4] Centro de Investigacion en Sanidad Animal Instituto Nacional de Investigacion y Technologia Agraria Alimentaria, Valdeolmos, Madrid, Spain
[5] Institute for Animal Health, Pirbright Laboratory, Woking, Surrey, U.K.

Summary. Two groups, comprising 4 donkeys and 4 mules (group 1) and 4 donkeys and 3 mules (group 2), were used to determine the duration of viraemia and to monitor the development of antibodies following inoculation with African horse sickness virus (AHSV). One group of animals was given a single dose of attenuated AHSV serotype 4 (AHSV 4) vaccine. The second group was inoculated with a virulent field strain of AHSV 4. Both groups were subsequently challenged with the virulent field strain of AHSV 4, 51 and 58 days, respectively, after their primary inoculation. Blood and serum samples, collected on alternate days after the primary inoculations and also after subsequent challenge, were assayed for virus and antibodies. Seven of the 8 AHSV vaccinated (group 1) and 7 of the 7 AHSV inoculated (group 2) animals showed humoral antibody responses after primary inoculation. Although no infectious virus could be isolated from either group for the duration of the study, reverse transcription-PCR data obtained for the second group did show the presence of AHSV viral RNA from as early as day 5 in mules and day 9 in donkeys after the primary inoculation. Viral RNA was detected consistently up to day 47 in some animals and intermittently thereafter. There was no evidence of a second viraemia in any of the animals after challenge. The detection of specific antibodies, against AHSV 4 NS3 protein, in all animals confirmed that both donkeys and mules were infected and that the virus had replicated.

Introduction

African horse sickness virus serotype 4 (AHSV 4) was first confirmed in the north of Morocco in October 1989 [5]. However, despite the stringent control measures implemented by the veterinary authorities, which included extensive and repeated monovalent AHSV 4 vaccination of horses, donkeys and mules, the disease continued to spread southward, re-occurring each year, the last cases being reported in October 1991 [1].

The epidemiology of African horse sickness (AHS) in mules and donkeys is poorly documented as are their responses to vaccination. Furthermore, the effectiveness of vaccination in these species is difficult to assess since they do not usually exhibit significant clinical signs after infection.

These studies were designed to assess viraemia and the development of antibodies in mules and donkeys following inoculation with AHSV 4 vaccine or wild-type AHSV 4, and subsequent to challenge with homologous wild-type virus.

Materials and methods

Animals

Two groups of 4 non-vaccinated donkeys and 4 non-vaccinated mules between 6 and 12 months of age were used. To ensure that these animals were sero-negative, blood samples were collected at weekly intervals for three weeks. Serum was decanted and assayed by virus neutralisation (VN) [4] tests and by competitive ELISA [2] for the presence of antibodies against AHSV. Only seronegative animals were used in the study. One mule that recorded AHSV positive by ELISA was therefore excluded from the studies.

Each group of animals was housed separately in insect proof isolation units at the Laboratoire Officiale de Control de Medicaments et Serum Veterinaire, Rabat, Morocco.

Viruses

The live, attenuated AHSV 4 monovalent vaccine, Lot No R117-4/94 was prepared and supplied by Biopharma, Rabat, Morocco.

The wild-type AHSV was isolated in embryonated hens eggs following intravenous inoculation of 20 μl/egg of homogenised spleen tissue, diluted 1 in 10 in phosphate buffered saline (PBS), that had been taken from a horse that died of AHS in Morocco. This virulent AHSV field strain was designated AHSV 4 H_1E_1 and was used at a concentration of $10^{4.9}$ ELD_{50}/ml.

Study A

The 4 donkeys and 4 mules were each inoculated by the sub-cutaneous route into the side of the neck with a single dose of vaccine according to the manufacturer's instructions. At 51 days post vaccination (dpv) each animal was challenged by sub-cutaneous inoculation into the same side of the neck with 1.0 ml of AHSV 4 H_1E_1. Animals were examined daily. Clinical appearance and rectal temperatures were recorded. Ten millilitres each of heparinised and non-heparinised blood were collected on alternate days until 92 dpv.

Sera were decanted from the non-heparinised blood samples and assayed by VN test [4] for the presence of antibodies against AHSV 4. The red blood cell (RBC) fractions were separated from the heparinised blood samples by centrifugation at 3000 rpm for 10 min, washed, lysed and stored in 50/50 v/v glycerol and distilled water containing 5% phenol and 5% potassium oxalate (OCG). A 1 in 10 dilution of each lysed RBC sample was prepared in PBS and was used for virus isolation. Virus isolation attempts were made by intravenous inoculation of 20 μl of each diluted RBC sample into 5, ten day old embryonated hens eggs, intracerebral inoculation of 20 μl into 2 day old suckling mice and inoculation of 1.0 ml onto monolayer cultures of Vero and BHK cells grown in 175 cm^3 flasks.

Study B

The 4 donkeys and 3 remaining sero-negative mules in this group were each inoculated by the sub-cutaneous route into the side of the neck with 1.0 ml of the virus, AHSV 4 H_1E_1. At 58 days post inoculation (dpi) each animal was challenged by sub-cutaneous inoculation into the same side of the neck with a further 1.0 ml of the same virus. Animals were examined daily. Clinical appearance and rectal temperatures were recorded. Ten millilitres each of ethylenediaminetetracetic acid (EDTA) and non-EDTA blood were collected on alternate days until 98 dpi.

Sera were decanted from the non-EDTA blood samples and assayed by competitive ELISA [2] and VN [4] for the presence of antibodies against AHSV 4 and by an indirect ELISA for specific antibodies against the AHSV non-structural protein 3 (NS3) [6]. The RBC fractions were separated from the EDTA blood samples by centrifugation at 3000 rpm for 10 mins, washed, lysed and stored in OCG. A 1 in 10 dilution of each lysed RBC sample was prepared in PBS and was used for virus isolation as described earlier. Each EDTA blood sample was also tested by reverse transcriptase PCR (RT-PCR) for the presence of AHSV viral RNA [8].

Results

Study A

The antibody responses recorded with sera from mules and donkeys are given in Figs. 1 and 2, respectively. Positive levels of neutralising antibodies were recorded in sera from 3 of the 4 mules from 14 dpv and these animals remained sero-

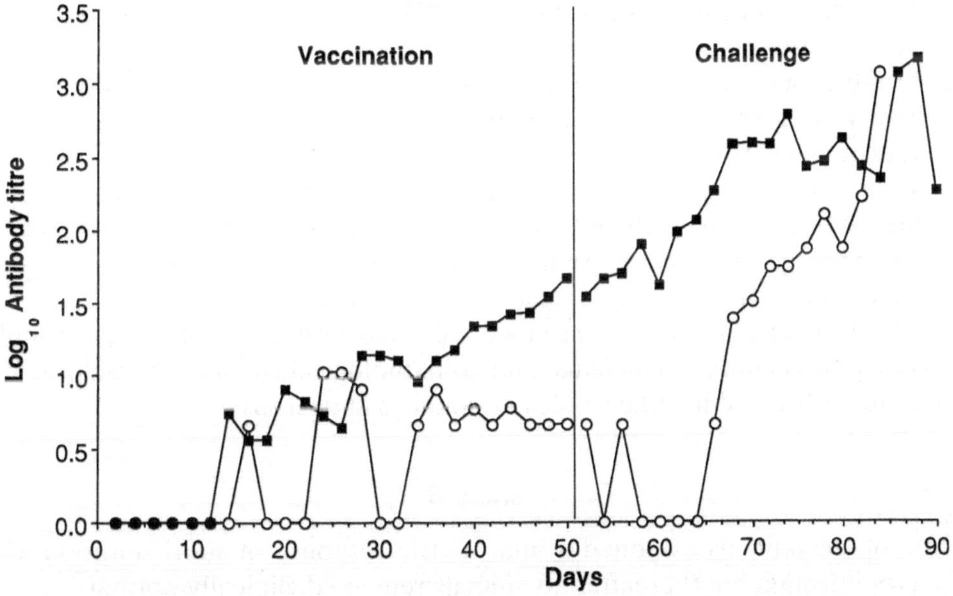

Fig. 1. The development of neutralising antibodies against AHSV 4 in sera taken from mules following vaccination and subsequent AHSV 4 challenge. ■—■ Mean values recorded for the animals that responded to primary vaccination. ○—○ Antibody titres recorded for the single mule that recorded a primary type antibody response after challenge

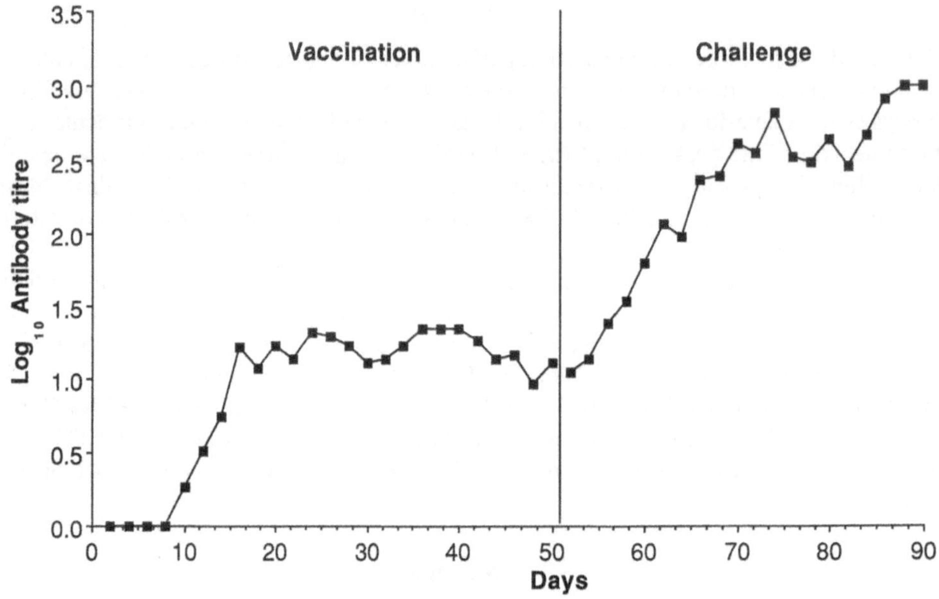

Fig. 2. The development of neutralising antibodies against AHSV 4 in sera taken from donkeys following vaccination and subsequent AHSV 4 challenge. ■—■ Mean value of the antibody titres recorded for the 4 donkeys

positive until the day of challenge (51 dpv). One of the VN positive mules died the day after challenge, however, the clinical signs and pathological changes observed were not consistent with death due to AHSV. Of the 2 surviving sero-positive animals one showed a boost in antibody titre at 7 days post challenge (dpc). No appreciable increase in antibody titre was recorded in sera from the other sero-positive mule after challenge. A primary type antibody response was recorded in the fourth mule at 17 dpc.

Virus neutralising antibodies, albeit low titres, were detected in the sera of all 4 donkeys from between 10 and 16 dpv. All 4 donkeys were sero-positive at the day of challenge (51 dpv). A rise in antibody titre was detected in the sera of all 4 animals between 3 and 9 dpc, achieving levels greater than 10^3.

AHSV was not isolated from blood collected from either donkeys or mules following vaccination or challenge and no pathological changes characteristic of AHS were observed in either mules or donkeys at necropsy.

Study B

Some of the animals exhibited a mild febrile response of short duration after primary infection but thereafter all animals remained clinically normal.

The serological responses in terms of ELISA antibody titre and the virological responses in terms of AHSV viral RNA, for the 4 donkeys and 3 mules, are given in Tables 1 and 2, respectively. ELISA-NS3 and VN antibody titres recorded at weekly intervals are also included. Positive titres of AHSV antibody were detected

Table 1. The immunological and virological responses recorded in serum and blood of 3 mules. Animals were inoculated with a horse virulent strain of AHSV 4 on day 0 and then challenged with homologous virus on day 58

Day	PCR			ELISA			NS3			VN		
	1	2	3	1	2	3	1	2	3	1	2	3
1	–	–	–	–	–	–						
3	–	–	–	–	–	–						
5	+	+	–	–	–	–						
7	+	+	+	–	–	–	–	–	–	–	–	–
9	+	+	–	–	0.7*	–						
11	+	+	+	0.7	0.7	0.7						
13	+	+	+	1.45	1.45	0.7						
15	+	+	+	1.45	1.45	0.7	2.3	2.0	1.4	1.98	1.98	1.74
21	+	+	+	1.45	1.45	1.15	2.3	2.0	1.7	1.98	1.98	2.94
29	+	–	+	1.45	1.45	1.75	2.3	2.0	1.7	2.46	2.34	2.46
31	+	+	+	1.75	2.05	1.75						
33	–	+	+	1.75	1.75	2.05						
35	–	+	+	2.05	2.05	1.75	2.3	2.0	1.7	2.58	2.94	3.18
37	+	+	+	2.05	2.05	1.9						
39	–	–	+	2.35	2.05	2.05						
41	–	+	–	2.35	2.05	2.35						
43	–	–	–	2.65	1.75	2.05	2.3	2.0	1.7	2.7	2.7	3.06
45	–	–	+	2.95	1.75	2.35						
47	+	+	+	2.65	2.35	2.35						
49	–	–	–	2.35	2.05	2.35	2.0	2.0	1.7	1.98	1.98	2.34
51	–	–	+	2.65	NT	2.65						
53	–	–	–	2.65	1.75	2.35						
55	+	–	–	2.5	1.75	2.35						
57	–	–	–	2.65	1.9	2.65	2.0	2.0	1.7	1.98	1.98	2.1
63	–	–	+	2.5	1.9	2.35				NT	1.62	NT
71	–	+	–	2.35	2.05	1.75				1.98	1.62	1.74
98	–	–	–	2.35	2.05	2.2	1.7	2.0	1.4	1.75	1.86	1.68

– Negative; + positive; * reciprocal Log_{10} titre; *NT* not tested

by competitive ELISA in sera from the donkeys between 9 and 13 dpi, and from the mules between 9 and 11 dpi. Thereafter antibodies persisted until the day of challenge (58 dpi). No significant changes in antibody titre were recorded by either ELISA or VN after challenge. AHSV viral RNA was consistently detected by RT-PCR from 11–13 dpi until 27–47 dpi in the RBC fractions from donkeys, and from 5–7 dpi until 31–47 dpi in the RBC fractions from mules, and intermittently thereafter. Positive RT-PCR reactions were also recorded on three separate occasions in RBC fractions collected after challenge from both species although these results did not appear to be related to a second cycle of AHSV replication.

Table 2. The immunological and virological responses recorded in serum and blood of 4 donkeys. Animals were inoculated with a horse virulent strain of AHSV 4 on day 0 and then challenged with homologous virus on day 58

Day	PCR				ELISA				NS3				VN			
	1	2	3	4	1	2	3	4	1	2	3	4	1	2	3	4
1	–	–	–	–	–	–	–	–								
3	–	–	–	–	–	–	–	–								
5	–	–	–	–	–	–	–	–								
7	–	–	–	–	–	–	–	–	–	–	–	–	–	–	–	–
9	–	–	–	+	–	–	0.7*	–								
11	+	+	–	–	–	0.7	0.7	0.7								
13	+	+	+	+	0.85	0.7	–	1.15								
15	+	+	+	+	0.85	0.7	0.7	1.15	1.4	1.4	1.4	1.4	1.14	1.38	1.5	1.14
21	+	+	+	+	1.0	0.7	0.7	0.7	1.7	1.4	1.7	2.0	1.71	1.62	2.1	1.86
23	+	–	+	+	1.15	0.7	0.7	0.7								
25	+	+	+	+	1.15	0.7	0.7	0.7								
27	+	+	+	+	1.45	1.15	1.15	0.7								
29	+	–	+	+	1.45	1.15	1.45	1.0	2.0	1.7	2.0	2.3	2.34	2.1	2.22	2.46
31	+	–	–	+	1.45	1.15	1.3	1.15								
33	–	+	–	+	1.45	1.45	1.45	1.15								
35	+	–	+	+	1.45	1.15	1.15	1.15	2.3	1.7	2.3	2.3	2.46	2.34	2.22	2.58
37	–	–	–	–	1.45	1.45	1.45	1.45								
39	–	–	–	–	1.75	1.3	1.45	1.15								
41	–	–	–	+	1.45	1.45	1.45	1.15								
43	–	+	+	+	1.45	1.45	1.45	1.45	2.3	1.7	2.3	2.0	2.46	2.58	2.22	3.18
45	–	+	+	+	1.75	1.45	1.6	1.45								
47	–	–	–	+	1.75	1.45	1.45	1.45								
49	–	–	–	–	1.75	1.45	1.45	2.05	2.3	2.0	2.3	2.0	2.22	1.86	1.86	2.1
57	–	–	–	–	1.75	1.45	1.15	2.05	2.3	1.7	2.3	2.0	2.22	1.86	1.86	2.1
63	–	–	–	–	1.6	1.45	1.45	1.45					NT	NT	1.86	NT
71	+	–	+	–	1.75	1.45	1.75	1.45					1.74	1.86	1.62	1.86
91	–	–	+	–	1.75	1.75	1.6	1.75					1.74	1.74	1.74	1.86
98	–	–	–	–	1.45	1.75	1.15	1.45	2.0	1.7	2.3	1.7	1.86	1.38	1.74	NT

– Negative; + positive; * reciprocal Log_{10} titre; *NT* not tested

Antibodies against AHSV NS3 were detected in sera from the donkeys and mules from 14 dpi but there were no apparent increases in the competitive ELISA, the VN or the ELISA-NS3 antibody titres after challenge.

AHSV was not isolated from blood collected from either donkeys or mules following primary inoculation or challenge and no pathological changes characteristic of AHS were observed in either mules or donkeys at necropsy.

Discussion

Donkeys and mules, together comprise approximately 80% of the equine population of Morocco. However, despite this preponderance in numbers and the fact that they only rarely exhibit clinical signs of disease, their importance in the perpetuation and spread of AHSV in Morocco and elsewhere remains uncertain. The present work, in conjunction with related studies reported elsewhere [3, 7], has been designed to help clarify the role of donkeys and mules in the epidemiology of AHS.

Neither pyrexia nor other clinical signs of infection were observed in any of 8 animals following vaccination or after subsequent challenge. A mild febrile response, however, was recorded in some individuals of a second group of 7 animals, after a primary inoculation of an AHSV strain virulent to horses but this was of short duration and in the absence of other clinical signs is unlikely to have been noticed under field conditions.

No infectious AHSV was isolated from any of the blood samples collected from either group of animals throughout the study. This negative result is an anomaly since other workers [3, 7] using similar isolation methods have previously demonstrated viraemia in donkeys after inoculation of AHSV strains virulent for horses. It is possible that in the present study the test samples contained only minimal amounts of virus, that they had been stored/transported under sub-optimal conditions or that, by chance, they just happened to be virus negative. In this context, the development of antibodies against AHSV NS3 in both mules and donkeys within the second group, is a significant finding since AHSV NS3 antibodies would have only been produced in sufficient amounts for detection by ELISA if the infecting virus had replicated in these animals [6]. Furthermore, the positive RT-PCR results obtained for RBC samples from both mules and donkeys confirms the presence of AHSV viral RNA in the blood of both equine species over an extended period of time, between 5 and 55 dpi, although this extended interval cannot be interpreted as being the duration of an infectious virus viraemia.

A specific humoral AHSV antibody response was recorded in 4 out of 4 donkeys and 3 out of 4 mules after primary vaccination with the live attenuated vaccine virus, however, the titres recorded were generally low. The validity of these titres was confirmed when all 7 animals showed a, secondary type, boost in antibodies after challenge. Differences in the timing of these secondary responses were possibly due to individual animal variation and/or a lower rate of virus replication in mules and donkeys than in horses [3]. The low variable levels of antibodies detected in one mule after vaccination and the subsequent primary antibody response on challenge is of concern and might suggest that AHSV vaccines which are designed specifically with the horse as a target species, are less efficacious in non-horse equines. Clearly further, larger scale, studies are necessary to assess the levels of efficacy attained by AHSV vaccines in non-horse equines and to determine the optimum dose required to stimulate an effective antibody response in these species. In contrast to the vaccinated group, no boost in antibody titres were detected in sera collected from the second group of mules and donkeys after challenge. This result would suggest that the challenge virus was rapidly neutralised and did not initiate a replicating infection at the site of inoculation or in surrounding tissues.

References

1. Baylis M, Hasnaoni H, Bouayoune H, Touti J, Mellor PS (1997) The spacial and seasonal distribution of African horse sickness and its potential *Culicoides* vectors in Morocco. Med Vet Entomol 11: 203–212
2. Hamblin C, Graham SD, Anderson EC, Crowther JC (1990) A competitive ELISA for the detection of group-specific antibodies to African horse sickness virus. Epidemiol Infect 104: 303–312
3. Hamblin C, Salt JS, Mellor PS, Graham SD, Smith PR, Wohlsein P (1998) Donkeys as a reservoir of African horse sickness. In: Mellor PS, Baylis M, Hamblin C, Calisher CH, Mertens PPC (eds) African horse sickness. Springer, Wien New York, pp 37–47 (Arch Virol [Suppl] 14)
4. Hasnaoui H, Harrak M, Tber A, Fikri A, Laghzaoui K, Bikour MH (1998) Application of an indirect fluorescent antibody assay for the detection of African horse sickness virus antibodies. In: Mellor PS, Baylis M, Hamblin C, Calisher CH, Mertens PPC (eds) African horse sickness. Springer, Wien New York, pp 29–36 (Arch Virol [Suppl] 14)
5. Lhafi A, Tber A, Fikri A, Laghzaoui K (1992) African horse sickness in Morocco: the epizootics of 1989 and 1990. In: Walton TE, Osburn BI (eds) Bluetongue, African horse sickness and related orbiviruses. Proc 2nd Int Symp. CRC Press, Boca Raton, pp 205–216
6. Laviada MD, Roy P, Vizcaino-Sanchez JM, Casal JI (1995) The use of African horse sickness virus NS3 protein, expressed in bacteria, as a marker to differentiate infected from vaccinated horses. Virus Res 38: 205–218
7. Venter GJ, Williams R, Nevill EM (1992) The value of donkeys as sentinels for African horse sickness. In: Walton TE, Osburn BI (eds) Bluetongue, African horse sickness and related orbiviruses. Proc 2nd Int Symp. CRC Press, Boca Raton, pp 825–831
8. Zientara S, Sailleau C, Moulay S, Wade-Evans A, Cruciere C (1995) Application of the polymerase chain reaction to the detection of African horse sickness viruses. J Virol Methods 53: 47–54

Authors' address: Dr. H. El Hasnaoui, Laboratoire Régional d'Analyses et de Recherches Vétérinaires, 43 Rue Nichokra Rahal, B.P. 13719, Casablanca, Morocco.

Donkeys as reservoirs of African horse sickness virus

C. Hamblin[1], **J. S. Salt**[1], **P. S. Mellor**[1], **S. D. Graham**[1], **P. R. Smith**[1], and **P. Wohlsein**[2]

[1] Institute for Animal Health, Pirbright Laboratory, Woking, Surrey, U.K.
[2] Institute of Pathology, School of Veterinary Medicine, Hannover, Germany

Summary. Investigations have been carried out to elucidate the possible role of the donkey in the epidemiology of African horse sickness (AHS). These studies have shown that despite the absence of pyrexia or other observable clinical signs, donkeys become infected with virulent AHS virus serotype 4 (AHSV 4) and that they develop a viraemia which can persist for at least 12 days, albeit at a comparatively lower titre than that recorded for similarly infected ponies. AHSV 4 showed a similar tissue tropism in the pony and donkey but the virus appeared to replicate less efficiently in donkey tissues. The only gross pathological changes observed in the donkeys *post mortem* were increased fluid accumulation in the serosal lined compartments, particularly the peritoneal cavity, and petechial and ecchymotic haemorrhages on the left hepatic ligament. The absence of infectious virus or viral antigens in any of the tissues collected at 14 and 19 days post inoculation (dpi) from 6 experimental donkeys suggest that, though susceptible to infection, the donkey is unlikely to be a long term reservoir for AHSV. Although AHSV 4 was detected in all 6 donkeys following the primary inoculation, no virus could be isolated from blood collected from two donkeys subsequently challenged with a second virulent virus, AHSV 5. Data generated from virus neutralisation tests showed a second primary antibody response, against AHSV 5, in these donkeys at 12 dpi. In contrast, the boost in antibody levels detected from 5 dpi, as measured by ELISA, was probably due to an anamnestic response against the AHSV group-specific viral proteins. Homogenised spleen tissue, collected *post mortem* from a donkey 7 dpi with AHSV 4, caused a lethal, cardiac form of AHS when inoculated into a susceptible pony.

Introduction

African horse sickness (AHS) is caused by a double stranded RNA virus belonging to the family *Reoviridae*, genus *Orbivirus* [11]. The virus is infectious but non contagious and is transmitted by biting arthropods, primarily *Culicoides* species. The virus is usually restricted to sub-Saharan Africa and the southernmost regions of the Middle East. However, the virus does make occasional excursions outside the endemic regions, the most recent being the un-precedented epizootic involving Spain, Portugal and Morocco, which occurred between 1987 and 1991 [14]. Despite repeated and widespread vaccination of equidae, African horse sickness virus (AHSV) persisted in one or more of these countries for 5 years.

The role of the donkey in the epidemiology of AHS during an epizootic is poorly defined. Previous workers [2, 3] have reported some clinical disease and deaths of donkeys during epizootics of AHS in the Middle East. Similarly, Theiler [15, 16] stated that although asses are reported to have died in Africa during 'bad horse sickness years' he had never seen an ass succumb to an 'injection of virus'. Despite these reports, it is generally accepted that donkeys are very resistant and most infections are sub-clinical [4]. Only two relevant reports of controlled studies involving AHSV and donkeys have been identified [2, 17]. The first paper [2] relates to the infection of a susceptible horse with blood collected from a donkey suspected of being infected with AHSV. The horse developed the fever form of AHS which was characterised by a febrile reaction and marked oedema of the supraorbital fossae, thereafter it made a complete recovery. The second group of workers [17] demonstrated the susceptibility of donkeys to infection with three different serotypes of AHSV. They failed to show any pyrexia or other clinical signs but they did detect viraemia of short duration in some of the infected animals.

Other authors have reported serological evidence of AHSV infection in donkeys and some have even speculated on the significance of this species in the epidemiology of AHS ([1], Hamblin, unpublished data from IAH, AHS Regional Reference Centre).

In countries like Morocco where there is a very high density of non-horse equines, donkeys may be important not only as silent maintenance hosts of AHSV but also in the epidemiology of these viruses. Investigations were therefore initiated to study AHSV serotype 4 (AHSV 4) infection in donkeys and to determine the possible role of this species in the epidemiology of AHS.

Materials and methods

Viruses

The AHSV 4 was originally isolated from spleen tissue taken from a horse which had died in Morocco during the 1991 epizootic. This virus was subsequently passaged on two occasions by intravenous inoculation of 11 day old embryonated hens eggs and once by subcutaneous (SC) inoculation into a pony. Spleen tissue collected *post mortem* from this pony was homogenised in an equal volume of Eagle's medium and lyophilised to provide the stock seed virus designated AHSV 4 $H_1E_2H_1$.

The AHSV 5 was originally obtained from Dr. B. Erasmus, Onderstepoort Veterinary Institute, S. Africa as lyophilised mouse brain passage 3 (AHSV 5 MB_3). This virus isolate was further passaged on three occasions in BHK tissue culture cells to provide AHSV 5 MB_3BHK_3. Abbreviations used in the passage history are H = horse, E = embryonated chicken egg, MB = mouse brain and BHK = baby hamster kidney cell line.

Animals and inoculation regime

Two 6 month and four 3–4 year old donkeys, and one 6 month, one 9 month and two 3–4 year old ponies were each housed in individual boxes of an insect proof isolation unit.

Study 1

Two 6 month old and two 3–4 year old donkeys numbered D1, D2 and D3, D4, respectively, and one 6 month old and two 3–4 year old ponies numbered P1 and P2, P3, respectively, were inoculated with a 2.0 ml dose of AHSV 4 $H_1E_2H_1$ containing $10^{5.8}$ ELD_{50}/ml. The four donkeys and the 6 month old pony were each given the inoculum by the intravenous (IV), intradermal (ID) and the SC routes (0.8 ml IV, 0.2 ml ID and 1.0 ml SC) whereas the two adult ponies were given the 2.0 ml dose by the IV route.

Study 2

Two 3–4 year old donkeys (D5 and D6) were inoculated with 2.0 ml of AHSV 4 $H_1E_2H_1$ containing $10^{5.8}$ ELD_{50}/ml into a single site on the side of the neck by the SC route. Fifty-four days after the primary inoculation these donkeys were re-inoculated with 2.0 ml of AHSV 5 MB_3BHK_3 containing $10^{7.5}$ ELD_{50}/ml by the IV route.

Study 3

One 9 month old pony (P4) was inoculated with a 2.0 ml dose of homogenised spleen suspension collected *post mortem* from donkey D2 (*Study 1*). The inoculum designated AHSV 4 $H_1E_2H_1D_1$, contained $10^{4.3}$ ELD_{50}/ml and was inoculated into three sites as detailed for the donkeys in Study 1.

Sampling

All equidae were examined daily and their clinical appearance and rectal temperatures recorded. Heparinised and non-heparinised blood samples were collected on three occasions each week. Sera were decanted from the clotted blood and stored at –20°C until required. The heparinised blood was washed, lysed and stored at 4°C [9].

Gross pathological changes observed for each animal were recorded *post mortem*. Tissue samples were collected from all animals for virological assay. A 20% suspension of each tissue was prepared by grinding 1.0 g of tissue in a mortar using a pestle with sterile sand and 4.0 ml of Eagle's medium. The homogenised tissue samples were clarified by centrifugation at 3000 **g** for 5 minutes.

Virological and serological assays

Titres of infectious AHSV in the lysed blood samples were determined by IV inoculations of 11 day old embryonated hens eggs [6]. Virus titres were calculated using the method of Spearmann and Kärber [5] and were expressed as ELD_{50}/ml of blood.

The titre of infectious virus in each homogenised tissue sample was determined by intracerebral inoculation of 2 day old suckling mice [9]. Virus titres were calculated using the method of Spearmann and Kärber [5] and were expressed as MLD_{50}/g of tissue.

Suspensions of homogenised tissue were assayed for viral antigen by indirect sandwich ELISA [8]. Samples which titrated and gave absorbance values at 492 nm above 0.15 for at least two consecutive dilutions were recorded as positive. Antigen titres were calculated using the method of Spearmann and Kärber [5] and were expressed as $\log_{10}$/g of tissue.

Serogroup specific antibodies against AHSV were detected by titration in competitive ELISAs [7]. End-point titres were expressed as reciprocal $\log_{10}$ of the last dilution of serum giving 50% reduction of the absorbance values recorded in the control wells [12]. Sera giving end-point titres above $\log_{10}$ 0.7 were considered positive.

Serotype specific antibodies against AHSV 4 and 5 were determined by virus neutralisation (VN) tests. Assays were performed in microtitre plates using BHK-21 tissue culture cells [10]. Antibody titres were expressed as reciprocal $\log_{10}$ of the final dilution of serum in the serum/virus mixtures at the 50% end-point [12].

Results

Study 1

No clinical signs were observed nor was pyrexia recorded in any of the donkeys (D1–D4) following inoculation of AHSV 4 into three different sites. Table 1 shows the development and titres of viraemia and antibodies recorded for each animal. The 6 month old donkeys (D1 and D2) were both euthanized 7 days post inoculation (dpi) while the older donkeys (D3 and D4) were euthanized separately on days 14 and 19, respectively, after infection. The gross pathological changes observed *post mortem* for the four donkeys were similar and included accumulation of fluid in the peritoneum (30–150 ml) with or without fibrin clots and a slight increase in fluid in the pericardial sac and thoracic cavity. Petechial and ecchymotic haemorrhages were observed on the left triangular hepatic ligament of D3 and D4. Petechial capsular/sub-serosal hepatic haemorrhages were also present in D3. No other pathological changes consistent with AHS were observed. The titres of infectious AHSV and viral antigen in representative tissues collected *post mortem* from the four donkeys euthanized sequentially after infection are given in Table 2.

Table 1. The detection of infectious virus and viral antibodies in donkeys following infection with AHSV serotype 4

	Donkey 1		Donkey 2		Donkey 3		Donkey 4	
DPI	virus ELD_{50}	antibody ELISA	virus ELD_{50}	antibody ELISA	virus ELD_{50}	antibody ELISA	virus ELD_{50}	antibody ELISA
1	NVD	–[a]	NVD	–	NVD	–	NVD	–
2	NVD	–	NVD	–	NVD	–	NVD	–
3	NVD	–	NVD	–	NS	NS	NS	NS
5	NS	NS	NS	NS	NVD	–	1.7	–
6	1.7	–	2.3	–	NS	NS	NS	NS
7	2.9	–	2.1	–	1.5	–	1.9	–
9					1.9	0.85	1.7	0.85
10					NS	NS	NS	NS
12					1.7	0.85	1.9	1.45
14					NVD	0.85	1.7	1.45
16							1.5	1.75
19							NVD	1.75

DPI Days post infection; *ELD_{50}* $\log_{10}$ Egg LD_{50}/ml; *ELISA* $\log_{10}$ end-point titre; *NVD* no virus detected at 10^{-1} dilution; *NS* no sample

[a] ELISA titres < 0.7

Pyrexia above 39.0°C was recorded from day 3 in the two older ponies (P1 and P2). Slight oedema of the supraorbital fossae was evident in these animals from days 4 and 5, respectively, after infection. The condition of the ponies declined rapidly and by day 8 both animals had congested conjunctival and oral membranes and were dyspnoeic. The temperatures of these animals dropped rapidly to 35.7°C and 37.5°C, respectively. Respiration was 60–70 per minute and the mucous membranes were cyanotic. Viraemia was detected in both animals from day 5 onward (Table 3). Both ponies were euthanized on welfare grounds 8 dpi at the terminal stage of the peracute, pulmonary form of disease. The gross pathological changes were similar in both ponies and included generalised oedema between the fascial planes particularly in the neck and around the oesophagus, straw coloured pleural and pericardial effusions, interlobular oedema and petechial haemorrhages on the epicardium and endocardium of the interventricular septum, the left ventricle and the atrio-ventricular valves. Additional findings in P2 included haemorrhages on the serosal surface of the small intestine and

Table 2. Distribution of infectious AHSV and viral antigens in tissues collected from donkeys following infection with AHSV 4

Tissues	Donkey 1 (7 dpi)		Donkey 2 (7 dpi)		Donkey 3 (14 dpi)		Donkey 4 (19 dpi)	
	ELISA	MLD_{50}	ELISA	MLD_{50}	ELISA	MLD_{50}	ELISA	MLD_{50}
Blood[a]	NT	2.9	NT	2.3	NT	NVD	NT	NVD
Respiratory system								
Lung	1.75	4.2	1.45	3.4	–	NVD	–	NVD
Cardio-vascular system								
Heart	–	NVD	–	NVD	–	NVD	–	NVD
Haemopoeitic system								
Bone marrow	–	NVD	–	NVD	NS	NS	NS	NS
Lymphoid system								
Spleen	1.45	4.6	2.65	5.2	–	NVD	–	NVD
Mediastinal L.N.	–	4.4	0.85	4.2	–	NVD	–	NVD
Tonsil	2.5	4.8	–	NVD	–	NVD	–	NVD
Digestive system								
Salivary gland	–	NVD	–	NVD	–	NVD		NVD
Liver	–	2.7	–	NVD	–	NVD	–	NVD
Stomach	–	NVD	–	NVD	–	NVD	–	NVD
Ileum		NVD	–	NVD	–	NVD	–	NVD
Colon	–	3.2	–	NVD	–	NVD	–	NVD
Endocrine system								
Adrenal gland	–	1.7	–	NVD	–	NVD	–	NVD
Thyroid	–	NVD	–	NVD	–	NVD	NS	NS

[a] Blood Log_{10} Egg LD_{50}/ml; *dpi* days post infection; *ELISA* titre Log_{10}/gm; *MLD_{50}* Log_{10} Mouse LD_{50}/gm; *NT* not tested; *NVD* no virus detected at 10^{-1} dilution; – no antigen detected in 20% homogenised tissue suspension; *NS* no sample

Table 3. Pyrexia, viraemia and antibody responses recorded in ponies following infection with AHSV serotype 4

	Pony 1		Pony 2		Pony 3		
DPI	Temp. °C	Virus ELD_{50}	Temp. °C	Virus ELD_{50}	Temp. °C	Virus ELD_{50}	Antibody ELISA
0	38.1	NVD	38.6	NVD	38.4	NVD	–[a]
1	38.4	NS	38.4	NS	38.5	NS	NS
2	38.6	NVD	38.5	NVD	38.7	NVD	–
3	39.1	NS	39.2	NS	38.6	NVD	–
4	39.3	NS	39.5	NS	38.4	NS	NS
5	40.4	4.1	40.3	4.1	39.2	NS	NS
6	37.5	NS	39.7	NS	39.0	2.3	–
7	35.7	4.3	37.5	4.5	39.2	NS	NS
8					39.5	2.9	–
9					39.5	NS	NS
10					39.6	3.5	0.85

DPI Days post infection; ELD_{50} Log_{10} Egg LD_{50}/ml; *NVD* no virus detected at 10^{-1} dilution; *NS* no sample

[a] ELISA titres < 0.7

increased articular fluid in the hip joints. No other pathological lesions were observed in these animals. Table 4 shows the quantities of AHSV and viral antigen present in representative tissues collected from each animal *post mortem*.

A persistent, mild pyrexia was detected in pony P3 from 5 dpi, however, no other clinical signs were observed. The pony was euthanized at 10 dpi while exhibiting the fever form of AHS. Viraemia was detected from day 6 until the day of euthanasia (Table 3). Gross pathological changes included an accumulation of straw coloured fluid in the peritoneal cavity (approximately 100 ml), slight oedema and haemorrhage in the lung, haemorrhage on the serosal surface of the ileum, oedema and haemorrhage of the mesenteric lymph nodes and increased articular fluid in the atlantoaxial and hip joints. The titres of infectious AHSV and AHS viral antigen present in representative tissues collected *post mortem* are given in Table 4.

Study 2

No clinical signs were observed or pyrexia recorded for donkeys D5 or D6 following infection with AHSV 4 or subsequent challenge with AHSV 5. Data expressed in Table 5 show that a viraemia was detected from days 5 and 10, respectively, which persisted in both animals until day 14. Table 5 also shows the comparison of the serological responses recorded for each donkey in terms of serogroup and serotype specificity against the two AHSV serotypes.

The donkeys were euthanized 14 and 19 days after the second virus inoculation. The gross pathological changes observed *post mortem* for donkeys D5 and D6 were similar and included straw coloured fluid in the peritoneum (approxi-

Table 4. Distribution of infectious AHSV and viral antigens in tissues collected from ponies following infection with AHSV 4

	Pony 1 (8 dpi)		Pony 2 (8 dpi)		Pony 3 (10 dpi)	
Tissues	ELISA	MLD_{50}	ELISA	MLD_{50}	ELISA	MLD_{50}
Blood[a]	NT	4.3	NT	4.5	NT	3.5
Respiratory system						
Lung	>2.95	5.6	>2.95	5.4	1.75	5.0
Cardio-vascular system						
Heart	2.8	NVD	>2.95	2.4	1.45	NVD
Haemopoeitic system						
Bone marrow	NS	NS	NS	NS	–	NVD
Lymphoid system						
Spleen	2.8	5.4	2.2	5.6	>2.95	6.6
Mediastinal L.N.	1.6	4.2	1.6	3.8	1.45	4.5
Tonsil	NS	NS	1.9	NT	–	4.8
Digestive system						
Salivary gland	1.45	4.0	1.6	5.0	–	3.6
Liver	2.65	2.8	2.2	4.4	1.75	3.4
Stomach	1.15	3.4	1.45	3.8	–	NVD
Ileum	2.2	3.8	1.75	3.4	–	3.4
Colon	1.9	3.4	1.6	NT	–	3.4
Endocrine system						
Adrenal gland	1.75	3.8	1.3	3.4	–	NVD
Thyroid	1.15	NT	1.9	3.8	–	2.8
Urinary system						
Bladder	NS	NS	NS	NS	–	NVD
Kidney	2.05	4.0	1.6	4.6	–	3.6

[a] Blood Log_{10} Egg LD_{50}/ml; *dpi* days post infection; *ELISA* titre Log_{10}/gm; *MLD_{50}* Log_{10} Mouse LD_{50}/gm; *NT* not tested; *NVD* no virus detected at 10^{-1} dilution; – no antigen detected in 20% homogenised tissue suspension; *NS* no sample

mately 100 ml) without fibrin clots and a slight increase in clear fluid in the pericardial sac. The greater omentum was also hyperaemic. No other pathological changes consistent with AHS were observed. Further, no infectious AHSV or viral antigens could be detected in any of the tissues collected *post mortem*.

Study 3

Pyrexia above 39.0°C was recorded in pony P4, 6 days after inoculation of homogenised spleen tissue collected 7 dpi from donkey D2. The rectal temperature peaked on the following day at 39.8°C, and subsequently fell on day 8 to 36.1°C (Table 6). Clinical signs on day 7 were limited to injection of the sclerae. Thereafter the condition of the animal declined rapidly and by the afternoon of

Table 5. The detection of infectious virus and viral antibodies in donkeys following sequential infection with AHSV serotypes 4 and 5

	Donkey 5				Donkey 6			
DPI	ELD_{50}	ELISA	AHSV 4 VN	AHSV 5 VN	ELD_{50}	ELISA	AHSV 4 VN	AHSV 5 VN
0*	NVD	–[a]	–[b]	–[b]	NVD	–[a]	–[b]	–[b]
3	NVD	–	–	–	NVD	–	–	–
5	1.7	–	–	–	NVD	–	–	–
7	1.7	–	–	–	NVD	–	–	–
10	4.1	–	–	–	1.9	–	–	–
12	2.3	0.85	–	–	2.0	–	1.05	–
14	1.75	0.85	1.65	–	2.5	0.85	1.2	–
17	NVD	0.7	1.65	–	NVD	0.85	1.2	–
19	NVD	1.0	1.65	–	NVD	1.15	1.05	–
21	NVD	1.15	1.8	–	NVD	1.6	1.05	–
24	NVD	1.9	1.8	–	NVD	1.45	1.8	–
33	NVD	1.75	1.65	–	NVD	1.45	1.8	–
35	NVD	1.75	1.65	–	NVD	1.45	1.8	–
54/0**	NVD	1.6	1.5	–	NVD	1.45	1.8	–
56/2	NVD	1.45	1.65	–	NVD	1.45	1.65	–
59/5	NVD	1.75	1.65	–	NVD	1.75	1.65	–
61/7	NVD	2.2	1.8	–	NVD	2.05	1.8	–
63/9	NVD	2.35	2.1	–	NVD	2.35	1.5	–
66/12	NVD	2.5	2.1	0.9	NVD	2.65	1.8	0.9
68/14	NVD	2.65	1.95	1.5	NVD	2.65	1.8	1.35
70/16					NVD	2.65	1.95	1.2
73/19					NVD	2.65	2.1	1.65

DPI Days post infection; *primary inoculation AHSV 4; **challenge virus AHSV 5; ELD_{50} Log_{10} Egg LD_{50}/ml; *ELISA* Log_{10} end-point titre; *VN* virus neutralisation Log_{10} end-point titre; *NVD* no virus detected at 10^{-1} dilution

[a] ELISA titre < 0.7

[b] VN titre < 0.75

day 8 the pony was depressed, anorexic and unsteady. Within four hours the pony became recumbent and was subsequently euthanized on humane grounds at the terminal stage of the peracute, cardiac form of disease. At the time of death the pony was tachypnoeic, the respiratory rate was approximately 60 per minute, and the congested mucous membranes were becoming cyanotic. Heart sounds were inaudible and peripheral pulses were not palpable.

Oedema was seen *post mortem* as straw coloured fluid between the fascial planes of the neck musculature and ligaments, around the oesophagus, in the axillary lymph nodes, in the hilus of the kidneys, in both the gastrophrenic and gastrosplenic ligaments and distributed widely in sub-cutaneous sites. Increased articular fluid was present in the atlantoaxial and hip joints. The pericardial sac contained

Table 6. Pyrexia, viraemia and antibody responses recorded in a pony following infection with AT-TSV serotype 4 derived from an infected donkey

DPI	Temperature °C	Virus ELD_{50}	Antibody ELISA
0	37.9	NVD	–[a]
1	37.9	1.5	–
2	38.3	NS	NS
3	38.8	3.1	–
4	38.8	NS	NS
5	38.3	NS	NS
6	39.4	5.1	–
7	39.8	NS	NS
8	36.1	5.1	–

DPI Days post infection; *ELD_{50}* Log_{10} Egg LD_{50}/ml; *ELISA* Log_{10} end-point titre; *NVD* no virus detected at 10^{-1} dilution; *NS* no sample

[a] ELISA titre < 0.7

approximately 100 ml of straw coloured fluid but there was no observed increase in peritoneal or pleural fluid, and the lungs appeared normal. Marked sub-epicardial haemorrhages were found over the left ventricle, which extended into the right ventricle. Haemorrhages were seen throughout the digestive tract, more particularly on the mucosae of the proventricular and pyloric regions of the stomach and the jejunum. Several mesenteric lymph nodes were haemorrhagic, often associated with haemorrhages in the mesenteric border of the bowel. Haemorrhages were also noted in the renal cortices and medullae, the gastrosplenic ligament and the peritoneal adipose tissue. There was inflamation of the greater omentum.

Discussion

The paucity of information relating to the potential role of donkeys in the geographical persistence and transmission of AHSV is probably due, at least in part, to the virtual absence of overt clinical signs following infection with the virus. These studies address that shortfall and provide new information on the possible role of donkeys in the epidemiology of AHS.

The AHSV 4 isolate used in these studies caused severe clinical disease and gross pathology in ponies, however, no overt disease at all was observed in any of the 6 donkeys inoculated similarly. The amounts of infectious virus recovered from the tissues collected *post mortem* from these ponies and donkeys shows that AHSV has a tropism for respiratory and lymphatic tissues. Despite the generally lower titre of viraemia in donkeys compared to ponies, similar high titres of AHSV and viral antigen were recorded in lung and lymphatic tissues collected from both species. Heart muscle collected from the ponies also contained large amounts of AHSV antigen but infectious virus was either absent or only present at low titre in this organ. It is likely that infectious virus in heart tissue is rapidly inactivated by lactic acid which is produced in this organ after death. In contrast to the ponies, no AHSV antigen could be detected in the heart tissue from donkeys. This might indicate a difference in the virus tropism for heart muscle between the 2 species but whether this a significant factor in the degree of pathology caused in each species is unknown.

Since donkeys rarely suffer from overt AHS and are considered to be of less value than horses, vaccination cover during an epizootic is likely to be less comprehensive in the donkey population than that for the horse population. Furthermore, the currently available vaccines are developed specifically for horses and therefore may be less efficacious for donkeys, thereby leaving some animals inadequately 'protected'. These factors could have important epidemiological implications, particularly since our studies have shown that viraemia does occur in donkeys and can persist for at least 12 days with virus titres up to $10^{4.1}$ ELD_{50}/ml. Mellor [13] estimated that a viraemia greater than $10^{4.0}$ of infectious bluetongue virus per ml of blood would be sufficient to infect a blood feeding vector *Culicoides* midge. If this value is similar for AHSV then donkeys must be considered a potential source of virus for insect vectors. It has also been shown here, that the AHSV 4 recovered from an infected, but apparently healthy donkey was lethal for a pony. Hence, infected donkeys could act as 'silent reservoirs' of AHSV, helping to maintain and spread the virus, even after vaccination.

Although a viraemia was detected in all 6 donkeys following inoculation with AHSV 4, no secondary viraemia could be detected when 2 of these animals were challenged with a second AHSV serotype, AHSV 5. The VN data confirm that AHSV 5 did replicate sufficiently to elicit a primary, AHSV 5 type-specific response at 12 to 14 dpi. However, ELISA and AHSV 4 VN data also record a boost in antibody titre from 5 and 9 dpi, respectively. This earlier boost in the antibody status of the 2 donkeys, even though apparently specific to AHSV 4, appears to have inhibited the development of a detectable viraemia to AHSV 5. Previous experience of one AHSV serotype, therefore, may not prevent initiation of infection with a second but the presence of sufficiently high levels of antibody to the first unrelated serotype may prevent or reduce the level of viraemia with the second serotype thereby precluding the animal from playing any further part in the epidemiology of the disease.

Acknowledgements

The authors would like to thank Christine Chisholm and Andrea Dearden for secretarial assistance, Sheila Shrigley for information services and Martin Broomfield for his assistance with animals in the isolation unit.

References

1. Abu El Zein EME, Mirghani ME, Ali BE (1989) Observations on African horse sickness in donkeys in the Sudan. Rev Sci Tech Off Int Epiz 8: 785–787
2. Alexander RA (1948) The 1944 epizootic of horse sickness in the Middle East. Onderstepoort J Vet Sci 23: 77–92

3. Carpano M (1931) African horse sickness as observed particularly in Egypt and Eritrea. Ministry of Agrig Tech and Science Service (Vet Service) Bulletin 115. Government Press, Cairo
4. Coetzer JAW, Erasmus BJ (1994) African horse sickness. In: Coetzer JAW, Thomson GR, Tustin RC (eds) Infectious diseases of livestock, vol 1. Oxford University Press, Oxford, pp 460–475
5. Finney DJ (1964) Statistical methods in biological assay, 2nd ed. Griffin, London, pp 524–553
6. Goldsmit L, Barzilai E, Tadmor A (1975) The comparative sensitivity of sheep and chicken embryos to bluetongue virus and observations on viraemia in experimentally infected sheep. Aust Vet J 151: 190–196
7. Hamblin C, Graham SD, Anderson EC, Crowther JR (1990) A competitive ELISA for the detection of group-specific antibodies to African horse sickness virus. Epidemiol Infect 104: 303–312
8. Hamblin C, Mertens PPC, Mellor PS, Burroughs NJ, Crowther JR (1991) A serogroup specific enzyme-linked immunosorbent assay (ELISA) for the detection and identification of African horse sickness viruses. J Virol Methods 31: 285–292
9. Hamblin C, Anderson EC, Mellor PS, Graham SD, Mertens PPC, Burroughs JN (1992) The detection of African horse sickness virus antigens and antibodies in young equidae. Epidemiol Infect 108: 193–201
10. Herniman KAJ, Gumm ID, Owen L, Taylor WP, Sellers RF (1980) Distribution of bluetongue virus and antibodies in some countries of the eastern hemisphere. Bull Off Int Epiz 92: 581–586
11. Holmes HI, Boccardo G, Estes MK, Furuichi MK, Hoshino Y, Joklik WK, McCrae M, Mertens PPC, Milne RG, Samal KSK, Shikata E, Wilton JR, Uyeda I, Nuss DL (1995) Family *Reoviridae*. In: Murphy FA, Fauquet CM, Bishop DHL, Ghabrial SA, Jarvis AW, Martelli GP, Mayo MA, Summers MD (eds) Virus Taxonomy. Classification and Nomenclature of Viruses. Sixth Report of the International Committee on Taxonomy of Viruses. Springer, Wien New York, pp 208–239 (Arch Virol [Suppl] 10)
12. Kärber G (1931) Beitrag zur kollektiven Behandlung pharmakologischer Reihenversuche. Arch Exp Path Pharmak 162: 480–483
13. Mellor PS (1990) The replication of bluetongue virus in *Culicoides* vectors. Curr Top Microbiol Immunol 162: 143–161
14. Mellor PS (1993) African horse sickness: transmission and epidemiology. Vet Res 24: 199–212
15. Theiler A (1921) African horse sickness (Pestis equorum). 5th Afr Dept Ag Sci Bull No 19
16. Theiler A (1930) African horse sickness. Medical Research Council. A system of bacteriology in relation to medicine, vol 7, chapter 31: Virus diseases – Bacteriophages. HSMO, London, pp 362–375
17. Venter GJ, Williams R, Nevill EM (1992) The value of donkeys as sentinels for African horse sickness. In: Walton TE, Osburn BI (eds) Bluetongue, African horse sickness and related orbiviruses. Proc 2nd Int Symp Paris 1991. CRC Press, Boca Raton, pp 825–831

Authors' address: Dr. C. Hamblin, Institute for Animal Health, Pirbright Laboratory, Ash Road, Pirbright, Woking, Surrey, GU24 ONF, U.K.

Clinical, virological and immune responses of normal and immunosuppressed donkeys *(Equus asinus africanus)* after inoculation with African horse sickness virus

O. Fassi-Fihri[1], **M. El Harrak**[2], and **M. M. Fassi-Fehri**[1]

[1] Département de Microbiologie, Immunologie et Maladies Contagieuses, Institut Agronomique et Vétérinaire Hassan, Rabat-Instituts, Rabat, Morocco
[2] Biopharma, Rabat Akkari, Rabat, Morocco

Summary. To elucidate the role that donkeys may play in African horse sickness virus (AHSV) persistence during inter-epizootic periods we looked for clinical signs of infection and studied the viraemia and neutralising antibody kinetics in 3 immunocompetent and 3 immunosuppressed donkeys inoculated with AHSV-4. None of the donkeys developed signs of AHS. However infectious AHSV was isolated from the blood of the immunocompetent donkeys for up to 17 days post infection (dpi) and viral antigens were detected for up to 28 dpi. Immune cells also increased significantly from 35 to 60 dpi. There was no evidence of a recrudescence of viraemia following immunosuppression of these donkeys at 90 dpi despite a decrease in the numbers of immune cells. Infectious virus was not isolated from the blood of donkeys that had been immunosuppressed, prior to AHSV inoculation. However viral antigens were detected for up to 35 dpi. The titres of AHSV-specific neutralising antibodies and the number of immune cells were also significantly lower than in immunocompetent animals. Our findings suggest that donkeys may be able to play a role in the epidemiology of AHS but the ability of vectors to become infected by feeding upon viraemic donkeys needs to be assessed before the significance of that role can be fully understood.

Introduction

During epizootic periods in Spain (1987–1990) and Morocco (1989–1991) cases of African horse sickness (AHS) were observed from June to early December, alternating with periods of apparent quiescence from December to May [9, 10].

These annual resurgences of AHS, over 4–5 consecutive years, imply either the re-introduction of infected vectors, principally *Culicoides imicola*, on the wind [2, 14] or the persistence of virus in the vector or in the host. Following entomological and bioclimatic studies, Mellor [9–8] formulated the hypothesis that long term persistence of African horse sickness virus (AHSV) is possible only in regions where populations of the vector "overwinter" as active adults. This author [10] also suggested that during the apparent inter-epizootic periods the virus may circulate "silently" between these surviving vectors and non-horse equines (mules, donkeys) since these species rarely show clinical signs of infection.

In Morocco the number of donkeys is very large (860,000) and these animals comprise over 40% of the total equine population. In the light of this information the present study was designed to investigate the role that donkeys may be able to play in the maintenance and spread of AHSV in Morocco.

Materials and methods

Donkeys

Six adult Moroccan donkeys (2 males, 4 females) weighing between 60 and 80 kgs were used in the study. These animals were deparasitised and then housed in 2 groups in insect proof accommodation for about one month before being tested by serum neutralisation test to confirm their sero-negativity for AHSV-4 antibodies.

Group 1: donkeys A1, A2, A3.

Group 2: donkeys A4, A5, A6.

Virus

AHSV serotype 4 was used. This virus had been isolated originally from a horse infected during the 1989 epizootic in Morocco and had been passaged 4 times in sucking mouse brain before lyophilisation and storage at –70°C. Prior to use in the present study the virus was passaged a further 3 times on Vero cells and the titre determined by the method of Kärber [5].

Inoculation of the donkeys

Each of the 3 donkeys in group 1 was inoculated intramuscularly with 10^6 $TCID_{50}$ of virus. After 90 days each of these donkeys was given 3 intravenous injections of 5 mg/kg of cyclophosphamide per injection (Endoxan A, ASTA Pharma, Frankfurt, Germany) at 3 day intervals. The animals were then observed for 35 days after the first injection of cyclophosphamide.

The 3 donkeys in the second group were given the same regime of cyclophosphamide at the same time as the first group and only subsequently were infected with a similar dose of AHSV-4. Observation also continued for a further 35 days.

Clinical examination

The temperature of each donkey was taken twice daily throughout the experiment and a full clinical examination was carried out each day.

Virus isolation from the donkeys

Group 1 donkeys: Whole heparinised blood samples were taken daily from each donkey for 30 days after infection and then at approximately weekly intervals for the duration of the experiment (125 days).

Group 2 donkeys: Whole heparinised blood samples were taken from each animal at approximately weekly intervals after infection until the end of the experiment.

Virus isolation was attempted from the blood samples using mouse kidney stable cells (MS) and Vero cells, grown in 25 cm^3 flasks on Eagles MEM supplemented with 0.25% lactalbumin, 10% foetal calf serum and antibiotics. One ml of whole blood was added to 2 flasks each of MS and Vero cells. After 1 hour at 37°C the cells were washed with PBS and were then maintained in serum-free medium at 37°C for 7 days. A positive result was indicated by the development of CPE after 4 to 7 days incubation. Presence of AHSV was confirmed by harvesting cell sheets showing CPE and applying the indirect-immunofluorescence (IFI) test. In the event of no CPE or of a negative IFI result then the supernatant from the cell sheet was passaged a second time in tissue culture and tested again by IFI.

The indirect-immunofluorescence test

Cells to be tested were centrifuged at 400 g for 10 min. The resultant pellet was resuspended in 15 ml of PBS and the suspension layered onto 5 microscope slides. The slides were dried, fixed in cold acetone (–20°C) for 30 min and then stored at –70°C until required. Donkey hyperimmune AHSV antiserum was layered onto the fixed cell suspensions and the slides were incubated at 37°C for 45 min. Slides were then washed in PBS, blotted dry and previously titrated, fluoescein isothiocyanate conjugated rabbit anti-donkey IgG (diluted to 1:80 in PBS) added, before incubation at 37°C for a further 30 min. Finally the slides were washed in PBS, and examined under a Leitz UV microscope at a magnification of 250 using 10% glycerine as a mountant.

Known uninfected and infected cell suspensions were processed in parallel with the test samples and were used as positive and negative controls.

RT-PCR

AHSV genomic dsRNA was extracted using a BIOPROBE (RNA) kit and then denatured with 0.02 M methyl mercuric hydroxide [15]. Primers designated as P1 and P2 were selected from a Spanish isolate of AHSV-4, segment 7 [13], and the RT-PCR conducted on the blood samples from the donkeys according to the method of Moulay et al. [11] using a digoxigenin labelled probe. The sequences and positions of the 2 primers are as follows:

P1: GTTAAAATTCGGTTAGGATG (bases 1–20)
P2: GTAAGTGTATTCGGTATTGA (bases 1159–1179)

Immunological tests

Plasma aliquots separated from the donkey whole blood samples were centrifuged at 1600 g for 15 min at 4°C. The resulting cell pellets were then resuspended to the original volume (4 ml) in Hank's saline solution. The cell suspensions were layered onto 5 ml of Ficoll-Paque (Lympho-preparation, Nyegaard), centrifuged at 400 g for a further 30 min at 4°C and the mononuclear cells harvested from the interphase and counted. Aliquots of these cells, each containing about 10^6 cells, were then examined by a direct immunofluorescence (IF) technique in order to enumerate those cells (mainly macrophages and lymphocytes) carrying donkey Ig on their surfaces. The cells were incubated for 30 min at 37°C in the presence of rabbit anti-donkey IgG conjugated with fluorescein isothiocynate. After 2 washes with Hank's solution the cell suspensions were examined under a UV microscope and the cells with fluorescing membranes counted.

An indirect IF test was also conducted on aliquots of the cell suspensions, according to the method of Bolin et al. [1], to identify B lymphocytes carrying AHSV specific antibodies on their membranes. This involved incubating the cell suspensions with AHSV-4 for 2 h at 37°C.

After 2 washes with Hank's solution, a previously titrated preparation of donkey anti-AHSV IgG conjugated with fluorescein isothiocynate was added and the mixture incubated for 30 minutes at 37°C. The fluorescing cells were counted as described previously.

Serology

AHSV-4 specific virus neutralisation tests were carried out on a 3-fold dilution series of donkey sera (1/3, 1/9, 1/27, 1/81, 1/243, 1/729) using a method similar to that of Herniman et al. [4].

Results

Clinical signs in donkeys infected with AHSV-4

Subsequent to their infection with AHSV-4 the 3 donkeys in group 1 exhibited a mild febrile response of between 0.6 and 1.9°C which peaked at about 24 h post inoculation. These animals also showed some evidence of depression and inappetence which persisted for 7–8 days, followed by complete recovery. Treatment with cyclophosphamide caused a transient increase in temperature but this quickly returned to normal. The 3 donkeys in group 2 responded similarly to those in group 1 to cyclophosphamide treatment and to infection with AHSV. At the end of treatment there was no obvious clinical differences between the immunosuppressed and immunocompetent donkeys and none of the clinical signs observed were consistent with those usually associated with AHS.

Virus isolation and detection

Group 1 donkeys – A viral agent was isolated using both the MS and Vero cell assay systems from blood samples taken from each of the 3 donkeys on day 1 after infection. The identity of the agent was confirmed as being AHSV by IF and RT-PCR assays. Infectious virus continued to be isolated from the group 1 donkeys using the MS cell system until as late as 17 days post infection (dpi).

IFI and RT-PCR assays for the detection of AHSV antigens carried out Vero and MS cells, and on blood samples, respectively, were positive from day 1 to day 28 post infection for each of the 3 donkeys in this group.

No infectious virus was isolated from any of 15 blood samples taken from the donkeys in group 1 subsequent to their treatment with cyclophosphamide at 90 dpi. IFI assays carried out on the MS and Vero cells, to which these blood samples had been added, were also negative.

Group 2 donkeys – No infectious virus was isolated via the MS and Vero cell assay systems at any stage. However the presence of viral antigen in the blood mononuclear cells of each of these animals from day 1 until at least 35 dpi infection, was confirmed by positive IFI assays carried out on Vero and MS cells.

Cellular responses

Figure 1a shows that initially there were between 4 and 6000 mononuclear cells/mm^3 of plasma in all 3 donkeys in group 1. Following infection with AHSV the count in the plasma of donkeys A2 and A3 increased significantly between 35 and 60 dpi to reach a peak of 4 times the initial concentration. In contrast there was no such variation in the cell counts of donkey A1 over the same time period. The concentrations of B lymphocytes in the 3 donkeys showed similar patterns of variation to those shown by the mononuclear cells (Fig. 1b).

B lymphocytes carrying specific antibodies against AHSV first appeared in the plasma of the group 1 donkeys at 28 dpi and the numbers of these cells peaked at 28–45 dpi (Fig. 1c). However by 60 dpi these cells had disappeared from the circulation.

In the group 2 (immunosuppressed) donkeys the number of mononuclear cells in general and of B lymphocytes in particular, decreased in parallel during the period of cycloposphamide treatment (Figs. 1d and 1e). Furthermore B lymphocytes carrying specific antibodies against AHSV were not detected in these animals.

Neutralising antibody responses

AHSV-4 specific antibodies were first detected in the group 1 donkeys 10 to 12 dpi. The titre in all 3 donkeys showed similar profiles and increased rapidly to reach similar maxima, after 21 days in donkeys A2 and A3, and somewhat later in donkey A1 (Fig. 1f). High titres of antibody then persisted in all 3 donkeys beyond 90 days. During cyclophosphamide treatment antibody titre decreased slightly for a period of about one week in donkeys A1 and A3 before returning to its former level. In donkey A2 the decrease in antibody titre persisted for the duration of the observational period (Fig. 1g).

Antibody titres in the group 2 donkeys increased more slowly than those in group 1 and the maximum titres attained were lower (Fig. 1h).

Discussion

No specific clinical signs of AHS were detected in any of the experimental donkeys after infection with AHSV-4, even in animals treated with the immunosuppressant, cyclophosphamide. This reflects the situation observed during the AHS epizootics in Morocco when donkeys were very rarely observed to develop disease. Nevertheless in the present study, AHS viraemias extending for as long as 16–17 days were recorded in some donkeys and infectious virus was isolated throughout this period. There was no recrudescence of viraemia, however, when at a later stage these donkeys were immunosuppressed. When donkeys were immunosuppressed prior to infection, viraemia extending for as long as 35 days was recorded by IFI. However no infectious virus was isolated at any stage during this 35 day period and therefore the significance of this apparently extended viraemia must be interpreted with caution.

In the present work MS cells were found to be more susceptible to AHSV infection and therefore more sensitive at detecting an AHS viraemia, than were Vero cells. Nevertheless even in situations where there was no obvious CPE, IFI

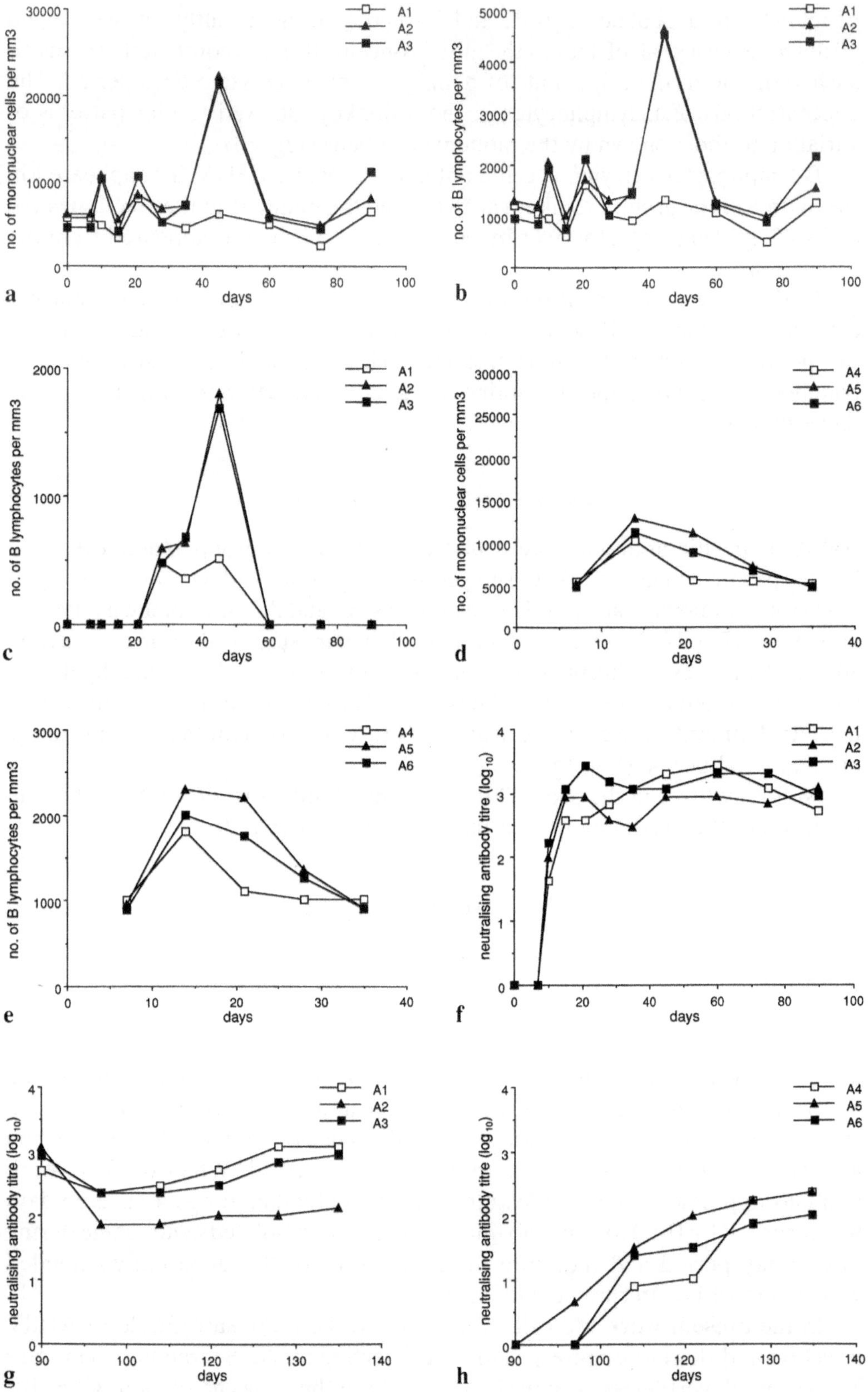
a
no. of mononuclear cells per mm3
30000
20000
10000
0
0 20 40 60 80 100
days
A1
A2
A3
b
no. of B lymphocytes per mm3
5000
4000
3000
2000
1000
0
0 20 40 60 80 100
days
A1
A2
A3
c
no. of B lymphocytes per mm3
2000
1000
0
0 20 40 60 80 100
days
A1
A2
A3
d
no. of mononuclear cells per mm3
30000
25000
20000
15000
10000
5000
0
0 10 20 30 40
days
A4
A5
A6
e
no. of B lymphocytes per mm3
3000
2000
1000
0
0 10 20 30 40
days
A4
A5
A6
f
neutralising antibody titre (log10)
4
3
2
1
0
0 20 40 60 80 100
days
A1
A2
A3
g
neutralising antibody titre (log10)
4
3
2
1
0
90 100 120 130 140
days
A1
A2
A3
h
neutralising antibody titre (log10)
4
3
2
1
0
90 100 120 130 140
days
A4
A5
A6

and RT-PCR techniques enabled the detection of viral antigens or of viral RNA, respectively.

In this context it is important to note that the AHSV present in some whole blood samples caused CPE in MS cells (2nd passage) until 16–17 dpi, during which time neutralising antibodies in the blood were either absent or only present at low titre. However, once antibody titre had reached a maximum (about 3 logs) CPE failed to occur, although viral antigen was detectable on the MS cells by IFI until 28 dpi (immunocompetent donkeys) and until 35 dpi (immunosuppressed donkeys). These findings suggest that the presence of high titres of antibody in the blood may prevent the establishment of a replicating infection in tissue culture cells but detectable amounts of viral antigen may persist for an extended time attached to these cells. It is not yet known whether this viral antigen also signifies the presence of infectious virus. However since it has long been known that AHSV circulates in association with the red blood cells (RBCs), where it is inaccessible to humeral antibody [6, 12], washing to remove this antibody prior to lysing the RBCs to release any infectious virus, should help resolve this question [3].

In relation to the role played by the donkey in the epidemiology of AHS our studies suggest that though these animals rarely show clinical signs of infection they are able to be infected with AHSV and can mount a prolonged viraemia, lasting for at least 17 days. It has not yet been shown whether this viraemia is of sufficiently high titre to infect significant numbers of vector *Culicoides*. If this should be the case then donkeys could well prove to be an efficient reservoir for AHSV during apparent inter-epizootic periods.

Acknowledgements

This study was supported by the Ministere de l'Agriculture et da la Mise en Valeur Agricole, Direction de l'Elevage and Réseau Thematique en Biotechnologie of AUPELF-UREF.

Fig. 1. a Mononuclear cells in the plasma of donkeys A1, A2 and A3 (group 1) following infection with African horse sickness virus serotype 4. **b** B lymphocytes in the plasma of donkeys A1, A2 and A3 (group 1) following infection with African horse sickness virus serotype 4. **c** B lymphocytes carrying African horse sickness virus specific antibodies in the plasma of donkeys A1, A2 and A3 (group 1) following infection with serotype 4 of the virus. **d** Mononuclear cells in the plasma of immunosuppressed donkeys A4, A5 and A6 (group 2) following infection with African horse sickness virus serotype 4. **e** B lymphocytes in the plasma of immunosuppressed donkeys A4, A5 and A6 (group 2) following infection with African horse sickness serotype 4. **f** African horse sickness virus specific, neutralising antibody responses of donkeys A1, A2 and A3 (group 1) following infection with serotype 4 of the virus. **g** African horse sickness virus specific, neutralising antibody responses of donkeys A1, A2 and A3 (group 1) after treatment with an immunosuppressant on days 90, 93 and 96. **h** African horse sickness virus specific, neutralising antibody responses of donkeys A4, A5 and A6 (group 2) after treatment with an immunosuppressant on days 90, 93 and 96 and infection with African horse sickness serotype 4 on day 96

References

1. Bolin SR, McClurkin AW, Coria MF (1985) Effects of bovine viral diarrhoea virus on the percentages and absolute numbers of circulating B and T lymphocytes in cattle. Am J Vet Res 46: 884–886
2. Fassi-Fehri MM (1993) Hypothèses sur le mode d'introduction de la peste équine au Maroc en 1989. Rev Med Vet 144: 615–619
3. Hamblin C, Anderson EC, Mellor PS, Graham SD, Mertens PPC, Burroughs JN (1992) The detection of African horse sickness virus antigens and antibodies in young equidae. Epidemiol Infect 108: 193–201
4. Herniman KAJ, Booman JPT, Taylor WP (1983) Bluetongue virus in a Nigerian dairy herd. 1. Serological studies and correlation of virus activity to vector population. J Hyg (Camb) 90: 177–193
5. Kärber G (1931) Beitrag zur kollektiven Behandlung pharmakologischer Reihenversuche. Arch Exp Pathol Pharmakol 162: 480–483
6. McIntosh BM (1953) The isolation of virus in mice from cases of horse sickness in immunised horses. Onderstepoort J Vet Res 27: 165–169
7. Mellor PS (1991) African horse sickness in Europe. Proc Soc Vet Epidem Prev Med, London 17th March–1st April 1991, pp 27–32
8. Mellor PS (1992) *Culicoides* as potential Orbivirus vectors in Europe. In: Walton TE, Osburn BI (eds) Bluetongue, African horse sickness and related orbiviruses. CRC Press, Boca Raton, pp 278–283
9. Mellor PS (1993) African horse sickness: transmission and epidemiology. Vet Res 24: 199–212
10. Mellor PS (1994) Epizootiology and vectors of African horse sickness. Comp Immunol Microbiol Infect Dis 17: 287–296
11. Moulay S, Zientara S, Sailleau C, Crucière C (1995) Detection of African horse sickness viruses by dot blot hybridization using a digoxygenin labelled probe. Mol Cell Probes 9: 233–237
12. Ozawa Y, Salama SA, Dardiri AH (1972) Methods for recovering African horse sickness virus from horse blood. Bull Off Int Epizoot 77: 1615
13. Roy P, Hirasawa T, Fernandez M, Blinov VM, Sanchez-Vizcaino JM (1991) The complete sequence of the group-specific antigen, VP7, of African horse sickness virus serotype 4 reveals a close relationship to bluetongue virus. J Gen Virol 72: 1237–1241
14. Sellers RF, Pedgley DE, Tucker MR (1977) Possible spread of African horse sickness on the wind. J Hyg (Camb) 79: 279–298
15. Wade-Evans AM, Mertens PPC, Bostock CJ (1990) Development of the polymerase chain reaction for detection of bluetongue virus in tissue samples. J Virol Methods 30: 15–24

Authors' address: Dr. O. Fassi-Fihri, Départment de Microbiologie, Immunologie et Maladies Contagieuses, Institut Agronomique et Vétérinaire Hassan II, B. P. 6202, Rabat-Instituts, Rabat, Morocco.

Immunohistochemical demonstration of African horse sickness viral antigen in tissues of experimentally infected equines

P. Wohlsein[1], **J. F. Pohlenz**[1], **J. S. Salt**[2], and **C. Hamblin**[2]

[1] Institute of Pathology, School of Veterinary Medicine, Hannover, Germany
[2] Institute for Animal Health, Pirbright Laboratory, Woking, Surrey, U.K.

Summary. African horse sickness virus (AHSV) antigen was demonstrated immunohistochemically in formalin-fixed, paraffin-embedded sections of tissues collected from three ponies suffering from the peracute form of the disease and from one pony affected by the fever form. The pattern of the antigen distribution indicated a particular organ tropism characterised by an accumulation of AHSV antigen in cardio-pulmonary tissues of the animals with the peracute disease and in the spleen of the pony with the fever form. AHSV antigen was identified in endothelial cells of small blood vessels, particularly capillaries and in large mononuclear cells resembling macrophages or reticular cells of lymphatic tissues. Occasional circulating mononuclear cells with the morphology of monocytes were also positively stained within the larger vessels. The immunohistochemical results confirm earlier work suggesting that AHSV may have different tropisms to particular organs during various forms of the disease and that different target cell populations exist in vivo. Immunohistochemistry may be an additional useful method for diagnostic and research purposes in AHS.

Introduction

African horse sickness (AHS) is an arthropod-borne viral disease of horses and other solipeds caused by African horse sickness virus (AHSV), a member of the genus *Orbivirus* of the family *Reoviridae* [7]. Infection of susceptible equines with AHSV may result in different forms of disease which traditionally have been defined by the predominant clinico-pathological findings as peracute or pulmonary, subacute or cardiac, acute or mixed and fever form [4, 5, 14]. Studies using plaque-purified virus isolates have indicated that the form of disease expressed by naive horses is a property of the virus inoculum [11]. However, the pathogenesis of the different disease forms is not completely understood. Virus particles have been identified in endothelial cells in tissue sections examined by electron microscopy [10]. Using in situ hybridization viral RNA has been detected in vascular endothelial cells of lung and heart tissues, and in large mononuclear cells of the spleen [2]. The only report concerning immunohistology and AHS was produced 25 years ago and was based on immunofluorescent staining of cryostat sections of spleen tissue but the antigen-containing cell populations were not defined [18]. This paper describes the topographic distribution and the target cell spectrum of AHSV in formalin-fixed, paraffin-embedded tissues collected from four ponies which had been inoculated with virulent AHSV serotype 4 (AHSV 4).

Materials and methods

Experimental design

Four ponies, approximately 3–4 years of age, were inoculated with AHSV 4. The virus was originally isolated from spleen tissue of a horse which had died during the epizootic in Morocco in 1991. This isolate had been passaged on two occasions by intravenous inoculation of 11 days old embryonated hens eggs (AHSV 4 H1E2). Spleen tissue collected from a pony which had been inoculated with this virus was homogenised with an equal volume of Eagles medium. The infectious supernatant fluid from the spleen homogenate was designated AHSV 4 H1E2H1.

Pony 1 was inoculated subcutaneously with AHSV 4 H1E2 as a 1.0 ml dose on each side of the neck ($10^{6.55}$ EID_{50}/pony). Ponies 2 and 3 were given a single 2.0 ml intravenous inoculation of AHSV 4 H1E2H1 ($10^{5.8}$ EID_{50}/pony). Pony 4 was inoculated with the same dose of AHSV 4 H1E2H1 as follows, 1.0 ml subcutaneously, 0.8 ml intravenously and 0.2 ml intradermally. One horse selected from the necropsy cases of the Institute of Pathology, School of Veterinary Medicine, Hannover, which had been euthanized due to severe orthopaedic lesions, was included as a non-infected control.

Tissue processing and immunohistochemistry

Full necropsies were carried out on all animals and the tissue specimens collected (Tables 1 and 2) were fixed in 5% neutral buffered formalin and embedded in paraffin wax according to a standard laboratory procedure. Tissue sections (3 μm) were cut and either stained with haematoxylin and eosin for histological examination or used for immunohistochemical processing. For immunohistochemistry, the deparaffinised tissue sections were immersed in 0.5% hydrogen peroxide (H_2O_2) in 85% ethanol for 30 min to inhibit any endogenous peroxidase activity. The tissue sections were then rinsed with three changes of 0.05 M TRIS-buffered saline pH 7.6 and incubated for 20 min at 37°C in a solution containing 0.05% pronase E (Merck Co., Darmstadt, Germany) and 0.1% $CaCl_2$. After rinsing three times with TRIS-buffered saline, the tissue sections were overlaid and incubated for 20 min at 37°C with normal goat serum, diluted 1:5 in phosphate-buffered saline (PBS), pH 7.1, to block any non-specific binding of immunoglobulins. The blocking serum was then drained and replaced with AHSV 9 polyclonal hyperimmune rabbit antiserum serum at a dilution of 1:1500 in PBS and incubated for 45 min at 37°C. This rabbit antisera had been raised following two inoculations of sucrose gradient purified virus particles [3], 28 days apart. The tissue sections were again rinsed in TRIS-buffered saline and then incubated for 30 min at 37°C with a biotin-labeled goat-anti-rabbit antibody (Vector Laboratories, Burlingame, CA, USA) which had been diluted 1:150 in PBS and contained 10% normal horse serum. After rinsing the tissue sections were overlaid with an avidin-biotin-peroxidase complex (ABC) (Vector Laboratories, Burlingame, CA, USA), prepared according to the manufacturer's instructions, and incubated for 1 h at

Table 1. Immunohistological results in the cardiovascular, respiratory, digestive, and urinary system of ponies Nos. 1–4 following infection with the African horse sickness virus serotype 4

Organ/tissue	Antigen staining[c]			
	1	2	3	4
Cardiovascular system				
Heart[a]	+++	+++	+	+/++
Aorta	NT	++	++	0
V. cava cranialis	NT	++	++	NT
Respiratory system				
Trachea	+	+	+	NT
Lung[b]	+++	+++	+++	++
Digestive system				
Tongue	NT	++	++	0
Stomach	0	+	++	0
Ileum	++	+	++	0
Caecum	NT	++	++	+
Colon	NT	+	++	0
Parotid gland	+	+	++	0
Pancreas, exocrine	+	0	+	0
Liver	0	+	+	0
Urinary system				
Kidney	+	0	+	0

[a] Identical findings in left and right ventricle, left and right atrium, and interventricular septum

[b] Identical findings in left apical, dorsal and ventral aspect of the left main lobe

[c] + Single antigen-containing cells
++ Moderate numbers of antigen-containing cells
+++ High numbers of antigen-containing cells
0 No detectable virus-specific staining
NT Not tested

ambient temperature. Peroxidase activity was visualized after rinsing with TRIS-buffered saline and incubating the tissue sections for 5 min at 37°C in a solution of 0.05% 3,3'-diaminobenzidine tetrahydrochloride (DAB) (Fluka Co., Buchs, Switzerland) containing 0.03% H_2O_2 for 5 min. The colour of the DAB chromogen deposit was changed into black by adding 2 ml of a 1% aqueous solution of cobalt chloride (Sigma Chemicals, Deisenhofen, Germany) to the chromogen [8]. This enhanced the staining intensity and produced a distinct contrast to the brownish tissue pigments. The immunoenzymatic reactions were stopped by rinsing the tissue sections in tap water. Finally, tissue sections were counter-stained with Mayer's haematoxylin, washed again in tap water, dehydrated and mounted. Tissues from the non-infected horse were treated similarly as a negative control. The specificity of the enzymatic reaction was assessed using duplicate sections where the first antibody (rabbit anti-AHSV) was omitted and substituted either with PBS or a rabbit antibody of a different specificity.

Table 2. Immunohistological results in the skin, supraorbital fat, lymphatic, endocrine and central nervous system of ponies Nos. 1–4 following infection with the African horse sickness virus serotype 4

Organ/tissue	Antigen staining[a]			
	1	2	3	4
Skin	+	NT	NT	0
Supraorbital fat	NT	0	+	+
Lymphatic system				
Spleen	++	+	++	+++
Mandibular lymph node	+	++	++	+
Retropharyngeal lymph node	0	+	++	NT
Axillar lymph node	NT	++	++	NT
Superficial cervical lymph node	NT	0	++	+
Mediastinal lymph node	+	+	+	+
Mesenteric lymph node	+	+	+	+
Popliteal lymph node	NT	0	+	0
Endocrine system				
Adrenal gland	+	NT	+	NT
Thyroid gland	0	0	+	0
Adenohypophysis	NT	NT	NT	0
Central nervous system	NT	0	0	+

[a] + Single antigen-containing cell
++ Moderate numbers of antigen-containing cells
+++ High numbers of antigen-containing cells
0 No detectable virus-specific staining
NT Not tested

Results

Clinically, ponies 1 to 3 developed peracute disease commencing at 4 to 6 days post inoculation (dpi) and characterised by pyrexia, nasal discharge, depression, dyspnoea and a terminal drop in body temperature which necessitated euthanasia on welfare grounds at 10 dpi for pony 1 and at 8 dpi for ponies 2 and 3. The only clinical sign observed in pony number 4 was a transient temperature rise. This animal was also euthanised at 10 dpi.

Grossly, the most striking pathological finding in ponies 1 to 3 was severe pulmonary oedema. Additionally in ponies 2 and 3, slight accumulations of serous fluid were observed between the fascial planes of the neck and in the pericardial and pleural cavity. No significant gross lesions were detected in pony 4. Microscopically, the haemotoxylin and eosin stained tissue sections revealed generalized activation of lymphatic organs in all 4 ponies. Acute, superficial glossitis was also seen in ponies 2 and 3.

AHSV-specific immunohistological staining was characterised by a diffuse, finely granular or dot-like, black chromogen deposit in the cytoplasm of infected cells. The number of antigen-containing cells was estimated semi-quantitatively and is summarized in Tables 1 and 2. For ponies 1 to 3, the highest number of positively stained cells was detected in cardio-pulmonary tissues but without an obvious predilection site among the different locations examined within these systems (Table 1). AHSV antigen was also found in these three animals in lesser amounts distributed widely throughout the body (Tables 1, 2). However in pony 4, the animal with the fever form of disease, the antigen was highly concentrated in the spleen, was present to a lesser extent in the cardio-pulmonary systems and was only sporadically present elsewhere (Tables 1, 2).

AHSV antigen was located primarily in cells with the morphology of endothelial cells of capillaries, arterioles, small veins and venules throughout the body. Antigen was not detected in cardiac muscle cells (Fig. 1). In the spleen, AHSV antigen was identified additionally in large cells with vesicular nuclei and abundant, occasionally branching cytoplasm, located in the red pulp. These cells resembled macrophages or reticular cells (Fig. 2). AHSV antigen was present also as a "rim" along the mantle zone of the lymphoid follicles of the mandibular, retropharyngeal, axillary, mediastinal, superficial cervical and popliteal lymph nodes of ponies 2 and 3, and appeared to be partially associated with long

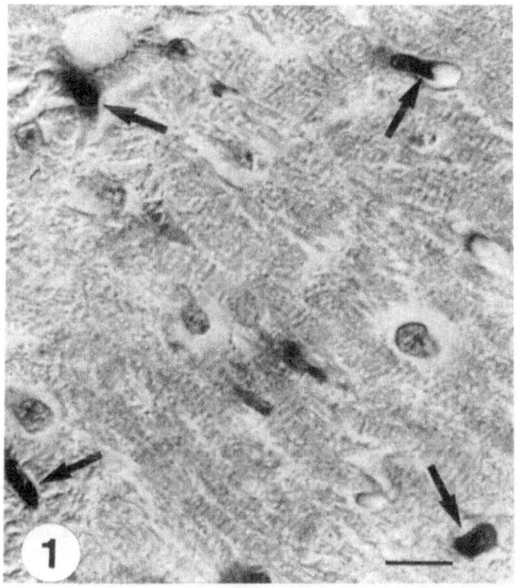

Fig. 1. Pony 1, heart; AHSV antigen in numerous endothelial cells of mycardial capillaries (arrows); ABC method, Mayer's haematoxylin counterstain, differential interference contrast microscopy. Bar: 10 μm

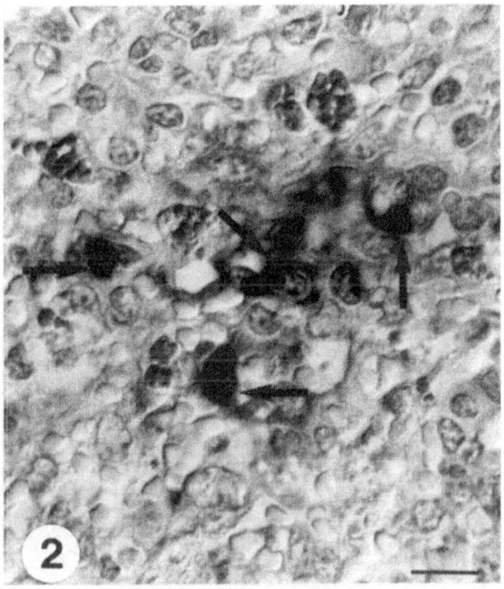

Fig. 2. Pony 1, spleen, red pulp; AHSV antigen localised in cells with large, vesicular nuclei and abundant cytoplasm, resembling macrophages and reticular cells (arrows); ABC method, Mayer's haematoxylin counterstain, differential interference contrast-microscopy. Bar: 10 μm

cytoplasmic processes. The morphology and anatomical location of these cells is suggestive of them being follicular dendritic cells. Occasionally, antigen was detected in the cytoplasm of large cells resembling macrophages or immunoblasts adjacent to, and in the centre of, lymphoid follicles. Additionally, spotted, dust-like staining was observed in follicle centres but without a clear association with individual cells (Fig. 3). Immunohistochemistry also revealed single or moderate numbers of antigen-containing cells in the digestive, urinary and endocrine systems. Individual endothelial cells in the supraorbital fat of ponies 3 and 4 were specifically stained. Although there was no specific staining along the endothelial lining of the larger blood vessels (e.g. aorta, cranial vena cava), AHSV antigen was detected in the endothelial cells of intramural capillaries, *vasa vasorum*. In some tissues, dot-like antigen staining was observed in the cytoplasm of occasional circulating cells which resembled monocytes (Fig. 4). Specific immunohistochemical staining was not detected in any of the tissues collected from the non-infected control animal.

Discussion

Ponies 1 to 3 developed a severe, rapidly progessive disease with marked pulmonary oedema, indicative of the peracute form of AHS. However, despite the rapid

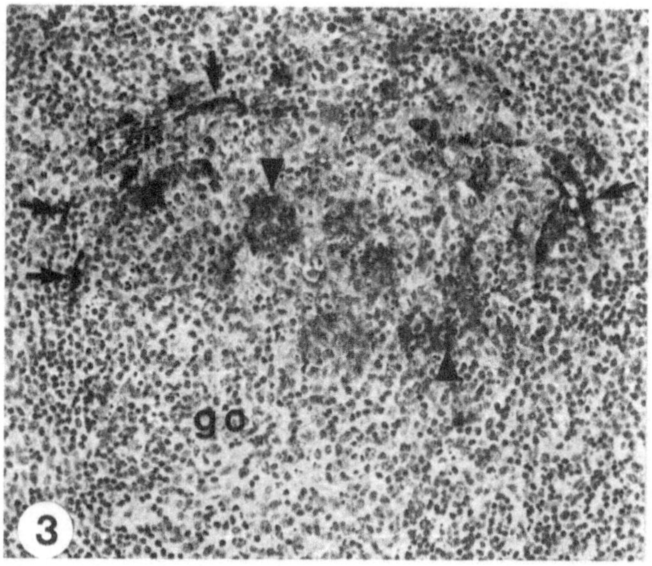

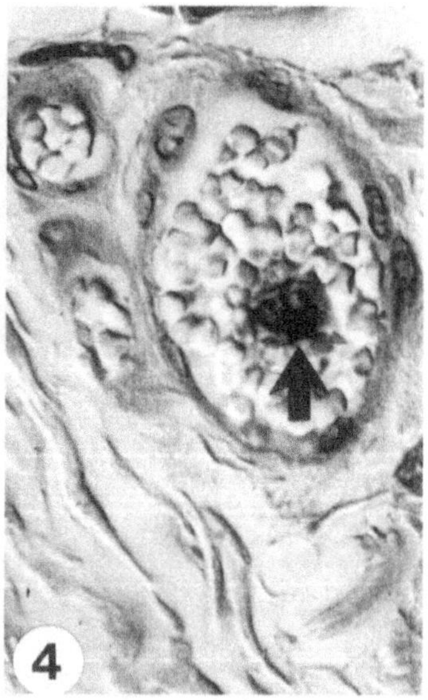

Fig. 3. Pony 2, mandibular lymph node; AHSV antigen associated with cells of the mantel zone (arrows) and the follicle centre (triangles), *gc* germinal centre. ABC method, Mayer's haematoxylin counterstain. Bar: 60 μm

Fig. 4. Pony 1, pancreas, venule of the interstitial connective tissue; dot-like staining of AHSV antigen in the cytoplasm of a large circulating mononuclear cell (arrow); ABC method, Mayer's haematoxylin counterstain, differential interference contrast microscopy. Bar: 10 μm course of disease in ponies 2 and 3, some of the gross lesions present in these animals resembled the more protracted, acute or subacute forms of AHS.

Immunohistochemical labelling of AHSV antigen in the tissues of the 4 experimental ponies appeared as diffuse, finely granular or dot-like chromogen deposits. In general the pattern of AHSV antigen labelling in all four animals was similar but differences did exist in topographical distribution. In those animals with the peracute form of the disease AHSV antigen was found primarily in the cardio-pulmonary systems, secondarily in the lymphatic system and to a lesser extent, distributed widely throughout the body. In the animal with the fever form of the disease the antigen was concentrated in the spleen with smaller amounts in the cardio-pulmonary systems and was generally less widespread. The immunohistochemical results of our work therefore support the suggestions of earlier workers [5, 10, 11] that there are variations in organ tropism among populations or sub-populations of AHSV particles which give rise to the different forms of AHS.

In all ponies examined in the present work, the predominant antigen-containing cells were those with the morphology of endothelial cells. The pattern of immunostaining in lung and heart tissues that we have recorded is similar to earlier findings obtained by in situ hybridization [2] and suggests that endothelial cells are a primary permissive target cell population for AHSV 4. However, the level of involvement of endothelial cells seems to be unrelated to the extent of the microvasculature of a particular organ. In man, marked diversity and heterogeneity of the microvasculature has been found, where endothelial cells lining the pulmonary vessels differ functionally from those of other vascular beds [1]. The factors determining the selectivity of the organ tropism of AHSV are currently not understood.

In our studies the endothelial lining of the large vessels was consistently negative for AHSV antigen in all animals examined. This finding is in marked contrast to earlier results obtained in vitro [12], which indicated that cultured endothelial cells of large vessels are permissive for AHSV.

Large mononuclear cells of the lymphatic organs appear to be a second target cell population for AHSV. The demonstration of AHSV antigen in these studies, and of AHSV RNA by in situ hybridization [2], in large cells of the red pulp of the spleen strongly supports the contention that these cells are permissive for AHSV replication. In lymph nodes, AHSV antigen was detected adjacent to germinal centres suggesting an association between the antigen and follicular dendritic cells. These cells are immune accessory cells which specialise in trapping immune complexes [16, 19]. The concentration of AHSV antigen at this site may be due to the trapping of immune complexes in the network of cytoplasmic processes or to an infection of follicular dendritic cells with or without subsequent virus replication or to the direct transmission of virus particles from mononuclear cells to follicular dendritic cells followed by possible replication at this site. Although all 4 animals remained serologically negative throughout the study period (data not shown), an early immune response is usually expected by 10 dpi with the formation of immune complexes [6]. The presence of AHSV antigen in large mononuclear cells, resembling phagocytic cells and surrounding lymphoid follicles suggests that these cells may be involved in the transport of viral proteins to lymphoid follicles. The detection of the antigen in large cells of the follicle centre may indicate phagocytosis by mononuclear cells or an association with immunoblasts during the germinal centre reaction of the immune response.

Viraemia in AHS has been shown to be cell-associated and virus can be recovered from horse erythrocytes [15]. The virus has also been shown to infect and replicate readily in monocytic cells of the peripheral blood in vitro [17]. Furthermore, monocytes and macrophages have been shown to be primary sites of replication for the related bluetongue virus [13]. In the present study, AHSV antigen was occasionally detected in circulating large mononuclear cells resembling monocytes. Therefore it is likely that circulating monocytes are an additional target cell population in AHS, contributing at least to a limited extent, to the cell associated viremia.

Ultrastructurally, virus inclusion bodies consisting mainly of AHSV VP3, VP7 and NS2 have been found in infected cells in vitro which appear to co-locate with the sites for the initiation of protein synthesis [9]. These structures may compare to the dot-like chromogen deposits observed in our work, indicating virus replication at these sites.

The results presented in this paper show that AHSV antigen can be detected in formalin-fixed, paraffin-embedded tissue sections. Consequently the method described here could be used as an additional diagnostic test and would be of particular value in countries where AHS is suspected but where it is difficult to refer fresh samples for diagnostic purposes. Furthermore, this method provides the ability to perform retrospective studies on archival material. Studies based on horses affected with other forms of the disease are now required to provide additional insights into the target cell spectrum of the virus in order to further our understanding of the pathogenesis of AHS.

Acknowledgements

The authors are grateful for financial contributions from the European Comission (Contract No. TS3-CT92-0151). We also wish to thank Peter Mertens for providing the purified AHSV and Paul Smith (Animal Technician, Pirbright Laboratory) for his assistance throughout the study.

References

1. Belloni PN, Carney DH, Nicolson GL (1992) Organ-derived microvessel endothelial cells exhibit differential responsiveness to thrombin and other growth factors. Microvasc Res 43: 20–45
2. Brown CC, Meyer RF, Grubman MJ (1994) Presence of African horse sickness virus in equine tissues, as determined by in situ hybridization. Vet Pathol 31: 689–694
3. Burroughs JN, O'Hara RS, Smale CJ, Hamblin C, Walton A, Armstrong R, Mertens PPC (1994) Purification and properties of the virus particles, infectious subviral particles, cores and VP7 crystals of African horsesickness virus serotype 9. J Gen Virol 75: 1849–1857

4. Coetzer JAW, Erasmus BJ (1994) African horse sickness. In: Coetzer JAW, Thomson GR, Tustin RC (eds) Infectious diseases of livestock. Oxford University Press, Oxford, pp 460–475
5. Erasmus BJ (1973) The pathogenesis of African horse sickness. In: Bryans JT, Gerber H (eds) Proceedings of the Third International Conference on Equine Infectious Diseases, vol III. Karger, Basel, pp 1–11
6. Hazrati A, Mirchamsy H, Bahrami S (1973) Comparative studies on the serological responses of horses to African horsesickness virus. In: Bryans JT, Gerber H (eds) Proceedings of the Third International Conference on Equine Infectious Diseases, vol III. Karger, Basel, pp 69–80
7. Holmes IH, Boccardo G, Estes MK, Furuichi MK, Hoshino Y, Joklik WK, McCrae M, Mertens PPC, Milne RG, Samal KSK, Shikata E, Winton JR, Uyeda I, Nuss DL (1995) Family *Reoviridae*. In: Murphy FA, Fauquet CM, Bishop DHL, Ghabrial SA, Jarvis AW, Martelli GP, Mayo MA, Summers MD (eds) Virus Taxonomy. Classification and Nomenclature of Viruses. Sixth Report of the International Committee of Taxonomy of Viruses. Springer, Wien New York, pp 208–239 (Arch Virol [Suppl] 10)
8. Hsu S-M, Soban E (1982) Color modification of diaminobenzidine (DAB). Precipitation by metallic ions and its application for double immunohistochemistry. J Histochem Cytochem 30: 1079–1082
9. Laegreid WW (1996) African horsesickness. In: Studdert MJ (ed) Virus infections of equines. Elsevier, Amsterdam, pp 101–123
10. Laegreid WW, Burrage TG, Stone-Marschat M, Skowronek A (1992) Electron microscopic evidence for endothelial infection by African horsesickness virus. Vet Pathol 29: 554–556
11. Laegreid WW, Skowronek A, Stone-Marschat M, Burrage T (1993) Characterization of virulence variants of African horsesickness virus. Virology 195: 836–839
12. Laegreid WW, Stone-Marschat M, Skowronek A, Burrage T (1992) Infection of endothelial cells by African horsesickness viruses. In: Walton TE, Osburn BI (eds) African Horsesickness and related orbiviruses. CRC Press, Boca Raton, pp 807–814
13. Lawman MJP (1979) Observations on the pathogenesis of bluetongue virus infections in sheep. PhD Thesis, University of Surrey
14. Maurer FD, McCully RM (1963) African horse-sickness – with emphasis on pathology. Am J Vet Res 24: 235–266
15. Ozawa Y, Salama SA, Dardiri AH (1973) Methods for recovering African horsesickness virus from horse blood. In: Bryans JT, Gerber H (eds) Proceedings of the third International Conference on Equine Infectious Diseases, vol III. Karger, Basel, pp 58–68
16. Schriever F, Nadler LM (1992) The central role of follicular dendritic cells in lymphoid tissues. Adv Immunol 51: 242–284
17. Stäuber N, McCullough KC, Kihm U (1992) Interaction of African horse sickness viruses with leukocytes and the capacity to induce an immune response in vitro. In: Walton TE, Osburn BI (eds) Bluetongue, African horse sickness, and related orbiviruses. CRC Press, Boca Raton, pp 815–824
18. Tessler J (1972) Detection of African horsesickness viral antigens in tissues by immunofluorescence. Can J Comp Med 36: 167–169
19. Van den Berg TK, Yoshida K, Dijkstra CD (1995) Mechanism of immune complex trapping by follicular dendritic cells. In: Kosco-Vilbois MH (ed) An antigen depository of the immune system: Follicular dendritic cells. Springer, Berlin Heidelberg New York Tokyo, pp 49–67

Authors' address: Dr. P. Wohlsein, Institute of Pathology, School of Veterinary Medicine, Bünteweg 17, D-30559 Hannover, Germany.

Entomology

The 1996 outbreak of African horse sickness in South Africa – the entomological perspective

R. Meiswinkel

Onderstepoort Veterinary Institute, Onderstepoort, South Africa

Summary. During the 1996 summer season (January–April) in South Africa an estimated 500 horses died of African horse sickness (AHS); 80% of deaths were due to AHS virus serotypes 2 and 4. Nearly all cases occurred in the northern, north-eastern and central parts of South Africa. This study reports the first attempt to verify the involvement of the biting midge *Culicoides imicola* in a field outbreak of AHS in southern Africa. In light-trap collections made at 47 sites over 12 weeks, *C. imicola* comprised 94.2% of 4.78 million *Culicoides. Culicoides imicola* was the most prevalent of 34 species captured and was the only species whose distribution matched that of the disease. Record catches of *C. imicola* were made, and reveal that in years of above average rainfall its numbers can show a 200-fold increase over those in dry years. Soil type appeared to determine strongly the distribution of *C. imicola*. The largest populations of *C. imicola* were found in areas with clayey, moisture-retentive soils whereas the lowest numbers, or none, occurred in areas where the soils were sandy and quick-draining. The deaths of two horses (confirmed AHS) in a sandy area were perplexing as they occurred in a region known to be free of *C. imicola*. The probable origin of these infections was established.

Introduction

Since the first recorded outbreak of AHS in South Africa in 1719, it has become accepted '... that the "horse sickness" is coeval with wet seasons succeeding drought, ... unquestionably proved by ... the mortality experienced during the years 1837, 1854 and 1863 ...' [16]. After researching AHS for 30 years Theiler [20] similarly concluded that AHS '... appears as a plague ... in such years that are characterized by unusual rainfalls, which lead to floods and are succeeded by warm weather.'

As regards outbreaks of AHS in South Africa, McIntosh reported that heavy losses occurred in 1780, 1801, 1839, 1854, 1862, 1891, 1914, 1918, 1923, 1940, 1946 and 1953 [11]. Except for those occurring in the years 1854/5, 1891/2 and 1922/3 these outbreaks have been poorly documented in the scientific literature. Most are very briefly referred to in agricultural reports or are known from newspaper clippings only.

In an analysis [4] of the 1922/3 outbreak in which 1700 horses died, it was noted that '... very heavy rains fell all over the country,' and that '... the insect transmitter of horse-sickness ... completes its life-cycle in little more than a week.' The latter deduction was remarkably accurate if one considers that it was not until 22 years later that a species of *Culicoides* biting midge was discovered to transmit AHS virus (*Reoviridae: Orbivirus*) [5, 23]. Although more than 50 years have lapsed since *C. imicola* was incriminated as the vector, no study has attempted to verify this species' involvement in field outbreaks of AHS in southern Africa.

During the 1996 outbreak such an attempt was made; the results from 66 of 75 *Culicoides* collections made at 47 sites over a 12-week period are reported here. All collections were made on farms which fell within the outbreak area and included stables at which horses had died. No attempt was made, however, to isolate AHS virus from pools of the various species of *Culicoides* collected.

The deaths of two horses (confirmed AHS) in the Port Elizabeth area of the south-eastern Cape were unexpected as they occurred in a part of South Africa recently reported to be free of *C. imicola* [13]. The deaths prompt the question as to whether a second, unknown, vector of the disease occurs in South Africa or whether the animals were infected elsewhere. These possibilities are examined. The conclusions have implications as regards the vehicular transport of horses, and *Culicoides*, in South Africa.

Finally, the supposed value that stabling has for protecting horses against AHS, is assessed.

Materials and methods

1996 outbreak of AHS in South Africa

The geographical extent of the outbreak is shown in Fig. 1 using cases confirmed by virus isolation and serotyping at the Onderstepoort Veterinary Institute (OVI), and cases rumoured. Mapping of the latter was anecdotal.

Trap sites

Forty-seven collecting sites (Table 1) were chosen on the basis of several criteria: a) there were calls from concerned owners/veterinarians; b) stables were experiencing deaths (marked with an asterisk in Table 1), and those in their vicinity; c) stables had been monitored previously in drier years (sites #18–#25 in Table 1); d) a site had been sampled weekly for an entire year (site #1); e) sites were on supersaturated, clayey soils in the Pretoria/Johannesburg outbreak area; f) sites were on sandy soils along the southern coast (sites #40–#47); g) sites were on the route of a horse transport float (sites #7, #29–#31, #36 and #40). Collecting sites are shown with an asterisk in Fig. 1.

Light trapping

Except for sites #1 and #31, all sites were sampled for one night only using one to four 220-volt down-draught suction light-traps each fitted with an 8-watt blacklight tube [22]. The

Table 1. *Culicoides* collecting sites and dates (1996), total number of *Culicoides*, percentage and log numbers of *C. imicola* in each catch. Asterisk denotes sites with confirmed AHS deaths. A grand total of 4.78 million *Culicoides* was collected. Region groups as given under Fig. 2

Site no.	Farm/site	Town/district (region group)		Date	Total *Culicoides*	% *imicola*	Log *imicola*
#1*	Kaalplaas (Onderstepoort)	Pretoria	(k)	26. II.	55366	97.3	4.7
	"	"		18. III.	1054620	99.1	6.0
	"	"		9. IV.	305520	99.2	5.5
	"	"		2. V.	49920	96.2	4.7
	"	"		28. V.	6657	90.4	3.8
#2	Bultfontein	Pretoria	(p)	28. III.	26664	61.7	4.2
#3	Vastfontein	Pretoria	(p)	15. III.	3373	93.5	3.5
#4	Cullen Horse'n Hound	Pretoria	(p)	14. III.	22609	89.6	4.3
#5*	Bieulieu	Johannesburg	(j)	27. II.	12678	86.8	4.0
#6	Gosforth Park	Johannesburg	(j)	11. III.	4721	90.7	3.6
#7*	Turffontein	Johannesburg	(j)	11. III.	966	74.8	2.9
#8	Diepsloot	Johannesburg	(j)	18. III.	55620	92.4	4.7
#9	Cellehof	Johannesburg	(j)	27. III.	57400	89.7	4.7
#10	Northlands	Johannesburg	(j)	27. III.	21464	89.7	4.3
#11	Burlington	Johannesburg	(j)	27. III.	149430	97.2	5.2
#12*	Brandbach stud	Cullinan	(m)	23. III.	94478	86.0	4.9
#13	Goedehoop	Middelburg	(m)	23. III.	65846	67.6	4.6
#14	Vaalbank	Middelburg	(m)	23. III.	10544	80.5	3.9
#15	Woestalleen	Middelburg	(m)	23. III.	48248	72.3	4.5
#16	Generaalsdraai	Belfast	(m)	23. III.	73634	68.6	4.7
#17	Leeubank	Belfast	(m)	23. III.	752	36.3	2.4
#18	M'bulwa	Sabie	(l)	20. III.	1024	79.4	2.9
#19	Malan	Sabie	(l)	20. III.	8064	89.3	3.9
#20	Weber	Sabie	(l)	20. III.	20335	95.2	4.3
#21	Ponieskrantz	Pilgrims Rest	(l)	20. III.	3200	91.6	3.5
#22	Skukuza	Kruger Park	(l)	20. III.	89376	94.2	4.9
#23	Sunrise	White River	(l)	20. III.	221878	83.7	5.3
#24	Kumula	White River	(l)	20. III.	486465	93.6	5.7
#25	Mataffin	Nelspruit	(l)	20. III.	477920	94 9	5.7
#26	Clairwood	Durban	(n)	3. IV.	19206	90.5	4.2
#27	Summerveld	Shongweni	(n)	1. IV.	19924	51.5	4.0
#28	Ashburton	Pietermaritzburg	(n)	2. IV.	38834	98.2	4.6
#29	Hillside	Springfontein	(s)	11. V.	1356	84.0	3.1
#30	Gelykfontein	Venterstad	(s)	11. V.	251	79.4	2.3
#31*	G. Player stud	Colesberg	(s)	30. IV.	5651	94.1	3.7
				1. V.	1946	91.3	3.2
#32	Wolwekop	Middelburg	(s)	27. IV.	367	17.5	1.8
#33	Rooispruit	Middelburg	(s)	27. IV.	303	46.2	2.1
#34	Redlands	Middelburg	(s)	27. IV.	473	19.8	2.0

(continued)

Table 1 *(continued)*

Site no.	Farm/site	Town/district (region group)		Date	Total *Culicoides*	% *imicola*	Log *imicola*
#35	Meersgeluk	Middelburg	(s)	27. IV.	410	30.7	2.1
#36	Duikerdorings	Graaff-Reinet	(g)	28. IV.	608	3.1	1.3
#37	The Island	Graaff-Reinet	(g)	28. IV.	208	9.5	0
#38	Tangey's View	Graaff-Reinet	(g)	28. IV.	327	6.7	1.3
#39	Syfergat	Uitenhage	(u)	21. II.	714	25.8	2.3
#40*	Fairview	Port Elizabeth	(c)	19. III.	83	1.2	0
#41	Arlington	Port Elizabeth	(c)	5. III.	639	0	0
#42	Danika stud	Port Elizabeth	(c)	20. II.	684	0	0
#43	Animal Welfare	Port Elizabeth	(c)	6. II.	26	0	0
#44	Yellow Sea dairy	Port Elizabeth	(c)	19. III.	78	0	0
#45	Ascot stud	Port Elizabeth	(c)	13. II.	1499	0	0
#46	Willows	Port Elizabeth	(c)	26. III.	25	0	0
#47	?	Bathurst	(c)	1. III.	479	0.2	0

Table 2. Average rainfall figures at Onderstepoort (1909–1995), and above-average figures for AHS outbreak years of 1922/3 and 1995/6

Month	Average (mm)		
	1909–1995	1922/3	1995/6
July	5.2	0.0	0.0
August	9.1	78.0	2.1
September	15.8	?	14.3
October	65.8	99.8	60 1
November	120.5	133.6	166.3
December	119.6	160.2	201.5
January	127.1	306.1	171.6
February	88.1	90.9	279.4
March	82.8	79.2	67.1
April	47.3	24.4	43.8
May	15.3	0.3	24.7
June	5.5	?	1.3
Total	702.1	> 972.5	1032.2

trap was hung from the eaves of any building forming part of a stable yard. The size of a *Culicoides* light-trap collection was estimated by subsampling [21]. Species were identified using the preliminary wing-picture atlas developed over the last 15 years at the OVI.

Other data

Rainfall data (Table 2) were obtained from the South Africa Weather Bureau. The route of the float (Fig. 1) that transported 12, later 10, horses between Johannesburg and Port Elizabeth was plotted based on telephone conversations with the float owners and the

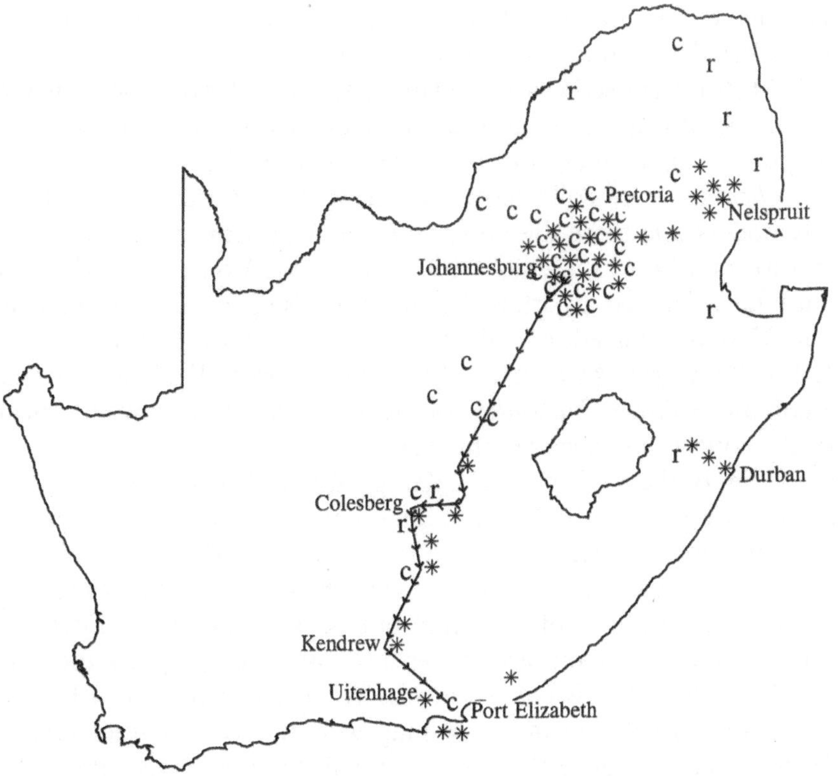

Fig. 1. The distribution of confirmed (*c*) and rumoured (*r*) cases of African horse sickness during the 1996 outbreak in South Africa. Asterisks show locations of light-trap sites in the survey. Arrowed line is the route between Johannesburg and Port Elizabeth taken by the float carrying horses a pair of which were later found to be infected with AHS

driver; exact times of loading and unloading were taken from the driver's logbook. The times given were confirmed by the owner or stable manager at whose premises the horses were either loaded or unloaded. All the sites visited at night by the float were sampled for *Culicoides*; only three of the five sites visited by the float during the day were sampled. As regards information and general interpretations of the outbreak numerous veterinarians, and colleagues at the OVI and the Faculty of the University of Pretoria, were consulted. Nearly all horse owners, stud managers, etc. were interviewed at stables sampled for *Culicoides*.

Results and discussion

Judging from the meagre published records available, the 1996 outbreak of AHS appears to be the third largest experienced in South Africa this century. The largest was the aforementioned outbreak of 1922/3 in which approximately 1700 equines died. As gleaned from newspaper clippings the second largest occurred in the western Cape in 1934 in which 1000 horses died in a broad swathe stretching from Porterville in the north-west to Robertson in the south-east of this region. Ominously, part of this area today lies within, and adjacent to, the eastern boundary that defines the AHS-free zone proclaimed in 1997 (Draft Commission Decision VI/2093/94 – Rev 11-1, dated 5/11/1996).

Table 1 lists the collecting sites and districts, and dates of collection during the 1996 outbreak; also given is the total number of *Culicoides* collected, and the percentage and log number of *C. imicola* at each site. The table shows that 4.78 million *Culicoides* were collected of which 94.2% were *C. imicola*, the species currently understood to be the only vector of AHS virus in Africa and beyond. The capture of 1054620 midges (99.1% *C. imicola*) on the 18 March 1996 at Kaalplaas, near Onderstepoort, is also the largest single collection made in South Africa in 35 years. Indeed, it ranks as one of the largest *Culicoides* collections reported in the world. As a rule < 3% of light-trap collections made in South Africa contain > 50000 *Culicoides*. In the 1996 outbreak, however, nearly one third of the collections exceeded this size.

Thirty-four of the approximately 120 species of *Culicoides* known to occur in South Africa were captured. The thirteen species most prevalent in the outbreak are ranked in Table 3; the first six species made up 99% of all *Culicoides* collected. *Culicoides imicola* was the most abundant and most prevalent species; its distribution closely matched that of the AHS outbreak. *Culicoides imicola* was only absent in the Port Elizabeth area (sites #41–#46); this supports the conclusion (given below) that the two AHS deaths reported from Port Elizabeth, and confirmed by virus isolation and serotyping, were the result of infections acquired outside the area. In Table 3 the current prevalence ranking is compared to that compiled during a three-year (1984–1986) survey of *Culicoides* found at 34 sites throughout South Africa [22]. While there is only moderate consensus, both studies show *C. imicola* to be dominant, widespread and abundant. This suggests that permanent populations of *C. imicola* have long been widely established in South Africa, and that only fluctuations in population size occur, these being between wet and dry seasons.

For an arthropod to be accepted as a vector of an arbovirus it must satisfy various criteria: a) its geographic distribution and seasonal prevalence must parallel that of the disease; b) after the arthropod takes an infective blood meal the virus must be able to replicate; and c) after an appropriate incubation period the arthropod must be able to transmit the virus. *Culicoides imicola* was the only species collected during the outbreak to fulfil the first criterion. AHS virus has been repeatedly isolated from *C. imicola* both in Africa and countries bordering the Mediterranean, fulfilling the second criterion to a great extent, and so no effort was made in this outbreak to isolate AHSV from it, or any other species of *Culicoides*. Evidence that *C. imicola* fulfils the third criterion is almost overwhelming ([5] and R. M. du Toit, pers. comm., cited in [23]) although it is remarkable that to date there have been no attempts, successful or unsuccessful, to obtain proof by infecting susceptible horses with AHS using pure samples of *C. imicola*. The extremely low numbers of species of *Culicoides* other than *C. imicola* that were caught in the present study, and their more restricted geographical distribution, suggest they played an extremely small, or no, role in the transmission of AHS virus during the 1996 outbreak.

Table 3. Prevalence and abundance of 13 *Culicoides* species most commonly encountered in 66 collections made at 47 sites during the 1996 outbreak of AHS in South Africa. The final column is the abundance rank of the same species according to [22]

Abundance rank	Species	% of all *Culicoides*	Prevalence in light-traps	Abundance rank [22]
1	*C. imicola*	94.2	60/66	1
2	*C. zuluensis*	3.0	44/66	5
3	*C. magnus*	0.7	25/66	10
4	*C. bolitinos*	0.6	34/66	15
5	*C. leucostictus*	0.4	33/66	7
6	*C. pycnostictus*	0.1	28/66	6
7	*C. nivosus*	< 0.1	15/66	8
8	*C. milnei*	< 0.1	13/66	17
9	*C. neavei*	< 0.1	11/66	16
10	*C. enderleini*	< 0.1	10/66	2
11	*C. subschultzei*	< 0.1	9/66	2
12	*C. onderstepoortensis*	< 0.1	9/66	20
13	*C. gulbenkiani*	< 0.1	9/66	11

If the data in Table 1 are examined as a whole, a significant decrease in log numbers of *C. imicola* is discernible as one moves from the top of the data set to the bottom (i.e. roughly from north to south). This trend is graphically depicted in Fig. 2 where the decline in log numbers of *C. imicola* parallels the decline in cases of AHS (upper curved arrow). It is suggested here that the sudden disappearance of *C. imicola*, especially in the Graaff-Reinet and Port Elizabeth areas (g and c in Fig. 2), may be due, principally, to a change in soil type. As illustrated by the diminishing scale at the bottom of Fig. 2 there is a general tendency for the clayey soils in the north to give way to sandier soils in the south. The largest populations of *C. imicola* were encountered at Kaalplaas (site #1). This farm is sited on dense cotton-soils; these are very clayey and thus highly moisture-retentive. At the opposite end of the scale is the absence of *C. imicola* on the nutrient-poor, quick-draining sandy soils found along the southern coast (sites #40–#46). This includes the greater Port Elizabeth area, a frost-free area which experiences a high, year-round average rainfall of > 600 mm/annum, but from which no cases of AHS had previously been reported this century.

The larvae of *C. imicola* live in the top one centimetre of the soil. Throughout their 7–10 day development period they require moisture and an ample supply of micro-organisms. It seems reasonable to conclude that *C. imicola* would therefore be unable to establish itself in both sandy and arid areas where moisture in the upper layer of the soil either drains out rapidly in a day or two, or is not regularly replenished. On rich clayey soils, however, intermittent rain (or irrigation) is sufficient to keep the soil saturated for long periods, and so enables *C. imicola* to develop large foci if animals, as sources of blood meals, are present. The general accumulation of clay in bottom-lands or depressions would thus suit *C. imicola*, and likely explains why the early colonists of the Cape had already observed more

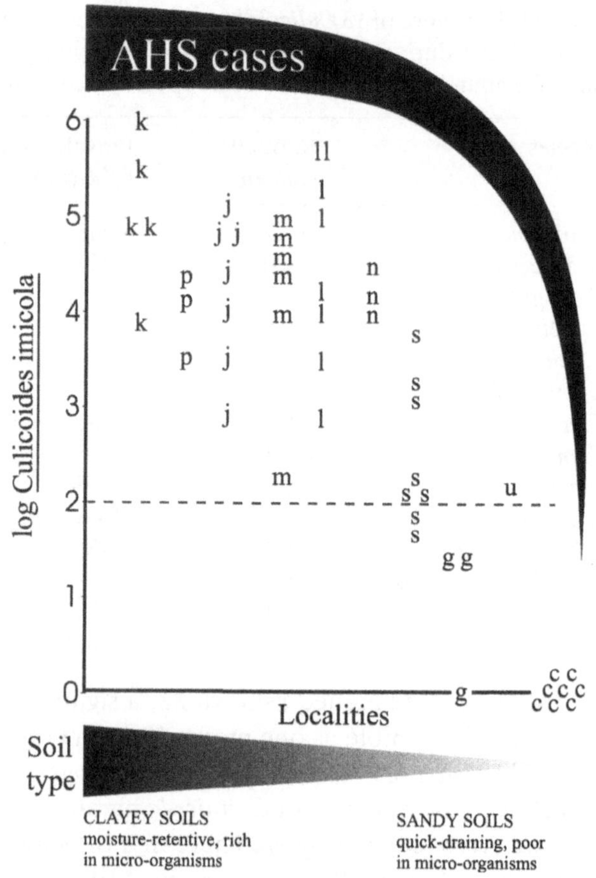

Fig. 2. A schematic representation of the proposed relationships between soil type, numbers of *Culicoides imicola* and the number of cases of African horse sickness during the 1996 outbreak in South Africa based on 52 insect collections at 47 sites (see Table 1). Sites were allocated to 10 region-groups which, moving from left to right, show a decrease in AHS cases and an increase in soil sand content. Region-groups are: Kaalplaas farm (*k*); Pretoria (*p*); Johannesburg (*j*); south-central Mpumalanga (*m*); eastern Mpumalanga lowveld (*l*); Natal (*n*); Free State and north-eastern Cape (*s*); Graaff-Reinet area of central Cape (*g*); Uitenhage, 30 km inland from the southern coast (*u*); and south-eastern Cape coast (*c*)

than 200 years ago, that '... low-lying parts ... are the permanent home ...' of AHS [16]. It may be from such foci that *C. imicola* spreads to colonize new areas during periods of excessive rain. In the 1996 outbreak it was noticed in the Johannesburg area that even the highest points of an undulating terrain would display seepage. As a result the high-lying Diepsloot stable (site #8) had the same numbers of *C. imicola* as found two km away at the low-lying Cellehof stable (site #9).

During the latter quarter of 1995 and the first quarter of 1996 most of South Africa experienced rainfall well above average (Table 2). Only one of a number of outbreaks of AHS to have occurred in South Africa this century has been well documented [4]; therein it was noted that prior to the 1922/3 outbreak '... very heavy rains fell throughout the country'. That outbreak commenced slowly in

October 1922 and reached its height in March/April 1923; 1700 horses, 114 mules and 100 donkeys died. The outbreak was described as '... one of the worst ... yet experienced' [4]. Of relevance to this study is that Du Toit includes observations on 42 (17%) of 242 horses that died on the Onderstepoort farm of 'Kaalplaats' which had the reputation of being '... undoubtedly one of the worst horse-sickness farms in the Transvaal'. It was on this farm that 12 (15%) of 80 horses died in 1996. Despite a 75-year gap in observations Kaalplaas retains its AHS notoriety. Or is it endemicity? Undoubtedly this notoriety is related to its capacity to produce exceptionally high numbers of *C. imicola*. Under such a massive field challenge from *C. imicola* it is clear that unvaccinated horses would stand little chance of survival as the various serotypes of AHS virus, if present, would be disseminated most efficiently. It remains unclear as to how the virus is introduced onto Kaalplaas each season.

What numbers of *C. imicola* can one expect in an average or dry season when few cases of AHS are reported? Between the years 1963–1970, *Culicoides* were collected almost nightly at Onderstepoort (± 3 km from Kaalplaas), and catch sizes charted against temperature and rainfall [14]. In a high rainfall season (> 1000 mm; July 1966–June 1967) the average number of *Culicoides* collected/night in the month of March was 124 267. Two seasons later (< 600 mm; July 1969–June 1970) the average catch size in March dropped to 598 *Culicoides*/night. This represents a 208-fold decrease. The same author also reported that the catches consisted chiefly of *C. pallidipennis* (= *C. imicola*), which at times accounted for up to 97.4% of the catch. The data from 1966/7 are very similar to those collected 30 years later at Kaalplaas during the 1996 outbreak of AHS. These two sets of data demonstrate the long-term stability of *C. imicola* as the dominant species in the area, and also that high rainfall (> 1000 mm) leads to an explosion in the numbers *of C. imicola.* It therefore seems reasonable to assume that similarly large numbers of *C. imicola* were active during the 1922/3 outbreak.

While there was not an outbreak of AHS during the high rainfall season of 1966/7, it was the year in which the "new" arboviral disease, equine encephalosis (EE), was recognized. It was found to be not only widespread but also the cause of deaths amongst horses [6]. In South Africa twelve isolations of EE virus (EEV) have been made from pools of unidentified *Culicoides* [15]; all isolations were made in areas where *C. imicola* can become abundant. It is interesting to note that three of the original isolates of EEV, i.e. Bryanston, Kaalplaas, and an unnamed third one from Colesberg, came from areas sampled during the 1996 outbreak of AHS, and all were found to possess very large populations of *C. imicola*. The sites were #1, #5, #8–#11 and #31; the owners of the last two sites reported that a large number of horses had fallen ill (> 20 at each stable) but that they had recovered. These may have been cases of EE, a disease recognised to have a low mortality rate.

As regards population sizes of *C. imicola* in wet and dry years similar, but fewer, comparative data are available for the Mpumalanga lowveld, the recognized AHS 'hotspot' of South Africa, which lies approximately 250 km east of Kaalplaas. Between August 1991 and August 1993, years of exceptional drought in this part of South Africa [24], monthly collections were made on 12 farms [12]. In March 1992 and March 1993, 15 collections yielded an average of 3980 and 12965 *Culicoides* respectively, of which 80% were *C. imicola*. During March 1996, eight of the same Mpumalanga farms (sites #18–#25) were resampled, and now yielded a higher average catch size of 147 869 *Culicoides* (90% *C. imicola*). This represents a 13–45-fold increase over dry-year catches. While there is no significant difference to be seen in Fig. 2 between the 1996 Kaalplaas (k) and Mpumalanga wet-season catch sizes, how is the marked differential between their dry-season catches explained? While the cause is likely multi-factorial it is important to note that in the dry years of 1991–1993 the pastures in the Mpumalanga paddocks were being irrigated at least twice weekly. No irrigation is practised at Kaalplaas, i.e. after the rainy season ends the soils over large areas dry out completely. This, and a lower winter minimum temperature, helps to 'knock-back' *C. imicola* to small foci in the lower-lying wet depressions at Kaalplaas. The data indicate that the higher numbers of *C. imicola* in Mpumalanga during the dry season were being maintained by irrigation; irrigation negates the suppressant effect that drying soils have on moisture-dependent *C. imicola* larvae.

Theiler [19] noted some 80 years ago that '... stabled animals can be protected against the disease effectively, provided the stable is insect proof.' However, the supposed value of stabling, and other remedies, have never been researched in Africa or any other part of the world where AHS occurs. Two observations merit brief discussion: (i) At site #31, Colesberg, all nine deaths were amongst unstabled horses; 60 stabled horses were unaffected. In this instance stabling appeared to be effective. However, horses out in the open, and surrounding a stable, may act as decoys deflecting *C. imicola* away from stabled ones. If these decoys are removed, or stabled, *Culicoides* may switch to feeding on stabled animals. The current data reveal that exceptionally large numbers of *C. imicola* were caught even where good quality stabling was given, indicating that *C. imicola*, which depends upon blood for the development of its eggs, was feeding successfully. Feeding may have been promoted by the common practice of leaving the upper half of a stable door open for ventilation. (ii) During the 1996 outbreak a few owners would, each night, encircle their stables with drums of smoking sawdust. The thick smoke was deemed 'bad for midges'; smoking of stables is a farmer's remedy that can be found in the literature dating back one hundred years [8]. However, light-traps operated in such palls of continuous smoke still yielded enormous catches of 50000–150000 *Culicoides* (> 92% *C. imicola*); these catches were as large as any made at smokeless stables.

Only in recent years was AHS declared a notifiable disease in South Africa. Despite this the 1996 outbreak witnessed, for the first time, a lack of centralisation in the current system of diagnosis. Case reporting remained at a low level; it is estimated that only 10–20% of the deaths were confirmed. One result of these breakdowns is that precise time-sequenced and age-linked information on the geographic spread of the outbreak are now lacking. Such data are needed to bear out, or counter, the oft-voiced fear or accusation, that the vaccine is not efficacious. Annual vaccination requires that two doses are given three to five weeks apart at the beginning of the season (September). Ideally, vaccination should be accompanied by a reduction in strenuous exercise, etc. However, because of this, many owners, especially those involved in horse-racing, do not vaccinate, saying that vaccination dulls the performance of the horse, and also, that the six-week 'go slow' period interferes too greatly with training and racing schedules. Fortunately, however, and also, curiously, AHS has to date not constituted a major problem in the racing industry in South Africa. The same observation was previously made in Zimbabwe [2, 3] and was ascribed to two facts: (i) most race horses are kept in urban and peri-urban locations, and (ii) being stabled at night the horses are not exposed outdoors when *Culicoides* are active. A third contributing factor may be that non-thoroughbred horses, i.e. ones that usually run free in the field and so are more likely to have been exposed to AHS-virus-infected *Culicoides*, are very rarely brought into contact with race horses. Whatever the reasons the racing industry, in the absence of broadscale vaccination, remains vulnerable to AHS. Sampling for *Culicoides* at two racing stables in peri-urban Johannesburg (sites #6 and #7) and one in Durban (#26), revealed *C. imicola* to be abundant enough to spread the virus if it were to be introduced into these stables and, indeed, a death from suspected AHS was reported at site #7. Another cogent reason for the race horse industry to comply with the new regulations on compulsory vaccination against AHS is that race horses are moved almost daily, and rapidly, to many centres throughout South Africa. Therefore, during one or other outbreak, it is only a matter of time before an infective horse is moved into a virus-free area. If the vector, *C. imicola*, is present a protracted and devastating outbreak can result amongst unvaccinated animals. As has been noted [10], vaccination of horses after the virus has been introduced rarely succeeds in preventing outbreaks.

At present there is little hope that *Culicoides* can be controlled chemically, especially if one has to deal with the enormous numbers reported above. Thus vaccination remains the major line of defence in the southern half of Africa where nine serotypes of the virus are present. It is worth repeating that in the two largest outbreaks of AHS to occur in South Africa 70 000 and 25 000 horses were lost within a season in 1854/5 and 1891/2 respectively. That South Africa no longer experiences such devastating losses, despite a larger horse population, is largely attributable to the widespread use of a polyvalent vaccine [7]. Shrinkage in range and numbers of zebras, which cycle the virus asymptomatically [1], is also likely to have played an important part in restricting the disease to the northern, north-eastern and central parts of the country.

The deaths of two horses in Port Elizabeth – case histories

On 20 April 1996 two horses died of suspected AHS at the Fairview racecourse stables 15 km west of Port Elizabeth (P.E) (site #40). The disease was later confirmed by virus isolation and serotyped as AHSV-3 cross-reacting with 7, and AHSV-4 (T. Gerdes, pers. comm.). The deaths were puzzling as they had occurred in the only area of South Africa reported to be free of *C. imicola* [13]. The fact that the horses died eight days after arrival in P.E. coincides with the known incubation period of the disease and so made it probable that the infections had been acquired outside the area. These two horses were amongst eight that had originated from Duikerdorings farm in the Kendrew district south of Graaff-Reinet (± 200 km north of P.E.). Within a week of the deaths three light-trap collections were made at Duikerdorings (site #36), and on two farms nearby (sites #37 and #38). Despite warm and calm conditions only a few individuals of *C. imicola* were caught amongst 20 species of *Culicoides* and a good diversity of other insects. The area has sandy soils and, at the time of collecting, conditions were dry, reflecting the fact that this part of South Africa experiences a low average rainfall of 300 mm/annum. Thus the ecological elements normally associated with large foci of *C. imicola* (i.e. moisture-retentive clayey soils, a higher rainfall and/or irrigation of pastures) were absent in the Kendrew area. This supported the owners' contention that he had never lost a horse to AHS in 47 years of breeding (and non-vaccination). These facts indicated that one had to seek elsewhere to answer the same owners' query: '... why do my horses, the minute they leave my farm, contract the illness?' This led to the route of the transport trailer being traced (Fig. 1). This revealed that the float had departed its Johannesburg base on 11 April to load first a horse at 16:20 at Lanseria, an hour later another at Randjiesfontein, and finally two horses at Turffontein racecourse (site #7), still within the Johannesburg area. Around nightfall the float departed southwards in the direction of Bloemfontein. At midnight (00:30, 12. IV. 1996) a horse was unloaded at Hillside farm, Springfontein (site #29); at 02:00 a second horse was unloaded at Gelykfontein stud, Venterstad (site #30). Both farms were subsequently sampled for *Culicoides*, and though the catches were moderately small, *C. imicola* was the dominant species. However, these owners reported no horses dying on their farms. After departing from Gelykfontein the float proceeded to Colesberg where the driver stopped, refuelled, and then rested until 06:00. He arrived in Kendrew at 09:20 to load the final eight horses. These, and the two Turffontein horses, were unloaded into two yards of the Fairview racing stables in the afternoon of 12 April 1996; eight days later two of the Kendrew horses died. At this time (April 1996) some 60 horses were reported to have died in the wider Colesberg area (G. Murdoch, pers. comm.); one stud (site #31) lost nine horses amongst many reported to have been ill. Two of four light-trap collections made at this stud were analysed and, in contrast to Kendrew, revealed *C. imicola* to comprise > 90% of unexpectedly large numbers of *Culicoides* captured. The abundance of *C. imicola* was surprising as Colesberg lies on the cold South African interior plateau, and is an area favoured by horse-breeders. However, discussions with breeders long resident in the Colesberg area confirmed that AHS is known to occur and that it always appears in April. No horse deaths were reported 200 km further south in the wider Graaff-Reinet/Kendrew area; this paralleled the significantly lower numbers of *C. imicola* collected there.

If one takes the various circumstances into consideration, i.e. widespread deaths, large numbers of *C. imicola*, and a horse-transport float visiting stables in the Colesberg area during the night (and remaining stationary for a number of hours with horses on board), it seems possible that some AHSV-infected

C. imicola could have entered the float and later, in the well-protected, and darkened, confines of the float infected two of the Kendrew horses.

An alternative hypothesis is that infective *Culicoides* entered the float in the Johannesburg area possibly at Turffontein (site #7) which was visited at dusk, where suspected AHS led to the death of one horse, and where *C. imicola* was captured.

Three aspects of the P.E. deaths merit discussion: (i) Once in P.E. the two infective horses spent their entire viraemic phase in the Fairview stables which held up to 400 horses (most unvaccinated). That no other horses died at these stables suggests that no local transmission of AHS virus took place. Indeed, these appear to be the first deaths from AHS reported in this part of South Africa this century. This is to be expected in an area where *C. imicola* is absent and also implies that no other vectors of the disease occur in the P.E. area, and also, in the absence of a biological vector, that mechanical transmission of AHS virus is not easily effected by other insects. (ii) As regards the historical presence of AHS in the south-eastern Cape this century an outbreak was reported in Uitenhage (u in Fig. 2) in the 1913/4 season. Uitenhage lies inland 30 km north of P.E. and is sited on the richer soils of the Kwa-Zungwa/Swartkops river valley. As can be seen in Table 1 (site #39) significant numbers of *C. imicola* were collected there following early season floods. Based on these, and historical, data the dotted line crossing Fig. 2 at log 2 *C. imicola* was chosen arbitrarily to separate areas at risk to AHS (above line) from those not at risk (below line). While this '100 *C. imicola*/light-trap risk line' may reflect the current and historical evidence it has never been calculated what percentage of *C. imicola* adults occurring at a site are actually captured in a light-trap. It may be as low as 1%. If so, it would be safer to lower this 'risk line' to log 1 *C. imicola*. This would leave all sites (#40–#47) in the sandy coastal zone, and only one inland site (#37), risk-free. (iii) While it has been argued that an outbreak of AHS will only occur after the introduction of an infective insect [9] the mode of introduction was not alluded to. Though it is well documented that mosquitoes can be transported by plane, train or road [18], no research has been done to ascertain whether *Culicoides* can be spread in a similar manner. If they are, as the data presented here suggest, it means that new outbreaks of AHS can be initiated by the vehicular transport of infected midges or equines.

Vehicular modes of dispersal would be rapid as infected equines and/or insects can be moved across large distances within 24 hours. Hitherto, on wind dispersal of infective *Culicoides* has been cited as the possible source of outbreaks of AHS or bluetongue in previously uninfected areas [17]. However, there remain conceptual problems with wind-aided dispersal of *Culicoides* over very long distances. For an outbreak to be initiated in an isolated locale requires that the timing and direction of the winds be precise for infected *Culicoides* to be deposited with pinpoint accuracy. It appears that the almost daily, and multidirectional, movements of vehicles transporting horses, and/or insects, creates better chances than winds do for spreading the virus from one stable to another. As far as the writer is aware the entry of vehicles by *Culicoides* has never been documented. Being chiefly nocturnal in their habits, and thus dependent on their powers of olfaction to locate hosts, it is probable that *Culicoides* will enter a stationary transport float especially if the midges are present in large numbers, and if horses are already on board. On the road these floats are not open to the elements, and so provide a protected environment for stowaway *Culicoides*. The possible transport of *Culicoides* needs to be researched as it remains an unrecognized threat in regulations surrounding AHS in South Africa. During periods of above average rains the insecticidal treatment of transport floats would be especially important as regards maintenance of the AHS-free zone of the south-western Cape.

Acknowledgements

Over the years I have benefitted greatly from my interactions with many colleagues. I should like to dedicate this small study to them: Banie Barnard, Matthew Baylis, Alan Dyce, Baltus Erasmus, Adele Faul, Truuske Gerdes, Alan Guthrie, Huck Jäger, Philip Mellor, Errol Nevill, Janusz Paweska, Albie van Dijk, Gert Venter and Bev Zietsman. Also my appreciation to Karien Labuschagne and Connie van Huyssteen for their disciplined help in the field and the laboratory. Finally, I thank the numerous members of the South African horse community who not only allowed me to collect *Culicoides* on their properties, but responded always positively to many questions posed.

References

1. Barnard BJH (1993) Circulation of African horsesickness virus in zebra (*Equus burchelli*) in the Kruger National Park, South Africa, as measured by the prevalence of type specific antibodies. Onderstepoort J Vet Res 60: 111–117
2. Blackburn NK, Swanepoel R (1988a) African horse sickness in Zimbabwe: 1972 to 1981. Trop Anim Health Prod 20: 169–176
3. Blackburn NK, Swanepoel R (1988b) Observations on antibody levels associated with active and passive immunity to African horse sickness. Trop Anim Health Prod 20: 203–210
4. Du Toit PJ (1924) Horse-sickness in 1923. J Dept Agric 8: 370–382
5. Du Toit RM (1944) The transmission of bluetongue and horse-sickness by *Culicoides*. Onderstepoort J Vet Sci Anim Ind 19: 7–16
6. Erasmus BJ, Boshoff ST, Pieterse LM (1976) The isolation and characterization of equine encephalosis and serologically related orbiviruses from horses. Proc 4th Int Conf Equine Inf Dis, Lyon, pp 447–450
7. Erasmus BJ (1978) A new approach to polyvalent immunization against African horse sickness. In: Bryans JT, Gerber H (eds) Equine Infectious Diseases: Proceedings of the Fourth International Conference on Equine Infectious Diseases, 24–27 September 1976. Veterinary Publications, Princeton, pp 401–403
8. Hayes MH (1896) South African horse sickness (Oedema mycosis). Vet J 42: 22–31
9. Howell PG (1996) South African horse sickness is with us again. S Afr Sport Horse 1: 24
10. Lord CC, Woolhouse MEJ, Mellor PS (1997) Simulation studies of vaccination strategies in African horse sickness. Vaccine 15: 519–524
11. McIntosh BM (1958) Immunological types of horsesickness virus and their significance in immunization. Onderstepoort J Vet Res 27: 465–538
12. Meiswinkel R (1995) Afrotropical *Culicoides*: biosystematics of the Imicola group, subgenus *Avaritia* (Diptera: Ceratopogonidae). With special reference to the epidemiology of African horse sickness. M.Sc. thesis, University of Pretoria

13. Meiswinkel R (1997) Discovery of a *Culicoides imicola*-free zone in South Africa: preliminary notes and potential significance. Onderstepoort J Vet Res 64: 81–86
14. Nevill EM (1971) Cattle and *Culicoides* biting midges as possible overwintering hosts of bluetongue virus. Onderstepoort J Vet Res 38: 65–72
15. Nevill EM, Erasmus BJ, Venter GJ (1992) A six-year survey of viruses associated with *Culicoides* biting midges throughout South Africa (Diptera: Ceratopogonidae). In: Walton EW, Osburn BI (eds) Bluetongue, African horse sickness, and related orbiviruses: proceedings of the 2nd international symposium. CRC Press, Boca Raton, pp 314–319
16. Paton T (1863) The "horse sickness" of the Cape of Good Hope. Veterinarian 36: 489–494
17. Sellers RF, Pedgley DE, Tucker MR (1977) Possible spread of African horse sickness on the wind. J Hyg Camb 79: 279–298
18. Steyn JJ, Schultz KH (1955) *Aëdes (Ochlerotatus) caballus* Theobald, the South African vector of Rift Valley fever. S Afr Med J 29: 1114–1120
19. Theiler A (1916) The problem of horse-sickness. Report of the 13th annual meeting of the South African Association for the Advancement of Science. Pretoria 1915 (July 5–10)
20. Theiler A (1921) African horse sickness (Pestis Equorum). Dep Agric Sci Bull 19
21. Van Ark H, Meiswinkel R (1992) Subsampling of large light trap catches of *Culicoides* (Diptera: Ceratopogonidae). Onderstepoort J Vet Res 59: 183–189
22. Venter GJ, Nevill EM, van der Linde TC de K (1996) Geographical distribution and relative abundance of stock-associated *Culicoides* species (Diptera: Ceratopogonidae) in southern Africa, in relation to their potential as viral vectors. Onderstepoort J Vet Res 63: 25–38
23. Wetzel H, Nevill EM, Erasmus BJ (1970) Studies on the transmission of African horsesickness. Onderstepoort J Vet Res 37: 165–168
24. Zambatis N, Biggs HC (1995) Rainfall and temperatures during the 1991/2 drought in the Kruger National Park. Koedoe 38: 1–16

Author's address: R. Meiswinkel, Onderstepoort Veterinary Institute, Private Bag X05, Onderstepoort, 0110 South Africa.

The seasonal and geographical distribution of *Culicoides imicola*, *C. pulicaris* group and *C. obsoletus* group biting midges in central and southern Spain

M. D. Ortega[1], **P. S. Mellor**[2], **P. Rawlings**[2], and **M. J. Pro**[3]

[1] Laboratorio de Sanidad y Producción Animal, Junta de Andalucía, Cordoba, Spain
[2] Institute for Animal Health, Pirbright, U.K.
[3] Laboratorio de Sanidad y Producción Animal, Ministerio de Agricultura Pesca y Alimentación, Algete, Spain

Summary. Pirbright-type light traps were used to collect *Culicoides* biting midges (Diptera: Ceratopogonidae) at fifteen sites in twelve provinces of central Spain and Andalusia. A total of 293 625 *Culicoides* were collected in 1 387 samples over a two year period. These comprised approximately 9.2% *Culicoides imicola*, 11.4% *C. pulicaris* group, 1.6% *C. obsoletus* group and 12.2% *C. circumscriptus*. *Culicoides imicola* was present at ten of the fifteen sites; the five sites from which it was absent were the most eastern of the fifteen. The greatest abundance of this species was at Navalmoral in Caceres province. *Culicoides pulicaris* group were present at all sites; *C. obsoletus* group were present at twelve sites. The annual peaks in abundance were: *C. imicola*, August–October; *C. pulicaris* group, May–June; and *C. obsoletus group*, March–June. The geographical and seasonal distributions of *C. imicola* are consistent with those of the outbreaks of African horse sickness (AHS) and bluetongue (BT) during epizootics in Spain, and support the contention that *C. imicola* was the major vector of AHS and BT viruses.

Introduction

Bluetongue (BT) and African horse sickness (AHS) are viral diseases of ruminants and equines, respectively. Both diseases are endemic to sub-Saharan Africa but they have a history of incursion into Iberia. In 1956 an epizootic of BT began in Portugal, and soon spread to Spain [2] where the first outbreaks were recorded in August in the provinces of Badajoz and Huelva, and thereafter Caceres, Cadiz, Seville, Cordoba, Ciudad Real, and Malaga. There was an epizootic of AHS in

Spain in 1966 [3]. The first cases were reported in October and only a small area of southern Cadiz province was affected; Cadiz is the province nearest to North Africa where the virus was also present at that time and which is therefore the probable source of the incursion [3]. A second epizootic of AHS in Spain started in 1987 and continued until 1990 [4–7]. Outbreaks began in July 1987, and were restricted to the Madrid, Toledo and Avila provinces of central Spain [4]. The following year (1988) the disease appeared in Cadiz province, and in the year after (1989) it spread to the provinces of Cordoba, Seville, Huelva, and Badajoz in southern Spain [6] and into Portugal [9] and Morocco [8]. The disease reappeared for a fourth year in Spain in 1990, with outbreaks reported in Malaga province only [7]. The single BT and two AHS epizootics all affected most heavily the southern region of Spain (Andalusia), although not all of the provinces in that region [13].

The viruses that cause BT and AHS are transmitted by the bites of *Culicoides* biting midges and, in the Mediterranean region, *C. imicola* is believed to be the most important vector species [11]. Other *Culicoides* species may also be vectors, however. BT virus has been isolated from *C. imicola* in Israel [1] and from *C. obsoletus* in Cyprus [12] while AHS virus has been isolated from *C. imicola* in Spain [10] and from mixed pools of *C. obsoletus* and *C. pulicaris* in the same study.

The occurrence of *C. imicola* in certain regions of Spain has been documented [11, 13–16]. We report here the seasonal and geographic distribution of *C. imicola*, *C. pulicaris* group and *C. obsoletus* group biting midges in central Spain and Andalusia. In addition, the relationship between the presence of certain species of *Culicoides* and the occurrence of AHS and BT is described.

Materials and methods

This study was conducted between May 1993 and August 1995. Fifteen Pirbright-type light traps were set up at farms in twelve provinces in central and southern Spain (Fig. 1) and were operated twice-weekly for most of the study period. Light traps were set outdoors at a height of 1.5–2 m and were operated from sunset until the next morning. Insects attracted to the light were sucked by a 12 v fan into a flask containing a 5% aqueous solution of formalin to which two drops of detergent had been added. Collections were returned to the laboratories in Algete (sites 1–10; Fig. 1) and Cordoba (sites 11–15), and transferred to 70% ethanol. *Culicoides imicola*, *C. pulicaris* group, and *C. obsoletus* group were sorted, counted and identified by their respective characteristic wing patterns using a stereo-zoom microscope. For sites 1–10, when catch sizes were very large, sorting and counting were limited to subsamples of the total collection; when this occurred sub-samples containing at least 500 *Culicoides* were sorted. For sites 11–15 a fourth species of *Culicoides*, *C. circumscriptus*, was also counted.

Results

The results are summarised in Table 1. A total of 293 625 *Culicoides* were collected in 1 387 samples (sites 1–10, 254 528 *Culicoides* in 665 samples; sites 11–15, 39 097 *Culicoides* in 722 samples). Of these, there were 33 463 *C. pulicaris* group (11.4%), 27 003 *C. imicola* (9.2%), 4 625 *C. obsoletus* group (1.6%) and, for sites 11–15, 4 755 *C. circumscriptus* (12.2%). *Culicoides* spp. were collected in all

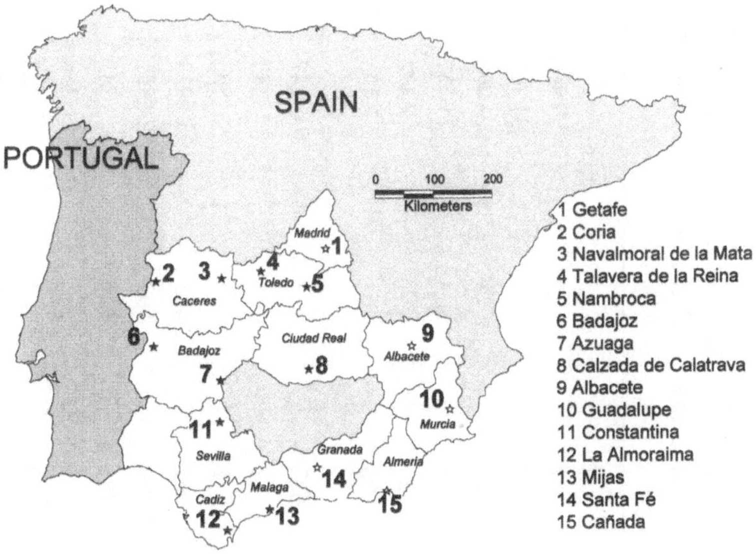

Fig. 1. The locations of the fifteen trapping sites in Spain. ★ Sites where *Culicoides imicola* is present; ☆ sites where *C. imicola* is absent. Province names are shown in italics

provinces but their numbers differed significantly from one province to another. The greatest catch of *Culicoides* was at Calzada de Calatrava in the province of Ciudad Real although, taking into account the number of samples as well, the largest relative density was at Navalmoral (Caceres). In contrast, at Cañada in Almeria only 25 *Culicoides* were collected in 105 samples. *Culicoides* were relatively scarce (< 1000 total) at Getafe (Madrid), Albacete (Albacete), Guadalupe (Murcia), Santa Fé (Granada) and Cañada (Almeria) but were abundant (> 2 000) at the remaining sites.

Those sites where *Culicoides* were scarce are also those where *C. imicola* appears to be absent (Table 1). *Culicoides* imicola was present at ten of the fifteen sites; the five from which it is absent are the most eastern of the fifteen (Fig. 1). The greatest catches of *C. imicola* were at Navalmoral (Caceres), followed by La Almoraima (Cadiz) and Constantina (Seville).

The other species of *Culicoides* that may act as vectors of AHS virus, *C. pulicaris* group, and *C. obsoletus* group [10], were more prevalent than *C. imicola* (Table 1). *Culicoides pulicaris* group were caught at all fifteen sites, and were more abundant than *C. imicola* at twelve. *Culicoides* obsoletus group were caught at thirteen of the fifteen sites, but were more abundant than *C. imicola* at only five.

The three species showed different seasonal distributions. The annual peak in the abundance of *C. imicola*, based on sites 11–13, was from August–October (Fig. 2), with lowest numbers after December, when temperatures were at their annual minimum [15]. The annual peak in the abundance of *C. pulicaris* group was generally May or June (Fig. 3) while that of *C. obsoletus* group was March–June (Fig. 4). This group was rarely collected after July when temperatures are at their annual maximum [15].

Table 1. Total numbers of *Culicoides* spp. caught using Pirbright-type light traps at 15 sites in central Spain and Andalusia over an approximately two-year period

Site	Number (Fig. 1)	Number of samples	Sampling period	*C. imicola*	*C. pulicaris* grp	*C. obsoletus* grp	*C. circum-scriptus*	All *Culicoides*
Getafe	1	87	5/93–12/94	0	37	0	[a]	771
Coria	2	41	6/93–11/94	158	2 405	61	[a]	9 640
Navalmoral	3	40	8/93–9/94	18 993	5 882	42	[a]	94 703
Talavera	4	73	7/93–11/94	31	147	27	[a]	2 963
Nambroca	5	48	6/93–12/94	5	2 051	64	[a]	24 437
Badajoz	6	89	7/93–12/94	535	2 360	29	[a]	1 332
Azuaga	7	17	10/93–11/94	64	1 656	20	[a]	2 438
Calzada de Calatrava	8	92	7/93–12/94	697	10 436	642	[a]	105 247
Albacete	9	92	5/93–12/94	0	39	1	[a]	143
Guadalupe	10	105	5/93–12/94	0	47	1	[a]	865
Constantina	11	112	4/94–8/95	1 524	1 019	961	2 463	10 922
La Almoraima	12	263	8/93–7/95	4 221	7 114	1 994	599	24 151
Mijas	13	82	9/93–8/95	775	243	781	1 441	3 741
Santa Fé	14	151	8/93–12/94	0	4	1	250	258
Cañada	15	105	5/94–6/95	0	23	0	2	25
Total		1 397		27 003	33 463	4 624	4 755	293 625

[a] Not recorded

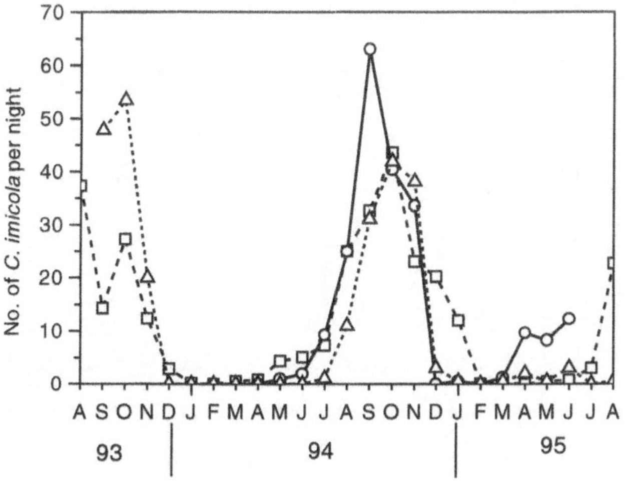

Fig. 2. Monthly mean catch of *Culicoides imicola* collected per night at the sites of La Almoraima (□), Constantina (○), and Mijas (△) from August 1993 through August 1995

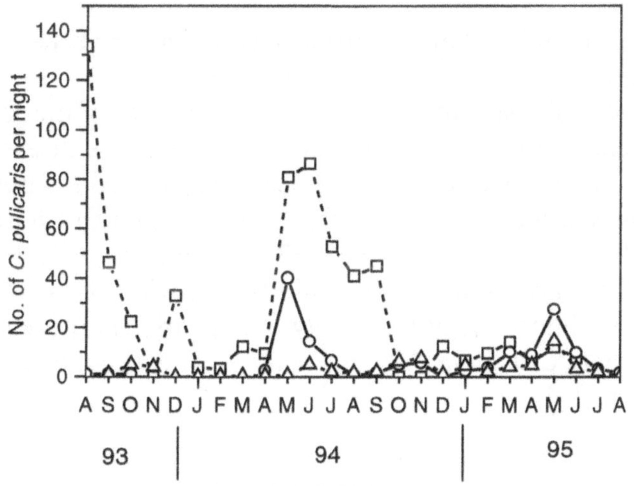

Fig. 3. Monthly mean catch of the *Culicoides pulicaris* group collected per night at the sites of La Almoraima (□), Constantina (○), and Mijas (△) from August 1993 through August 1995

Discussion

All outbreaks of BT and AHS in Spain have been reported between July and December. However, relatively few individuals of the *C. obsoletus* and the *C. pulicaris* groups appear to be present at this time of year, although they may be abundant earlier in the year. Therefore, even though AHS virus has been isolated from these two species, they may be relatively unimportant vectors of the virus. In contrast, there is considerable similarity between the seasonal distributions of BT and AHS in Spain, and the abundance of *C. imicola*. This study therefore lends

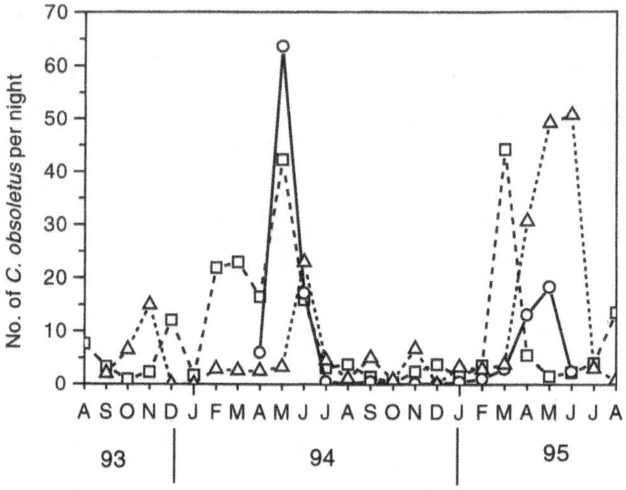

Fig. 4. Monthly mean catch of the *Culicoides obsoletus* group collected per night at the sites of La Almoraima (□), Constantina (○), and Mijas (△) from August 1993 through August 1995

support to the contention that *C. imicola* was the major vector or BT and AHS viruses during the three epizootics in the region.

The geographical distribution of *C. imicola* and that of outbreaks of BT and AHS are also consistent with this contention. Cases of the diseases were limited to the central and south-western areas of Spain [16]. *Culicoides* imicola appears to be absent in the south-east of the country, and in the north where no cases were reported [16]. Interestingly, during the 1987–1990 epizootic of AHS there were no cases reported in Caceres although within this province is Navalmoral, the site with the greatest abundance of *C. imicola*. It appears, therefore, that while AHS did not occur where *C. imicola* is absent, it also did not occur in all areas where the vector is present [13].

Acknowledgements

This work was supported by the European Commission (8001-CT91-0211) and the Junta de Andalucía. We thank those people who supplied the farms J. Fernández (Almería), J. M. Gómez (La Almoraima, Cádiz), M. Martín (Yeguada Indalo, Málaga), F. Garrido (LSPA, Granada) and L. González (Dehesa Frías, Sevilla), the field agents F. Calandria, F. Ramírez, M. A. Cornejo, P. Ramírez, J. L. Cordero, A. Perales and J. Calvo, and those who helped sorting the samples: M. D. Gasca and J. L. Humanes. We thank Dr. W. L. Jr Grogan for support and comments on the manuscript. In addition, M. J. P. wishes to express her thanks to M. D. O. for the taxonomic knowledge she has conveyed and without which much of the work presented here would have been impossible.

References

1. Braverman Y, Barzilai E, Frish K, Rubina M (1985) Bluetongue virus isolation from pools of *Culicoides* spp. in Israel during the years 1981 to 1983. In: Barber TL,

Jochim MM (eds) Bluetongue and related orbiviruses. Alan R. Liss, New York, pp 191–193 (Progress in Clinical and Biological Research, vol 178)
2. Campano Lopez A, Sánchez Botija C (1955) L'epizootie de fièvre catarrhale ovine en Espagne (Blue Tongue). Bull Off Int Epiz 50: 65–93
3. Diaz Montilla R, Paños Martí P (1967) Epizootiología de la Peste Equina en España. Bull Off Int Epiz 68: 705–714
4. Diaz-Yubero MA (1988) Study of the African horse sickness outbreak in Spain. Inf Epiz Esp 88/1/16: 2–3
5. Diaz-Yubero MA (1988) African horse sickness in Spain. Off Int Epiz 1: 34
6. Escribano Mora C (1989) African horse sickness in Spain. Off Int Epiz 2: 99–100
7. Escribano Mora C (1990) African horse sickness in Spain. Off Int Epiz 3: 100–103
8. Lhafi A, Tber A, Fikri, A, Laghzaoui K (1992) African horse sickness in Morocco: the epizootics of 1989 and 1990. In: Walton TE, Osburn BI (eds) Bluetongue, African horse sickness, and related orbiviruses. CRC Press, Boca Raton, pp 205–216
9. Machado Gouveia JM (1989) African horse sickness in Portugal. Off Int Epiz 2: 119–137
10. Mellor PS, Boned J, Hamblin C, Graham S (1990) Isolations of African horse sickness virus from vector insects made during the 1988 epizootic in Spain. Epidemiol Infect 105: 447–454
11. Mellor PS, Jennings DM, Wilkinson PJ, Boorman JPT (1985) *Culicoides imicola*: a bluetongue virus vector in Spain and Portugal. Vet Rec 116: 589–590
12. Mellor PS, Pitzolis G (1979) Observations on breeding sites and light-trap collections of *Culicoides* during an outbreak of bluetongue in Cyprus. Bull Entomol Res 69: 229–234
13. Ortega MD, Holbrook FE (1994) Presence of *C. imicola* (Diptera: Ceratopogonidae) in Jaen, Spain. J Am Mosq Control Assoc 10: 463
14. Ortega MD, Holbrook FE, Lloyd J (1997) Seasonal and geographical distribution of *Culicoides imicola* Kieffer (Diptera, Ceratopogonidae) in Southwestern Spain. J Am Mosq Control Assoc 13: 227–232
15. Ortega MD, Holbrook FE, Lloyd J (1998) Seasonal distribution and relationship to temperature and precipitation of the most abundant species of *Culicoides* in the five provinces of Andalusia, Spain. J Am Mosq Control Assoc (in press)
16. Rawlings P, Pro MJ, Pena I, Ortega MD, Capela R (1997) Spatial and seasonal distribution of *Culicoides imicola* in Iberia in relation to the transmission of African horse sickness virus. Med Vet Entomol 11: 49–57

Authors' address: Dr. M. D. Ortega, Delegación Provincial de Medio Ambiente, Tomás de Aquino s/n 7ª Pta, 14071 Córdoba, Spain.

The relationship between climate and the distribution of *Culicoides imicola* in Iberia

P. Rawlings[1], **R. Capela**[2], **M. J. Pro**[3], **M. D. Ortega**[4], **I. Pena**[2], **C. Rubio**[3], **A. Gasca**[4], and **P. S. Mellor**[1]

[1] Institute for Animal Health, Pirbright, U.K.
[2] Universidade da Madeíra, Funchal, Madeíra, Portugal
[3] Laboratorio de Sanidad y Producción Animal, Ministerio de Agricultura Pesca y Alimentación, Madrid, Spain,
[4] Laboratorio de Sanidad y Producción Animal, Junta da Andalucía, Cordoba, Spain

Summary. The biting midge *Culicoides imicola* was captured at 17 of 27 farms in Spain and Portugal during a survey of its distribution following outbreaks of African horse sickness in Iberia that occurred between 1987 and 1990. Farms were sampled approximately twice weekly from October 1992 to February 1995. Farms were widely spaced apart (maximum 850 km) and had considerable variation in climate. Across sites, summer temperatures ranged from 18.3°C–27.2°C; in winter the range was 4.4°C–11.6°C. Relative humidities in summer ranged from 37.2% to 90.1%. Proximity to southern Spain (Seville) was the most significant predictor of the presence/absence of *C. imicola*, but high summer temperatures and possibly dry summer conditions, were also important. Vila Nova de Milfontes in Portugal, where *C. imicola* was abundant and the climate is relatively cool, was an exception to the climatic trends at the other 26 sites. This exception points to a lack of knowledge of climatic requirements for immature development of *C. imicola.* The absence of *C. imicola* from the three most easterly sites, which have apparently favourable climates, suggests a relatively recent invasion by this species into Iberia.

Introduction

Outbreaks of African horse sickness (AHS) occurred in Spain and Portugal between 1987 and 1990 [4, 10, 11]. A routine importation of zebras (the wildlife reservoir of the virus causing AHS) from Namibia to a safari park near Madrid in 1987 turned into a significant event costing the lives of more than 1500 horses, mules and donkeys. At the time information was lacking on where *Culicoides imicola*, an efficient insect vector of African horse sickness virus (AHSV) and bluetongue virus (BTV), is distributed in Iberia. This Afro-Asian tropical and subtropical species of biting midge was found for the first time in southern Spain in 1982 [7] and in Portugal in 1984 [8]. These findings followed the earlier discovery of *C. imicola* at the margins of the eastern Mediterranean where outbreaks of bluetongue and AHS [1, 6] had also occurred.

The distribution of *C. imicola* in Iberia is now known to be extensive and to cover all those areas affected by the 1987–1990 outbreaks of AHS [2, 10]. *Culicoides imicola* is known to occur further north and west of the outbreak areas in Portugal [2], but it has not yet been recorded from the east or north of Spain [10]. This distribution of *C. imicola* in Iberia may be determined by certain, as yet unidentified, climatic factors such as winter temperature, which is a major factor controlling the survival and distribution of many insect species in Europe. In southern Europe, specifically within the Mediterranean climatic zone [3], understanding which climatic factors promote the establishment of viable and persistent populations of virus vectors, such as *C. imicola*, is important if we are to evaluate the effects of global climate change on disease [13]. As a first analysis of the relationship between environmental factors and Iberian collections of *C. imicola*, the effects of air temperature, relative humidity and geography on the presence and abundance of this vector are given in this article.

Materials and methods

Insect collection

Collection of biting midges, together with farm-based climate recording, started at most of the 11 sites in Portugal and 16 in Spain (Fig. 1) in October 1992 and continued through to February 1995. The minimum number of months of collection at any site was five. At all sites *Culicoides* were captured with Pirbright-Miniature light/suction traps (PMLTs) that operated from 1 h before sunset to between 06:00 and 09:00 h the following morning. Midges were attracted to the light of a 12 V, 25 W car bulb, then were pulled by a 12 V suction fan into plastic containers containing 20 ml of water and surfactant screwed to the base of the PMLT, and the midges were later preserved in either 70% ethanol or 2% formalin. Traps were intended to be operated twice weekly by a local agent. Samples of insects returned to the laboratory were sorted in a petri dish using fine forceps under a binocular microscope. In samples where there were fewer than 500 *Culicoides*, the entire catch was sorted and all major species were identified. Where samples were estimated to contain substantially more than 500 *Culicoides* a measured fraction of the sample was sorted and identified and estimated totals were then calculated. A decision chart or key was available to help identify the 10 *Culicoides* species or species groups of special interest as potential vectors.

Climate monitoring

Climate in the vicinity of 25 of the PMLTs was recorded using automatic weather stations, which measured air temperature, soil temperature (at 5 cm depth), relative humidity and wind speed (at 1–2 m above ground level). Air temperature and humidity probes were housed in a solar radiation shield at approximately 1.5 m above ground level. A datalogger was programmed to take readings every 5 min and record the average of the 12 readings per hour. Datalogger memory was sufficient to allow downloading of data at six monthly intervals. At the two sites without automatic weather stations (Castelo Branco and Talayuela) data were obtained from manual readings of standard instruments housed in Stevenson screens.

Fig. 1. Map of Iberia showing the location of the farms used in the study of *Culicoides imicola* distribution, population abundance and relationship with climatic and geographic factors

Data analysis

The capture of at least one *C. imicola* was taken to indicate the presence of the species at a site. Where *C. imicola* was present, relative abundance was inferred on the basis of whether *C. imicola* was present in < 5% of samples, from 5–20% of samples or > 20% of samples. Sites were ranked on the basis of certain climatic and geographic variables and the ranking compared with that of the relative abundance of *C. imicola*. Mean monthly air temperature was calculated by averaging the daily means which in turn were the mean of 24 hourly air temperature readings. Similar calculations were made for relative humidity. At Castelo Branco and Talayuela mean daily air temperature is the average of the daily maxima and minima.

Results

Culicoides imicola was captured at 17 of the 27 sites (Table 1). The site with the hottest mean monthly air temperature in summer was Barrancos (27.2°C), in southern Portugal, the coolest was Oviedo (18.3°C), north of the Cantabrian Mountains in northern Spain. The site with the coldest mean monthly temperature in winter was Alijó (4.4°C), in northern Portugal; the warmest was Cañada del Rosal (11.6°C), on the coast of eastern Andalusia. Across sites, the ranges of mean monthly air temperatures in summer and winter were only 9.0°C and 7.2°C, respectively, despite the large distances between certain sites (e.g. 850 km between Oviedo and Almoraima). The difference between winter and summer mean monthly air temperatures ranged from 9.9°C at Oviedo to 21.7°C at Freixo de Espada à Cinta and also at Calatrava, with an average of 17.2°C.

Table 1. Rank order of monitoring sites according to temperatures (°C): mean monthly air temperature during the hottest summer month and coldest winter month (for the period October 1992 to February 1995)

Site name	*C. imicola* presence[b]	Rank	Summer	Rank	Winter
Barrancos	+++	1	27.2	19	6.1
Constantina	+++	2.5	27.0	15.5	7.1
Coria	+++	2.5	27.0	12	7.7
Cañada del Rosal	–	4.5	26.9	1	11.6
Azuaga	+++	4.5	26.9	17	6.5
Guadalupe	–	6.5	26.8	3	11.0
Talayuela[a]	+++	6.5	26.8	14	7.2
Freixo de Espada à Cinta	+++	8	26.7	23	5.0
Calatrava	+++	10	26.6	24.5	4.9
Sagrajas	++	10	26.6	11	7.9
Getafe	–	10	26.6	21.5	5.5
Mijas	+++	12	26.5	2	11.4
Almoraima	+++	13	26.4	4	10.9
Talavera	++	14	25.9	18	6.3
Castro Marim	+++	15	25.8	6	9.9
Castelo Branco[a]	+	16	25.2	24.5	4.9
Nambroca	+	17	24.7	21.5	5.5
Santa Fé	–	18	24.2	20	5.6
Albacete	–	19	23.6	26	4.6
Evora	+++	20	23.0	9.5	8.2
Muge	+	21	22.8	13	7.4
Alijó	–	22	22.4	27	4.4
Monçao	–	23.5	20.8	7	9.5
Carvalhos	–	23.5	20.8	15.5	7.1
Vila Nova de Milfontes	+++	25	20.6	5	10.2
Vagos	–	26	20.0	9.5	8.2
Oviedo	–	27	18.3	8	8.4

[a] Air temperatures taken from Stevenson screen measurements

[b] "–" No *C. imicola* captured at monitoring site; "+" *C. imicola* present but captured in < 5% of samples; "+ +" *C. imicola* present but captured in > 5 and < 20% of samples; "+++" *C. imicola* present and captured in > 20% of samples

The 27 sites were ranked according to the air temperature of the hottest summer month and coldest winter month (Table 1). The summer temperature rankings corresponded well with the presence or absence of *C. imicola*: the vector was present at 11 of the 14 (top half of the table) and 4 of the 5 sites with the hottest summers, but was also present (and abundant) at the site with the third coolest summer (rank 25, Vila Nova de Milfontes). The winter temperature rankings gave a less clear pattern. *Culicoides imicola* was present at only 9 of the 14 and 3 of the 5 sites with the least cold winters.

The 27 sites were then ranked according to the mean relative humidity in summer and in winter, and by the number of days of low relative humidity

(< 60%) in summer (Table 2). *Culicoides imicola* was present at 10 of the 14 and 4 of the 5 sites with the lowest relative humidity in summer, and at 9 of the 14 and 3 of the 5 sites with the lowest relative humidity in winter. The vector was present at 10 of the 14 and 4 of the 5 sites with the greatest number of summer days of relative humidity < 60%. Contrary to these patterns, Vila Nova de Milfontes, where *C. imicola* was abundant, had a very high summer relative humidity and very few summer days of low relative humidity.

The topography of southern Iberia is complex, with several ranges of mountains and large river systems. The influence of topography and geography on the

Table 2. Rank order of monitoring sites according to relative humidity: mean summer relative humidity (%RH), number of dry days (%RH < 60) during the summer and mean winter relative humidity

Site name	*C. imicola* presence[b]	Rank	Summer %RH	Rank	No. dry days[c]	Rank	Winter %RH
Talayuela[a]	+++	1	37.2	2.5	87	24	86.6
Azuaga	+++	2	38.5	5.5	84	4	78.1
Getafe	–	3	39.3	1	91	20	84.8
Nambroca	+	4	41.1	8	82	22	85.5
Constantina	+++	5	41.8	4	86	6.5	79.2
Calatrava	+++	6	42.0	8	82	8	79.6
Coria	+++	7	42.6	10	79	13	81.6
Barrancos	+++	8	44.9	2.5	87	10.5	81.2
Freixo de Espada à Cinta	+++	9	45.8	5.5	84	25	87.2
Talavera	++	10	46.4	8	82	17	83.4
Sagrajas	++	11	47.9	11	75	15.5	83.1
Albacete	–	12	50.5	12	72	15.5	83.1
Guadalupe	–	13	53.7	14.5	59	1	54.0
Santa Fé	–	14	55.0	13	62	3	75.1
Alijó	–	15	56.6	14.5	59	18	84.1
Castro Marim	+++	16	57.9	16	51	6.5	79.2
Mijas	+++	17	58.5	17.5	46	2	67.9
Evora	+++	18.5	59.3	17.5	46	21	85.2
Castelo Branco[a]	+	18.5	59.3	21	20	19	84.6
Almoraima	+++	20	62.1	19	45	5	78.7
Cañada del Rosal	–	21	67.4	20	27	23	86.3
Monçao	–	22	70.1	24	6	26	88.2
Muge	+	23	70.2	23	7	27	88.5
Carvalhos	–	24	74.9	22	12	10.5	81.2
Vagos	–	25	78.4	25	5	14	82.5
Vila Nova de Milfontes	+++	26	79.2	26	1	12	81.3
Oviedo	–	27	90.1	27	0	9	80.6

[a] Relative humidities taken from Stevenson screen measurements
[b] See abundance key at the foot of Table 1
[c] See parameter labels in Table 4

sustainability of immigrant insect populations, such as *C. imicola*, can be difficult to assess. The 27 sites were ranked by latitude, longitude and by straight-line distance from Seville (Table 3), which is close to the presumed incursion route of *C. imicola* into Iberia from North Africa. *Culicoides* imicola was present at 10 of the 14 and 3 of the 5 most southerly sites, and at 10 of the 14 and 2 of the 5 most westerly sites. The vector was present at 12 of the 14 and at all 5 of the sites closest to Seville. This degree of correspondence is greater than that of any of the climatic variables examined so far.

The degree of correspondence between the presence/absence of *C. imicola* and the various climatic and geographic parameters considered so far is summarised in Table 4. Analyses of differences between the median values of each

Table 3. Rank order of monitoring sites according to geographic position: by latitude, longitude and straight-line distance (km) to Seville (37° 23' N, 6° 0' W)

Site name	*C. imicola* presence[a]	Rank	Latitude (N)	Rank	Longitude (W)	Rank	Distance to Seville
Almoraima	+++	1	36° 17'	18	5° 27'	5	138
Mijas	+++	2	36° 33'	20	4° 38'	6	146
Cañada del Rosal	–	3	36° 50'	25	2° 24'	12	285
Castro Marim	+++	4	37° 10'	8.5	7° 30'	3	119
Santa Fé	–	5	37° 12'	23	3° 43'	7	179
Vila Nova de Milfontes	+ + +	6	37° 42'	1	8° 47'	10.5	220
Constantina	+++	7	37° 53'	15	5° 38'	1	66
Guadalupe	–	8	38° 1'	27	1° 11'	20	382
Barrancos	+++	9	38° 7'	10	7° 1'	2	118
Azuaga	+++	10	38° 26'	16.5	5° 34'	4	130
Evora	+++	11	38° 30'	6	8° 2'	9	207
Calatrava	+++	12	38° 36'	22	3° 53'	10.5	220
Albacete	–	13.5	38° 56'	26	1° 55'	19	368
Sagrajas	++	13.5	38° 56'	12	6° 51'	8	197
Muge	+	15	39° 5'	2	8° 41'	13	294
Nambroca	+	16	39° 47'	21	3° 56'	18	330
Castelo Branco	+	17	39° 50'	8.5	7° 30'	16	316
Coria	+++	18	39° 51'	13	6° 49'	14	303
Talayuela	+++	19	39° 54'	16.5	5° 34'	15	304
Talavera	++	20	39° 58'	19	4° 51'	17	323
Getafe	–	21	40° 19'	24	3° 34'	21	400
Vagos	–	22	40° 33'	3	8° 40'	22	433
Carvalhos	–	23.5	41° 5'	5	8° 5'	24	473
Freixo de Espada à Cinta	+++	23.5	41° 5'	11	6° 54'	23	449
Alijó	–	25	41° 17'	7	7° 32'	25	483
Monçao	–	26	42° 4'	4	8° 28'	26	594
Oviedo	–	27	43° 18'	14	5° 55'	27	710

[a] See abundance key in the caption of Table 1

parameter for sites where *C. imicola* was present versus those where it was absent, using the Mann-Whitney non-parametric statistic U, are summarised in Table 5. There were significant differences between sites where the vector is present or absent for only 2 of the parameters. The straight-line distance from Seville to sites where *C. imicola* was present (median = 220 km) was significantly smaller (U = 22, P < 0.01) than to sites where *C. imicola* was absent (median = 416.5 km). The mean air temperature of the hottest summer month was significantly greater (U = 45,

Table 4. Summary of the fit between ranks of sites according to five climatic and three geographic parameters, and the presence (+) or absence (–) of *Culicoides imicola*

	Top 14 ranks		Mid 3 ranks		Last 10 ranks	
	No. +	No. –	No. +	No. –	No. +	No. –
Perfect fit	14	0	3	0	0	10
Mean temperature of hottest month	11	3	3	0	3	7
Mean temperature of coldest month	9	5	2	1	6	4
Mean relative humidity of driest Summer month	10	4	2	1	5	5
No. dry days in summer (mean %RH < 60%)	10	4	2	1	5	5
Mean relative humidity of driest Winter month	9	5	2	1	6	4
Latitude	10	4	3	0	4	6
Longitude	9	5	3	0	5	5
Straight-line distance from Seville	12	2	3	0	2	8

Table 5. Summary of Mann-Whitney U-tests (S1 = sites where *C. imicola* was present, S2 = sites where *C. imicola* was absent)

Variable	Median		Mann-Whitney		S.s.[a]
	S1	S2	U1	U2	
Mean temperature of hottest month	26.50	23.00	45	124	0.05
Mean temperature of coldest month	7.20	7.65	87	83	ns
Mean relative humidity of driest summer month	46.40	62.00	49	121	ns
No. dry days in summer (mean %RH < 60%)	79	43	52	118	ns
Mean relative humidity of driest winter month	81.60	82.80	86	84	ns
Latitude	38° 36'	40° 26'	51	119	ns
Longitude	6° 49'	4° 49'	67	103	ns
Straight-line distance from Seville	220	416.5	22	148	0.01

[a] *S.s.* Statistic significance. For sample sizes S1 = 17 and S2 = 10 either U1 or U2 must be equal to or less than 45 (5% significance level) or equal to or less than 34 (1% significance level) for the difference in medians to be significant

$P < 0.05$) at sites where the vector was present (median = 26.5°C) than at sites where it was absent (23.0°C). Mean summer relative humidity was lower (median = 46.4%) at sites where *C. imicola* was present than at sites where it was absent (median = 62.0%) but the difference was not significant (U = 49, ns). Similarly, sites where *C. imicola* was present tended to be more southern, more western and have more dry days in summer than sites where *C. imicola* was absent, but not significantly so (U = 51, U = 67, U = 52 respectively). Mean temperature during the warmest winter month and mean winter relative humidity were similar for sites where *C. imicola* was present or absent (U = 83, U = 84, respectively).

Discussion

Across the southwestern half of the Iberian peninsula *C. imicola* was captured on more than one occasion at 17 of the 27 sites studied. The distance of a site from Seville was a better predictor of the presence or absence of *C. imicola* than any of the climatic parameters examined. The median distance from Seville of the 17 sites with *C. imicola* present was 220 km, while for those sites where the vector was absent it was nearly twice as distant (416.5 km). Nevertheless, *C. imicola* was also present at one site about 450 km from Seville (Freixo de Espada à Cinta) and absent from one site only 179 km from Seville (Santa Fé). Only one other relationship was statistically significant: the mean air temperature of the hottest summer month. There were no significant relationships between the presence of this Afro-Asian tropical and subtropical vector at sites in Iberia and winter temperatures and relative humidity.

These analyses suggest that the presence of *C. imicola* on farms in Iberia is favoured by high summer temperatures, possibly in conjunction with low summer relative humidity. Although such climates occur in inland areas close to the Mediterranean Sea, east of the Sierra Nevada, and across central Spain, *C. imicola* appears most likely to be found in hot and dry areas close to Seville. In this context, Seville is simply representative of the most southerly provinces of Spain and Portugal which are, in all likelihood, the first areas of Iberia encountered by *C. imicola* upon its presumed incursion into the peninsula from North Africa on favourable winds [12].

The absence of *C. imicola* from three Spanish sites in particular – Cañada del Rosal, Santa Fé and Guadalupe – suggests that the geographic spread of *C. imicola* across Iberia may not be complete. All three sites are in the southeast of Spain and are in relatively hot, dry climatic zones. Along the Mediterranean shore, east of Gibraltar, *C. imicola* has only been found as far east as Velez-Malaga (36° 52' N 3° 43' W) [9]. The eastward spread of *C. imicola* may be constrained by having to move through the narrow, coastal corridor between the Sierra Contraviesa – Sierra de Gador and the Mediterranean Sea. Movement westward is much less constrained by topography.

Another interesting exception is the presence of *C. imicola* at Freixo de Espada à Cinta, despite its northern location and considerable distance from Seville (449 km). Interestingly, it has a summer climate typical of more southern sites [2, 10]. However, the apparent disappearance of *C. imicola* from the site of Alijó, where it was recorded in 1990 [2], may indicate how fragile some populations can be at these higher latitudes.

In general, the establishment of viable populations of *C. imicola* in Iberia appears to be favoured by hot and dry climates. More specifically, a mean monthly air temperature in the hottest month of 25°C or more and, possibly, a mean relative humidity in summer of less than 60%, promote sustainable populations of *C. imicola*. Thirty year, long-term average climate measurements show that these conditions are already found from Andalusia to central Spain [5]. In Portugal, however, the influence of Atlantic weather systems creates a broad front of maritime climate near the seaboard in which summer temperatures are moderated. The presence of a thriving population of *C. imicola* at Vila Nova de Milfontes indicates that hot and dry summer days alone are not the prerequisite for viable populations of *C. imicola*. This notable exception to the trends observable in most of the other 26 sites points to a crucial lack of knowledge concerning the breeding requirements of *C. imicola* and local climates which favour larval, pupal and adult development.

Predicting changes to the distribution of *C. imicola* in the next 10 or 20 years is difficult. There is some evidence that this species has not been on the Iberian peninsula for much more than 50 years and possibly less [10]. If true, then the timing of this incursion coincides with a general warming of the western part of the Mediterranean basin over the past 20 years [13]. Different general circulation models (GCMs) of the future climate in the Mediterranean basin, assuming a doubling of CO_2 concentration from 1990 levels, agree that the basin will warm on average by 3–6°C, but these changes will probably not be evenly distributed over each season. This degree of warming may be attained by the year 2030. The GCMs cannot predict what rainfall changes might occur in the Mediterranean basin as a whole, or in the Iberian peninsula in particular. Even without significant warming and drying of the peninsula it seems likely that *C. imicola* will expand its distribution and will spread east of the Sierra Nevada and north of Madrid in Spain. In Portugal, *C. imicola* will continue to exploit locally favourable sites in maritime climates, like that found at Vila Nova de Milfontes, as indeed it may also do in Spain. The difficulty in identifying the location of these favourable sites continues to make AHS and BT disease risk management very difficult. Detailed studies to identify the factors which have enabled the establishment of very large *C. imicola* populations at sites like Vila Nova de Milfontes would reduce the imprecision of our current knowledge and would assist in predicting further extensions in its European range.

Acknowledgements

This work was carried out under the European Commission research contract 8001-CT91-0211 under the project title "African Horse Sickness in Europe". The authors are very grateful to a wide range of people who assisted in the collection of insect samples and climatic data from the field.

References

1. Boorman J, Wilkinson PJ (1983) Potential vectors of bluetongue in Lesbos, Greece. Vet Rec 113: 395–396
2. Capela R, Sousa C, Pena I, Caeiro V (1993) Preliminary note on the distribution and ecology of *Culicoides imicola* in Portugal. Med Vet Entomol 7: 23–26
3. Le Houérou HN (1992) Vegetation and land-use in the Mediterranean Basin by the year 2050: A prospective study. In: Jeftic L, Milliman JD, Sestini G (eds) Climatic change and the Mediterranean. Edward Arnold, London, pp 173–232
4. Lubroth J (1988) African horse sickness and the epizootic in Spain 1987. Equine Prac 10: 26–33
5. MOPT (1992) Atlas Nacional de España. Sección II. Climatología. Ministerio de Obras Públicas y Transportes, Dirección General del Instituto Geográfico Nacional, Madrid, España, p 29
6. Mellor PS, Pitzolis G (1979) Observation on breeding sites and light trap collection of *Culicoides* during an outbreak of bluetongue in Cyprus. Bull Entomol Res 69: 229–238
7. Mellor PS, Boorman JPT, Wilkinson PJ, Martinez-Gomez F (1983) Potential vectors of bluetongue and African horse sickness viruses in Spain. Vet Rec 112: 229–230
8. Mellor PS, Jennings DM, Wilkinson PJ, Boorman JPT (1985) *Culicoides imicola*: a bluetongue virus vector in Spain and Portugal. Vet Rec 116: 589–590
9. Ortega MD (1995) Distribution and seasonal fluctuations of *Culicoides* spp. in Occidental Andalucia, Spain. MSc thesis, University of Wyoming
10. Rawlings P, Pro MJ, Pena I, Ortega MD, Capela R (1997) Spatial and seasonal distribution of *Culicoides imicola* in Iberia in relation to the transmission of African horse sickness virus. Med Vet Entomol 11: 49–57
11. Rodriguez M, Castaño M, Escolar E, Flores JM, Toni P, Gonzalez M, Jimenez F, Gonzalez JL, Montoya A (1987) Peste equina africana: descripción del brote en España, 1987. Med Vet 4: 537–557
12. Sellers RF, Pedgley DE, Tucker MR (1979) Possible windborne spread of bluetongue to Portugal, June–July 1956. J Hyg Camb 81: 189–196
13. Wigley TML (1992) Future climate of the Mediterranean Basin with particular emphasis on changes in precipitation. In: Jeftic L, Milliman JD, Sestini G (eds) Climatic change and the Mediterranean. Edward Arnold, London, pp 15–44

Authors' address: Dr. P. Rawlings, Faculty of Computer Studies and Mathematics, University of the West of England, Coldharbour Lane, Bristol BS16 1QY, U.K.

Simulation studies of African horse sickness in Spain

C. C. Lord[1,*], **M. E. J. Woolhouse**[1,**], and **P. S. Mellor**[2]

[1] Department of Zoology, University of Oxford, Oxford, U.K.
[2] Institute for Animal Health, Pirbright Laboratory, Pirbright, U.K.

Summary. Factors affecting epidemics of African horse sickness in Spain were studied using a mathematical model. The model examined the likelihood of an epidemic after the introduction of the virus, and the effectiveness of vaccination strategies. Two host species (horses and donkeys) and one vector species (the biting midge *Culicoides imicola*) were included. A stratified random sampling method (Latin hypercube sampling) was used for sensitivity analysis of the likelihood of an epidemic. Systematic variation of vaccination parameters was used to consider alternative control strategies. In general, when an epidemic occurred most potential hosts became infected. The peak prevalence in *C. imicola* was low, and never exceeded 3%. The most significant factors in the likelihood of an epidemic were vector population size, the recovery rate in horses and the time of year when the virus was introduced. The lag between virus introduction and protection, the proportion of hosts vaccinated, and including donkeys in vaccination programmes where the factors that most strongly affected the success of different vaccination strategies. These factors should be priorities for empirical research, and should be considered in the design of control strategies in areas at risk of virus introduction.

Introduction

African horse sickness (AHS) virus (*Reoviridae: Orbivirus*) has caused severe epidemics several times in the past when introduced to naive host populations [18, 24]. The virus infects most equids, although severe disease is generally found only in horses. It is vector-borne, primarily by the biting midge *Culicoides imicola*. The role of equines other than horses (such as donkeys) in maintaining these epidemics is poorly understood, as is the importance of including these animals in vaccination programmes.

Current addresses: *Florida Medical Entomology Laboratory, University of Florida, U.S.A.; ** Centre for Tropical Veterinary Medicine, University of Edinburgh, U.K.

Identifying factors that affect the ability of the virus to invade is critical in predicting areas at risk and in the design of control programmes. Beyond the presence of appropriate vector and host populations, little is understood about factors that affect the likelihood of an epidemic. Mathematical models can be used both to identify those factors where more information is needed to predict the outcome of a virus introduction and to examine potential control strategies.

Uncertainty in parameter estimates may affect predictions or recommendations based on models, and so should be included in any modelling study. Biological variability (e.g. variation between locations, virus strains, host types or vector populations), measurement error, and/or gaps in knowledge all contribute to uncertainty in parameter estimates. Formal sensitivity analysis is one method of incorporating uncertainty into a model, with the model output then interpreted taking into consideration the parameter variation included.

Here, we report the results of a simulation study for AHS in Spain, examining the factors affecting the likelihood of epidemics and the effectiveness of different vaccination strategies.

Materials and methods

Model development

The model used was compartmental and deterministic with one vector species (*Culicoides imicola*) and two host species (horses and donkeys). The details of the model and parameter estimation can be found in Lord et al. [14, 15]; only a brief description will be given here. Seasonal variation in some parameters was included with a temperature-based driving function.

The model tracks the proportion of hosts which are susceptible, infective or immune, and the number of midges which are susceptible, latently infected or infective. No natural host mortality or reproduction is included for the hosts, as their natural lifespan is much longer than the duration of an epidemic. Disease-induced mortality was included for both types of hosts; mortality in horses was assumed to be greater than in donkeys. The shorter lifespan of midges, however, requires that both a latent period and mortality in uninfected midges are included. As transovarial transmission is not known for viruses in *Culicoides* [3, 13, 20], all newly emerged midges enter the susceptible class. The spatial scale of the model was defined by the interaction of host and vector populations; one vector population may interact with hosts on one or more than one farm, depending on breeding sites and other factors.

Temperature-dependent midge reproduction and mortality were used to generate seasonal midge populations. Because of the high variability in midge numbers that has been observed [2, 8, 9, 11, 21, 26, 29], a wide range of values for the maximum midge population were used. The degree of seasonality in the midge population was varied, as was the time of year when the virus was introduced. The peak of the temperature curve was defined as June 21; all other dates are determined from this.

C. imicola feeds on other mammals in addition to equines [10, 19, 28]; therefore, the estimates of biting rates included the possibility of feeding on non-equines. However, no host preference between types of equid (horses or donkeys) was included.

By definition, transmission rates range between 0 and 1 (no infection or all bites lead to an infected host or vector). Little information was available beyond this, although most data suggest low transmission rates [16, 17, 25]. We assumed that 1) transmission rates do not vary between species of host and 2) the same distributions apply to transmission from host to vector as vector to host. Donkeys appear to have longer infectious periods than horses [18, 22, 27], although the frequent asymptomatic cases make this difficult to estimate. We assumed that the infective periods in different hosts are correlated, with the period in donkeys up to twice as long as that in horses.

For the simulation runs which were categorized as epidemic (see below), three vaccination strategies were tested. In strategy 1, 50–95% of the horses were protected before virus introduction. In the second strategy, the same range of protection was used, but both donkeys and horses were protected. Finally, the third strategy protected 90% of horses either before virus introduction or from 10 to 50 days after introduction. Note that here, the fraction protected includes both the coverage (proportion of animals which are vaccinated) and the efficacy of the vaccine. Similarly, the lag before protection includes both the time from virus introduction until vaccination is implemented and the time lag from inoculation until protection against challenge.

Sensitivity analysis

To examine how the uncertainty in the parameter estimates affected the likelihood of an epidemic, Latin hypercube sampling was used. This technique allows exploration of the full parameter space with a greatly reduced number of simulation runs [4, 5, 12]. Each parameter distribution was sampled 100 times. These samples were then randomly reordered to create 100 sets of parameters used to run the model. Thus, the full range of each parameter was used while requiring only 100 simulation runs. A systematic sensitivity analysis was used for the vaccination strategies; each set of parameter values from the initial simulations which resulted in an epidemic was tested with all levels of the vaccination parameters.

Programming

The simulation model was implemented in the differential equation simulation package SOLVER (version 4 [6]). The Latin hypercube sampling scheme was programmed in Turbo Pascal [7].

Outcome variables

We formally defined an epidemic to have occurred if 1) at least 2 horses were infected at the time of the peak prevalence; and 2) there were less than 2 days where the virus was not present in at least 1 host or vector between the start of the simulation run and the time of peak prevalence in horses. In practice, there was a clear division between epidemic and not epidemic runs. Four outcome variables were used for the final analysis: whether an epidemic occurred, the peak prevalence in horses, the time of the peak prevalence in horses, and the peak prevalence in midges. The same definitions were used to study vaccination strategies; however, we concentrated on whether an epidemic occurred and did not examine the other outcome variables in detail.

Statistical analysis

Logistic stepwise regression was used to analyse the effect of each of the independent parameters on the presence or absence of epidemics [1]. All statistics were calculated using SAS [23].

Results

Presence of epidemics

There were 27 runs which met the criteria for an epidemic. Considering only those runs classed as epidemic, the peak prevalence in midges tended to be very low (Fig. 1a), whereas the peak prevalence in horses was much higher (Fig. 1b). Five parameters were included in the logistic regression on the presence or absence of an epidemic. The uncertainty in the maximum vector to host ratio had the most influence on the variation in occurrence of epidemics (Fig. 2a), and appeared to have a threshold below which epidemics did not occur. The likelihood of an epidemic was affected by the time of year when the virus was introduced (Fig. 3). Both the proportion of infected horses which died and the recovery rate of horses were significant factors in the regression. The total removal rate of infective horses is the sum of these two parameters, which also appears to have a threshold effect on the presence of epidemics (Fig. 2b).

Vaccination strategies

A very nonlinear relationship was apparent between the fraction of animals protected and the number of epidemics prevented when prophylactic vaccination is considered (Fig. 4). Prevention of epidemics is substantially enhanced by

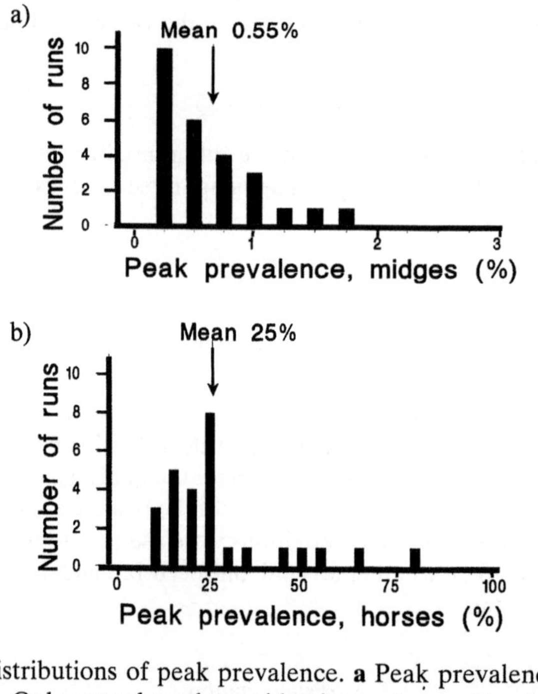

Fig. 1. Frequency distributions of peak prevalence. **a** Peak prevalence in midges; **b** peak prevalence in horses. Only runs classed as epidemic are shown. Reprinted with permission from Lord CC, Woolhouse MEJ, Rawlings P, Mellor PS, Simulation studies of African horse sickness and *Culicoides imicola* (Diptera, Ceratopogonidae). J Med Entomol 33: 328–338, © 1996, Entomological Society of America

vaccinating donkeys as well as horses (Fig. 4), especially when a high fraction of hosts are vaccinated. The proportion of epidemics provented declines with the time lag until protection is effective (Fig. 5); such responsive vaccination rarely prevented epidemics in the simulations.

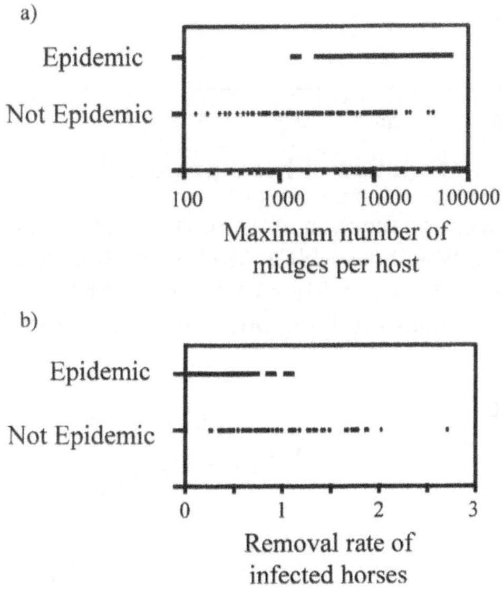

Fig. 2. Effect of two parameters on the likelihood of an epidemic. **a** Midge population size; **b** removal rate of horses (recovery and mortality). Reprinted with permission from Lord CC, Woolhouse MEJ, Rawlings P, Mellor PS, Simulation studies of African horse sickness and *Culicoides imicola* (Diptera, Ceratopogonidae). J Med Entomol 33: 328–338, © 1996, Entomological Society of America

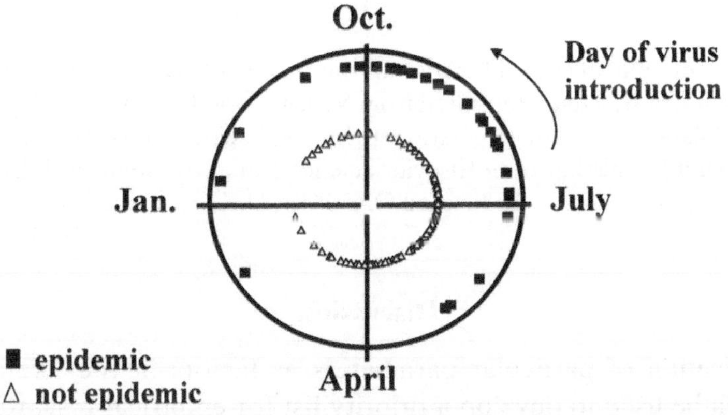

Fig. 3. Effect of the time of year of virus introduction on the likelihood of an epidemic. Dates based on the temperature curve; June 21 was defined as the day of maximum temperature. Reprinted with permission from Lord CC, Woolhouse MEJ, Rawlings P, Mellor PS, Simulation studies of African horse sickness and *Culicoides imicola* (Diptera, Ceratopogonidae). J Med Entomol 33: 328–338, © 1996, Entomological Society of America

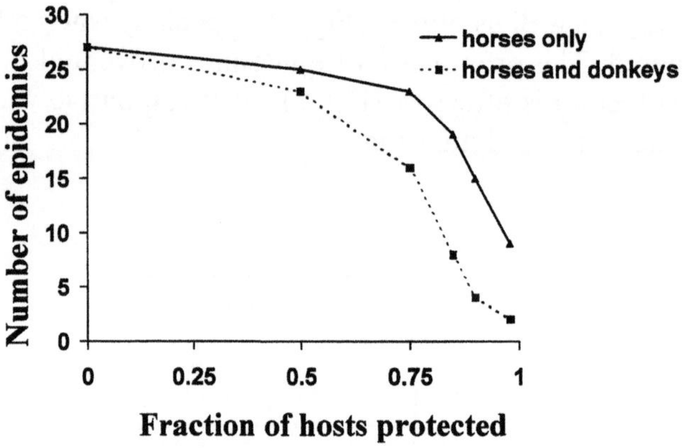

Fig. 4. Number of epidemics as a function of the fraction of hosts protected. Reprinted from Vaccine, vol. 15, Lord CC, Woolhouse MEJ, Mellor PS, Simulation studies of vaccination strategies in African horse sickness, pp. 519–524, © 1997, with kind permission from Elsevier Science Ltd., The Boulevard, Langford Lane, Kidlington OX5 1GB, U.K.

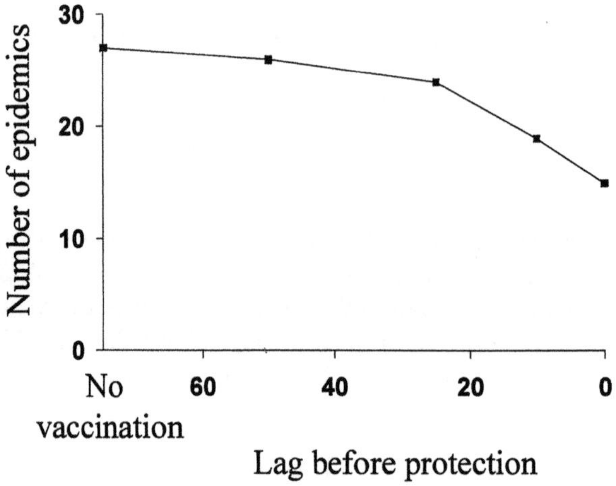

Fig. 5. Number of epidemics as a function of the time lag between virus introduction and when protection was effective. Reprinted from Vaccine, vol. 15, Lord CC, Woolhouse MEJ, Mellor PS, Simulation studies of vaccination strategies in African horse sickness, pp. 519–524, © 1997, with kind permission from Elsevier Science Ltd., The Boulevard, Langford Lane, Kidlington OX5 1GB, U.K.

Discussion

The identification of particular parameters as factors in the likelihood of an epidemic can be used to develop a priority list for empirical research (Table 1). Reducing the uncertainty in these parameter estimates will lead to more precise predictions of the outcome. High priority parameters are those where further information will contribute the most to our ability to predict epidemics. Low priority parameters are those where further information will be less valuable until further data are available on other parameters.

Table 1. Prioritised list for research based on the results of the simulation model

Higher priority	Lower priority
Population size of vectors	Transmission rates
When the infection is introduced	Horse and donkey population sizes
Recovery and mortality rates of infected horses	Recovery and mortality rates of infected donkeys
Interbloodmeal interval	Virus development rates in vectors
	Seasonality in vector
	Lifespan of vector

Regional differences in light trap catches were used to generate the uncertainty included in the population size of midges. Thus, the high priority accorded this parameter indicates that midge population dynamics need to be studied in each location of interest. In addition, relating the numerical values used here (based on trap catches) to actual population sizes in the field will require further research, as the relationship between light trap catches and actual population size is unknown.

The uncertainty in some of the parameters identified as important was derived from a combination of several aspects of the biology and epidemiology of AHS. The mortality rate, the duration of viraemia and the slaughter of infected animals are all factors in the rate of removal of horses from the infective class. The interbloodmeal interval in the model was defined as the time period between blood meals on equids, and depends on several factors. These factors include the length of the gonotrophic cycle (time nèeded for clutch development) and host preferences, both between types of equid and between equids and other animals. For both of these parameters, more information on the components will assist in developing more precise predictions of the likelihood of epidemics.

Further research is unlikely to change the uncertainty in the time of year of virus introduction, as there are several ways the virus can be introduced into a new area. Since no time of year guaranteed either the success or failure of an introduction, the results from this analysis should be interpreted only in terms of the utility of increased vigilance during periods of increasing or peak midge populations.

Several features of the epidemics of AHS in Spain were observed in the simulations. The timing of epidemics was similar in that they were more likely in late summer and autumn. In the absence of control measures, most horses in a herd were affected by the virus following its introduction.

The addition of vaccination strategies to the basic model identified two crucial features of successful vaccination programmes: protection of equines prior to the introduction of the virus and the inclusion of donkeys in the control strategies. More data on the time course of the development of protective immunity following inoculation will aid in the development of effective control strategies, as will further information on the relative efficacy of vaccines in horses and donkeys.

This study prioritises areas for further research that will improve our ability to identify areas at risk, predict epidemics and develop control programmes. Many areas are at risk of the introduction of the virus, either from infected midges carried on the wind or midges or vertebrates moved by humans. To predict where a virus introduction may lead to an epidemic, more information is needed most on vector population dynamics and on the removal rate of infected hosts. Designing improved control strategies requires more information on the time lag between inoculation and the development of protective immunity, and on the problems associated with locating and immunising donkeys. With this information, continuing development of the model could determine with greater precision the locations and seasons of greatest risk, and be used in the design of specific control strategies.

Acknowledgements

Support for this research was provided by a BBSRC Link Grant and CEC (Contract Nos. 8001-CT91-0211 and TS3-CT92-0151). M.E.J.W. is supported by the Royal Society.

References

1. Agresti A (1990) Categorical data analysis. Wiley, New York
2. Al-busaidy SM, Mellor PS (1991) Epidemiology of bluetongue and related orbiviruses in the Sultanate of Oman. Epidemiol Infect 106: 167–178
3. Allingham PG, Standfast HA (1990) An investigation of transovarial transmission of Akabane virus in *Culicoides brevitarsis*. Aust Vet J 67: 273–274
4. Blower SM, Dowlatabadi H (1994) Sensitivity and uncertainty analysis of complex models of disease transmission: an HIV model, as an example. Int Stat Rev 62: 229–243
5. Blower SM, Hartel D, Dowlatabadi H, Anderson RM, May RM (1991) Drugs, sex and HIV: a mathematical model for New York City. Phil Trans R Soc Lond B 321: 171–187
6. Blythe SP, Gurney WSC, Maas P, Nisbet RM (1990) User guide for SOLVER, revision 4.0. APIC, Glasgow
7. Borland International (1992) Turbo Pascal user's guide. Borland International, Scotts Valley, CA
8. Braverman Y, Linley JR (1988) Parity and voltinism of several *Culicoides* spp. (Diptera: Ceratopogonidae) in Israel, as determined by two trapping methods. J Med Entomol 25: 121–126
9. Braverman Y, Linley JR, Marcus R, Frish K (1985) Seasonal survival and expectation of infective-life of *Culicoides* spp. (Diptera: Ceratopogonidae) in Israel, with implications for bluetongue virus transmission and a comparison of the parous rate in *C. imicola* from Israel and Zimbabwe. J Med Entomol 22: 476–484
10. Braverman Y, Boreham PFL, Galun R (1971) The origin of bloodmeals of female *Culicoides pallidipennis* trapped in a sheepfold in Israel. J Med Entomol 8: 379
11. Capela R, Sousa C, Pena I, Caeiro V (1993) Preliminary note on the distribution and ecology of *Culicoides imicola* in Portugal. Med Vet Entomol 7: 23–26
12. Iman RL, Shortencarrier MJ (1984) A Fortran 77 program & users guide for the generation of Latin Hypercubes & random samples for use with computer models. NUREG/CR-3624, SAND83-365. Sandia National Laboratories, Albuquerque, NM
13. Jones RH, Foster NM (1971) Transovarian transmission of bluetongue virus unlikely for *Culicoides variipennis*. Mosq News 31: 434–437

14. Lord CC, Woolhouse MEJ, Mellor PS (1997) Simulation studies of vaccination strategies in African horse sickness. Vaccine 15: 519–524
15. Lord CC, Woolhouse MEJ, Rawlings P, Mellor PS (1996) Simulation studies of African horse sickness and *Culicoides imicola* (Diptera, Ceratopogonidae). J Med Entomol 33: 328–338
16. Meiswinkel R, Nevill EM, Venter GJ (1994) Vectors: *Culicoides* spp. In: Coetzer JAW, Thomson GR, Tustin RC (eds) Infectious diseases of livestock with special reference to southern Africa, vol 1. Oxford University Press, Oxford, pp 68–89
17. Mellor PS (1990) The replication of bluetongue virus in *Culicoides* vectors. In: Walton TE, Osburn BI (eds) Bluetongue, African horse sickness, and related orbiviruses: Proceedings of the Second Symposium. Curr Top Microbiol Immunol 162: 143–161
18. Mellor PS (1993) African horse sickness: transmission and epidemiology. Vet Res 24: 199–212
19. Nevill E, Anderson D (1972) Host preferences of *Culicoides midges*: (Diptera: Ceratopogonidae) in South Africa as determined by precipitin tests and light trap catches. Onderstepoort J Vet Res 39: 147–152
20. Nunamaker RA, Sieburth PJ, Dean VC, Wigington JG, Nunamaker CE, Mecham JO (1990) Absence of transovarial transmission of bluetongue virus in *Culicoides variipennis*: immunogold labelling of bluetongue virus antigen in developing oocytes from *Culicoides variipennis* (Coquillett). Comp Biochem Physiol 96A: 19–31
21. Rawlings P, Mellor PS (1994) African horse sickness and the overwintering of *Culicoides* spp. in the Iberian peninsula. Rev Sci Tech Off Int Epiz 13: 753–761
22. Rodriguez M, Hooghuis H, Castaño M (1992) African horse sickness in Spain. Vet Microbiol 33: 129–142
23. SAS Institute (1988) SAS user's guide. SAS Institute, Cary, NC
24. Sellers RF, Pedgley DE, Tucker MR (1977) Possible spread of African horse sickness on the wind. J Hyg (Camb) 79: 279–298
25. Venter GJ, Hill E, Pajor ITP, Nevill EM (1991) The use of a membrane feeding technique to determine the infection rate of *Culicoides imicola* (Diptera, Ceratopogonidae) for 2 bluetongue virus serotypes in South Africa. Onderstepoort J Vet Res 58: 5–10
26. Venter GJ, Sweatman GK (1989) Seasonal abundance and parity of *Culicoides* biting *midges* associated with livestock at Roma, Lesotho (Diptera: Ceratopogonidae). Onderstepoort J Vet Res 56: 173–177
27. Venter GJ, Williams R, Nevill EM (1992) The value of donkeys as sentinels for African horse sickness. In: Walton TE, Osburn BI (eds) Bluetongue, African horse sickness and related orbiviruses: proceedings of the second symposium, Paris 1991. CRC Press, Boca Raton, pp 825–831
28. Walker AR, Boreham PFL (1976) Blood feeding of *Culicoides* (Diptera: Ceratopogonidae) in Kenya in relation to the epidemiology of bluetongue and ephemeral fever. Bull Entomol Res 66: 181–188
29. Walker AR, Davies FG (1971) A preliminary survey of the epidemiology of bluetongue in Kenya. J Hyg (Camb) 69: 47–60

Authors' address: Dr. C. C. Lord, Florida Medical Entomology Laboratory, University of Florida, 200 9th St. SE, Vero Beach, FL 32692, U.S.A.

The *Culicoides* vectors of African horse sickness virus in Morocco: distribution and epidemiological implications

H. Bouayoune[1], **J. Touti**[2], **H. El Hasnaoui**[3], **M. Baylis**[4], and **P. S. Mellor**[4]

Laboratoire Régional d'Analyses et de Recherches Vétérinaires de Fès[1], d'Agadir[2], de Casablanca[3], Morocco
[4] Institute for Animal Health, Pirbright, U.K.

Summary. African horse sickness (AHS) is a vector-borne, infectious disease of equids caused by African horse sickness virus. The only proven field vector of the virus is the biting midge *Culicoides imicola*, although *C. obsoletus* and *C. pulicaris* are suspected vectors. In 1994–5 a total of 3887 light trap samples were collected from 22 sites distributed over most of Morocco. *Culicoides imicola* was found to be very widely distributed with the greatest catches in the low-lying north-western areas (between Tangier and Rabat) and at Marrakech. *Culicoides imicola* was absent at one site only, near Settat. In general, the catch of *C. imicola* peaked in late summer and autumn, with a smaller peak in spring. Catches of *C. obsoletus* were greatest in the north-western provinces of Morocco and in the south, while catches of *C. pulicaris* were greatest in the north. Although both species were widely distributed, trap catches were much lower than those of *C. imicola*. Peak catches were in spring or late summer and autumn. In general, the findings for *C. imicola* correspond well with the seasonal and spatial distribution of disease outbreaks during the 1989–1991 epizootic of AHS in Morocco. It is suggested that *C. obsoletus* and *C. pulicaris* were probably of little significance in the epidemiology of AHS in Morocco in 1989–91.

Introduction

African horse sickness (AHS) is an infectious, non-contagious, insect-borne disease of equids. In susceptible equine populations the disease can be devastating, with mortality rates exceeding 90% in horses and 70% in mules [12] although donkeys are much less affected [28]. The disease is caused by African horse sickness virus (AHSV), a double-stranded RNA virus (*Reoviridae: Orbivirus*), of which there are 9 known serotypes. AHS is enzootic to sub-Saharan Africa where zebra may also be infected by the virus although they are not known to display any clinical signs of the disease.

In the present century there have been several epizootics of AHS outside sub-Saharan Africa (e.g. Egypt, 1928; Middle East, 1943; Middle East, Pakistan and Afghanistan, 1959–60; North Africa and Spain, 1965–66). The most recent was an epizootic (caused by AHSV serotype 4) in Iberia and Morocco (1987–91). The first outbreaks were reported in Spain in 1987; the virus reappeared in 1988 and again in 1989, when it also spread to Portugal and Morocco. The epizootic persisted in Morocco until 1991. The number of cases of AHS in Spain, Portugal and Morocco was estimated to be 1385 [10, 23, 24], 94 [24] and 1244 [1] respectively. Between reports of the last cases of AHS in one year and the first in the next, there were periods of 6–12 months in which no cases of AHS were reported.

The only proven field vector of AHSV is the biting midge *Culicoides imicola*. This species is widely distributed in sub-Saharan Africa [17], and has been found in Asia as far east as Laos and Vietnam [13, 30], in North Africa [2, 16, 27] and in Europe, where it has been reported from Spain [21], Portugal [9, 22], Cyprus [6] and several Greek islands [7, 8].

During the 1987–1991 epizootic of AHS in Iberia and Morocco there were several isolations of the virus from parous, non-engorged, field-caught *C. imicola*, demonstrating that it was a vector. However, other species of *Culicoides* have been shown, under laboratory conditions, to be capable of supporting AHSV replication [20] and AHSV serotype 4 was successfully isolated from mixed pools of parous, non-engorged *C. pulicaris* and *C. obsoletus* collected from Spain in 1988 [19]. These species must therefore also be considered to be potential vectors of the virus.

The recent epizootic prompted an investigation of the seasonal and spatial distributions of *C. imicola, C. obsoletus* and *C. pulicaris* in Morocco, and the effects of climate on these distributions, based on the operation of a network of 20 light traps and automatic weather stations spread over the country. The results of these investigations have been reported elsewhere [4, 5]. Here we describe the distribution and seasonalities of the 3 species derived from data collected in 1994–5 and compare the findings with the distribution and timing of the AHS outbreaks in Morocco in 1989–91.

Materials and methods

Study area – Morocco

Morocco, situated in the northwest of Africa, is a country of great geographical and climatological diversity which arises largely from the interaction of three features: the Mediterranean and Atlantic coasts to the north and west; areas bordering the Sahara desert to the south and east; and two mountain ranges, the Rif in the far north and the Atlas which run through the middle of the country in an approximately southwest to northeast direction. The Atlas are further subdivided into three chains, the Middle Atlas, the High Atlas and the Anti-Atlas. In general, moving from the north and west to the south and east, there is a trend towards greater dryness and hotter temperatures; this trend is interrupted by the Atlas range, in which humid to subhumid and cool to cold conditions are found. The climate is Mediterranean, with a wet season from October/November to April/May, and a dry season from May to September. Mean annual rainfall ranges from 800–900 mm in the north and 400–500 mm in the west, to 100–150 mm in the south and east [11].

Trap sites

In 1993 light traps were installed at 20 sites distributed over much of the territory of Morocco. In early 1995 two sites were transferred (from M'diq to Chefchaouen and Bouizakarn to Goulimime) so that a total of 22 sites were sampled. *Culicoides imicola* is caught usually only in the presence of livestock and therefore all sites were livestock holdings: either stables, equestrian clubs or cavalries where the main livestock were horses, or farms and smallholdings where the main livestock were cattle, sheep, goats or poultry. The number of livestock at each site ranged from 10 to 3900 and in some cases there were additional livestock in the surrounding areas. Ideally, all trap sites would have had similar numbers of livestock; however, the choice of trap site was constrained in some areas of Morocco by the concomitant needs for electricity (to power the trap) and security. The precise geographical coordinates (± 50 m) and altitudes (± 100 m) of all sites were recorded with a Garmin GPS 55. The positions of all sites are shown in Fig. 1.

Methodology

The traps used were Pirbright-type miniature light traps consisting of a fan driven by a 12 V motor and using a 12 V, 21 W car tail light bulb as the light source [6]. Traps were powered from the mains supply via a 12 V transformer and rectifier unit or, where necessary, directly from a 12 V, 105 Ah sealed lead-acid battery. At each site traps were installed either outside an animal enclosure or inside, near the entrance. Traps were operated twice weekly, from dusk to dawn. At some sites traps were operated less frequently during certain periods; at

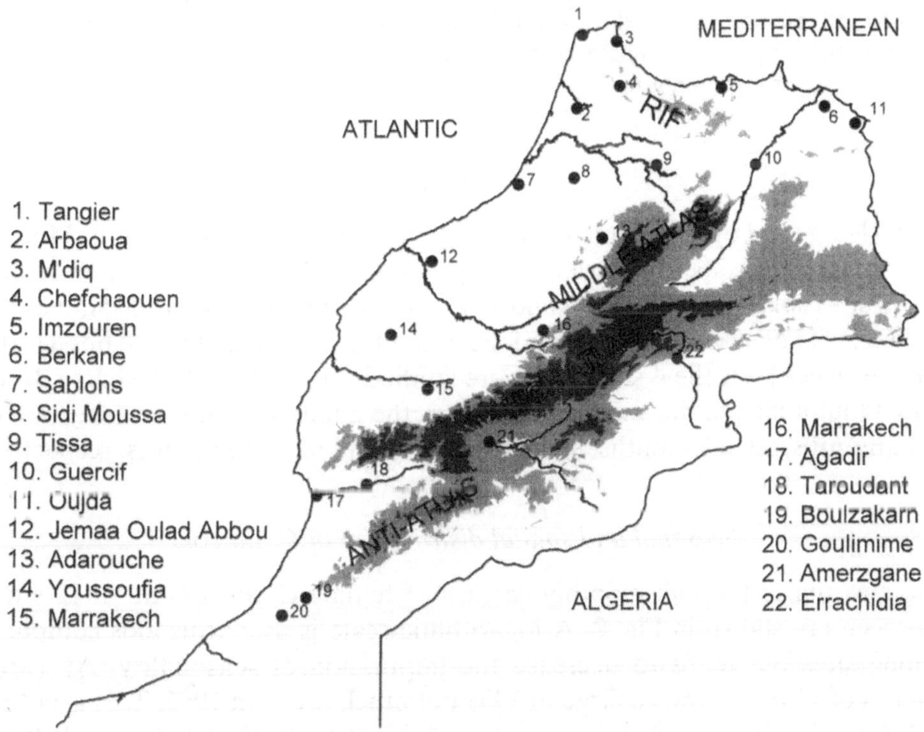

Fig. 1. Geographical locations of the 22 sites in Morocco where light traps for *Culicoides* were operated in 1994–1995. Major rivers are shown as dark lines; altitude, light grey > 1200 m, dark grey > 1800 m

others they were operated more frequently, even daily. At certain sites trap-breakdown led to lost data for several weeks.

Insects attracted to the light source were blown by the fan into a collecting jar containing water with a little detergent, and were preserved in 1–2% formalin. Catches were returned periodically to regional laboratories in Morocco (Agadir, Casablanca, Fès, Oujda, Marrakech and Tangier) where *C. imicola, C. obsoletus* and *C. pulicaris* were sorted, sexed and counted by trained local personnel. Male *Culicoides* do not feed on blood and do not transmit arboviruses; they are therefore ignored in the present work.

Data analysis

Baylis et al. [4] demonstrated that there is large day-to-day variation in the catch of *C. imicola* over relatively short time periods. This variation may result from changes in the *population size*, changes in the *activity rate* of the population (the proportion of the population that are active on a given night) or changes in *trap efficiency* (the relative efficiency with which traps attract and capture the insects). The objective of the present study was to obtain estimates of population size and it was necessary, therefore, to attempt to remove the effects of a varying activity rate and/or trap efficiency. Following [4], we argue that the maximum trap catch in a certain period is a better measure of population size than the mean, and that the submaximal samples in the same period result from a reduced activity rate or trap efficiency (perhaps because of adverse weather conditions). Baylis et al. [4] arbitrarily selected a half-monthly period, giving 48 estimates of population size in 1994–5. These estimates were averaged to give the mean annual trap catch of *C. imicola* at each site, which is used here as a comparative measure of abundance of the vector. Trap catch data were not normally distributed, with a skew towards low values, and therefore the 48 half-monthly maxima were log (n + 1) transformed prior to taking means. Transformed means were detransformed for presentation.

Results

In 1994–5 a total of 3887 samples were taken from the 20 light traps in Morocco, an average of 194.5 per light trap, corresponding to nearly 2 samples per week. Of the 18 sites studied in both 1994 and 1995 the maximum number of samples were at Tangier (329 or 3.2 per week) and the minimum at Jemaa Oulad Abbou (102 or 0.98 per week). Of the 4 sites that were studied for less than two years (Bouizakarn, Goulimime, M'diq and Chefchaouen), the number of samples taken were 78 in 16 months, 60 in 8 months, 90 in 14 months and 63 in 10 months, respectively.

Seasonal and spatial distribution of C. imicola

The maxima of the half-monthly catches of female *C. imicola* at all 22 sites in Morocco are shown in Fig. 2. A logarithmic scale is used: this aids comparison among sites but tends to decrease the impression of seasonality. At Tangier catches of *C. imicola* were large in 1994 but much lower in 1995. This is believed to be due to the trap developing a mechanical fault; therefore, in all further analyses data for Tangier in 1995 are excluded. At Beni Mellal the catches of *C. imicola* were considerably greater in 1995 than 1994. The reasons for this are unclear and no data are excluded in analyses.

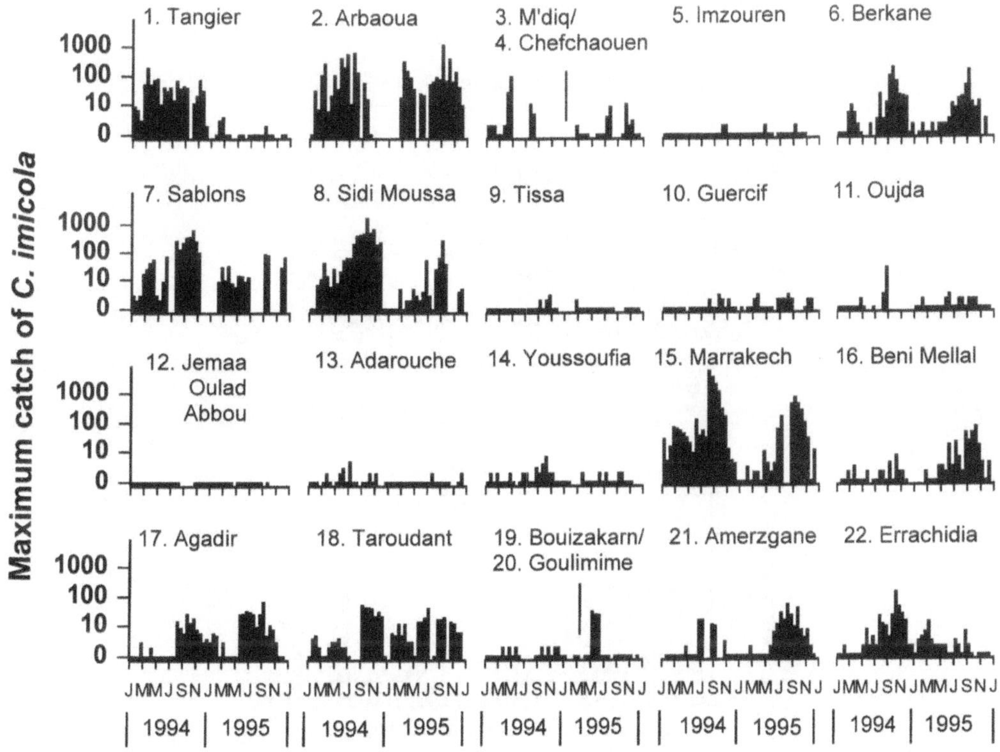

Fig. 2. The light trap catches of female *Culicoides imicola* at 22 sites in Morocco in 1994–1995. Each year was divided into 24 half-monthly periods; solid bars are the maximum catch in each period. The x-axes are displaced downwards slightly; catches of zero show as low bars (e.g., at Jemaa Oulad Abbou). The absence of bars indicates that traps were not operated. Two sites were transferred in 1995: M'diq to Chefchaouen, and Bouizakarn to Goulimime. Vertical black lines indicate when these transfers occurred

In general (Fig. 2), *C. imicola* appears most abundant at the 4 sites in the north of Morocco (Tangier, Arbaoua, Sablons and Sidi Moussa) and at Marrakech; there were intermediate trap catches at Berkane, Beni Mellal, Agadir, Taroudant, Amerzgane and Errachidia; and *C. imicola* appears rare at the remaining sites. There is only one site (Jemaa Oulad Abbou, near Settat) at which *C. imicola* appears to be absent. *Culicoides imicola* was found at altitudes ranging from 4 to 1275 m and in climatic conditions ranging from subhumid to Saharan.

At three sites (Marrakech, Sablons and Tangier) at least 1 female *C. imicola* was caught during every half-monthly period of 1994 in which traps were operated; at two sites (Arbaoua and Sablons) this was also the case in 1995.

Figure 2 demonstrates that at most sites there was a peak in the catch of *C. imicola* in the late summer and early autumn. This is confirmed in Fig. 3 which presents the mean of the catches of 21 sites for each half-monthly period of the year. The exception to this pattern was at Tangier where, in 1994, catches were large throughout most of the year (excluding January and February) and the greatest catch was in March (Fig. 2). Scrutiny of Fig. 2 also indicates that at many

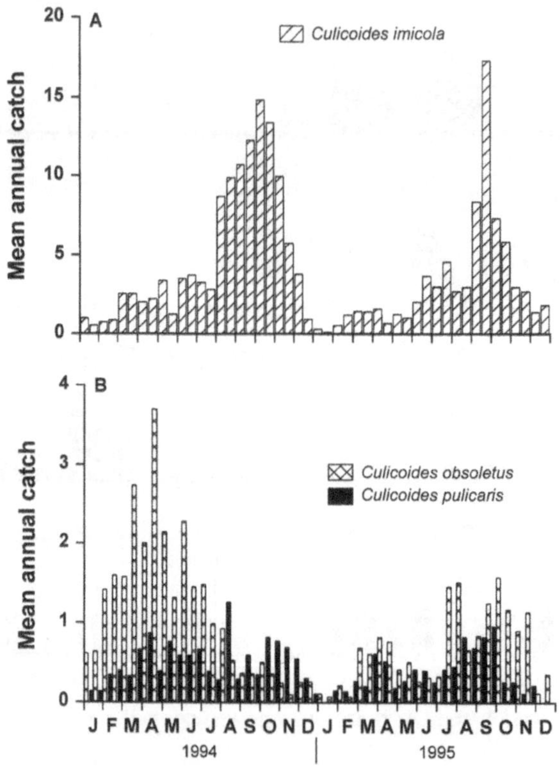

Fig. 3. The seasonal distribution of **A** *Culicoides imicola* and **B** *C. obsoletus* and *C. pulicaris* in Morocco in 1994–1995. Bars are the means across all 22 sites, excluding those where a species appears to be absent. Maximum catches were log (n + 1) transformed, averaged for each time period, and detransformed for presentation

sites there was a lesser peak earlier in the year (e.g., March/April). This is particularly clear at Marrakech, Sidi Moussa, Arbaoua, Sablons and Berkane.

The mean annual catches of *C. imicola* at the 22 sites are shown on a map of Morocco in Fig. 4. *Culicoides imicola* is very widely dispersed in Morocco, with consistently large catches in the northwest and at Marrakech.

Seasonal and spatial distribution of C. obsoletus *and* C. pulicaris

Catch data for *C. obsoletus* and *C. pulicaris* were analysed in the manner described above for *C. imicola*. In general, the mean maximum catches of *C. obsoletus* and *C. pulicaris* were lower than that of *C. imicola*, although they are

Fig. 4. The spatial distribution and catch sizes of **A** *Culicoides imicola*, **B** *C. obsoletus* and **C** *C. pulicaris* in Morocco, based on light trap catches at 22 sites in 1994–1995. Areas of solid black circles are proportional to the mean annual catch at each site. Note that the scales differ for the 3 species. Where not occluded, trap sites are shown as open circles

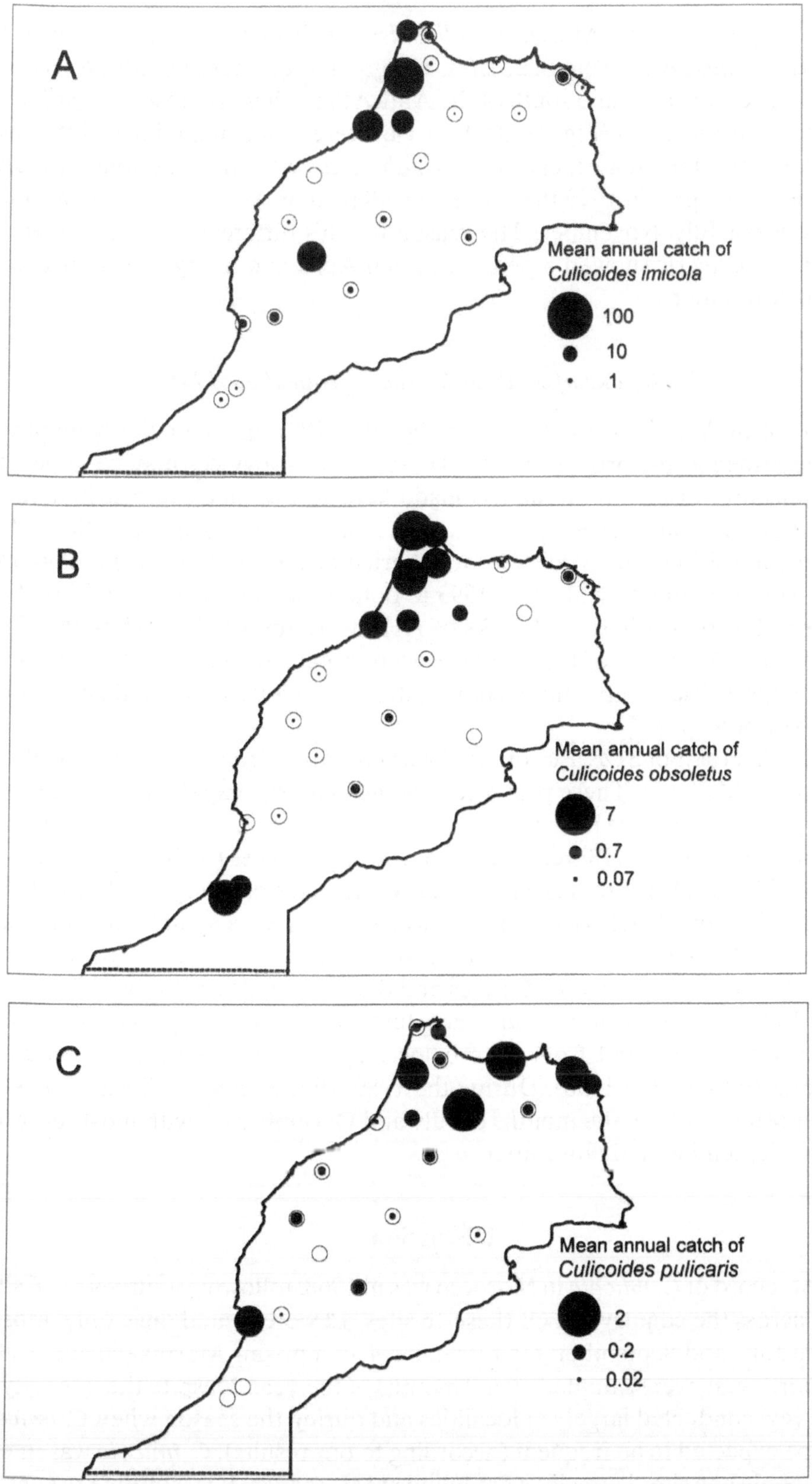
A
Mean annual catch of
Culicoides imicola
100
10
1
B
Mean annual catch of
Culicoides obsoletus
7
0.7
0.07
C
Mean annual catch of
Culicoides pulicaris
2
0.2
0.02

both widely distributed in Morocco. *Culicoides obsoletus* was found at 20 out of 22 sites and *C. pulicaris* at 19 out of 22 sites. The greatest catches of *C. obsoletus* were in the north of Morocco and south of the Anti-Atlas, whereas those of *C. pulicaris* were more northeastern (Fig. 4). Both species were found at all times of the year (Fig. 3). In 1994 the greatest catches of *C. obsoletus* were from February to July, with a peak in April; in 1995 there was a small peak in March/April and a larger peak between July–November. The reason for this difference between years is unclear. *C. pulicaris* showed a peak in March/April and a larger peak between August–November.

Outbreaks of AHS in Morocco from 1989–1991

The spread of AHS in Morocco during the 1989–1991 epizootic has been previously described at the provincial level [1, 15]. Since provinces may be large and geographically varied, an attempt is made here to examine the distribution of disease outbreaks at a larger spatial scale. Records of the AHS epizootic held at the Direction d'Elevage, Ministère de l'Agriculture et de la Reforme Agraire, Rabat, were examined by M. B. in 1993 to obtain the names and localities of all villages in Morocco where outbreaks of AHS were reported in 1989–1991. The latitudes and longitudes of these villages were obtained from 1:50,000 Morocco survey maps; subsequently, these coordinates were plotted upon a digitised outline of Morocco (Fig. 5).

In 1989, a total of 512 cases of AHS were reported from the 3 most northern provinces of Morocco. These provinces are the nearest to Spain, from where the virus is presumed to have been introduced. In 1990 the epizootic spread south and east, and 555 cases of AHS were reported from 6 provinces (including 2 previously affected in 1989). There was a particularly large cluster of reported cases (463) in the province of Chefchaouen, most of which (81%) were in mules and donkeys while only 19% were in horses. In contrast, elsewhere, from 1989 to 1991, only 13.3% of reported cases were in mules and donkeys while 86.7% were in horses [1]. In 1991 the epizootic spread much further than in the previous 2 years: outbreaks were reported from 20 provinces spread over most of the country (including 1 affected in 1990). During the 1989–1991 epizootic all cases of AHS were reported between the months of July and December [1] with most reported between September and November.

Discussion

The first record of *C. imicola* in Morocco was in 1968, following a survey of 16 sites spread across the country [2]. Of these 16 sites, 12 were visited once only, mostly between June and September; the remaining 4, comprising Marrakech and 3 sites in the northwest, were sampled over 9 months of the year. Despite the intensity of this survey, conducted largely at localities and during the season when *C. imicola* would be expected to be frequent (according to our results), *C. imicola* was in fact found at only one site: 4 females were found at Marrakech. It is tempting to suggest

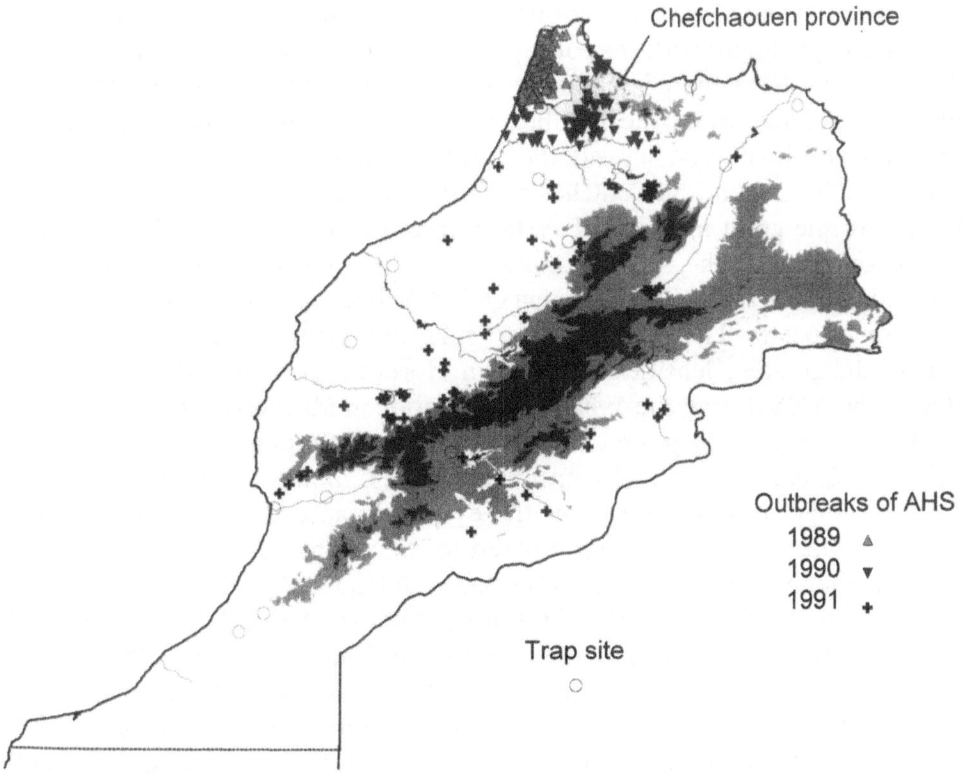

Fig. 5. The distribution of outbreaks of AHS in Morocco from 1989–1991. Because of incomplete recording it was not possible to locate the positions of all of the outbreaks reported in Morocco by [1]. The rates at which outbreaks were successfully located were: 465 out of 512 (90.8%) for 1989; 302 out of 555 (54.4%) in 1990; and 129 out of 177 (72.9%) in 1991. The particularly low success rate in 1990 is attributable to the low number of reports found for the province of Chefchaouen: in the remaining 5 provinces affected by AHS in that year 92.4% of the outbreaks were successfully located. For 1989–1991 a total of 348 reported outbreaks were not located, but only one of these was reported from a 'commune rurale' (a subdivision of a province) in which no other outbreaks were located. In general, therefore, the distribution of AHS outbreaks shown here is probably a good approximation to the distribution of all reported outbreaks

that in the last 25 years the distribution of *C. imicola* may have expanded dramatically but two observations argue against this interpretation. First, in 1965–1966 there was an epizootic of AHS in Morocco, with outbreaks reported from much of the national territory [14]. Assuming that *C. imicola* was the vector, it seems likely that the distribution of this insect in 1968 was far more widespread than is suggested by [2]. Second, most of the catches of other *Culicoides* species listed by [2] were small, with rarely over 10 of any one species. It is possible that the collecting methods of [2] (battery-operated CDC traps or aspiration of *Culicoides* from illuminated white sheets) were not sufficiently powerful or not used over enough time to capture many species at sites where they were present. Alternatively, it is possible that following the 1965–1966 epizootic insecticides were widely used around animal holdings and this affected the survey results in 1968.

The spatial and seasonal distribution of *C. imicola* in Morocco reported here is in broad agreement with that of reported outbreaks of AHS during the 1989–1991 epizootic. First, there is a large peak in the catch of *C. imicola* in late summer and early autumn while the number of reported cases of AHS peaked from September to November. Second, the northwest of Morocco, between Rabat and Tangier, is the area in which catches of *C. imicola* are most consistently large; it is also where the great majority of AHS outbreaks were reported. Third, a large area to the west of the Atlas mountains in which no outbreaks of AHS were reported contains the only site (Jemaa Oulad Abbou) in which no *C. imicola* have been found to date. This result is of particular interest because this area contains some of the greatest densities of equids in Morocco [1]. This region also was not affected by AHS during the 1965–1966 epizootic in Morocco [14].

Despite this broad agreement, there are some differences between the distributions of *C. imicola* and AHS outbreaks. First, *C. imicola* appears to be particularly abundant in Marrakech, yet there was not an especially high density of disease outbreaks in the area (compared to the north). It is interesting to note, however, that of the 20 provinces affected by AHS in 1991, Marrakech province had the second largest number of reported cases [1]. Second, surveying in 1995 revealed only small numbers of *C. imicola* at Chefchaouen yet, in 1990, there was a very large number of reported cases of AHS in mules in Chefchaouen province. One possibility is that *C. obsoletus*, which was abundant at Chefchaouen in 1995, was a significant vector of AHSV in the province. More likely, perhaps, is that the site at Chefchaouen was badly chosen, and the catches of *C. imicola* are not representative of its abundance in the area. For logistical reasons (combining a mains electricity supply with the presence of livestock) it was difficult to find a trapping site in the area of Chefchaouen; the site eventually chosen was a small cheese factory at which there were many goats in the surrounding area. The trap itself was situated at some distance from the nearest livestock, and this may have led to the trap catches of *C. imicola* having been low.

An important question prompted by the 5 year epizootic of AHS in Iberia and Morocco was the mechanism by which the virus managed to overwinter for periods of 6–12 months between the report of the last outbreak in one year and the report of the first outbreak in the next; this question arises because, to date, the longest demonstrated duration of viraemia of AHSV in known hosts is only 40 days (demonstrated by virus isolation in cell culture from the blood of inoculated zebra foals [3]) and the maximum lifespan of adult *C. imicola* is about 70 days [26]. It was recently suggested that a possible contributing factor could be the extended longevity of recently-fed *Culicoides* at low temperatures (10°C) followed by maturation of the viral infection upon their return to higher temperatures [29]. While *Culicoides* at low temperatures may live for 3 months (J. Boorman, unpubl. data), such cold temperatures are rarely found for long periods in Morocco and this mechanism is unlikely to have played an important, if any, role in viral overwintering during the 1989–1991 AHS epizootic. Results presented here demonstrate that at certain sites, however, adult *C. imicola* may be caught on the wing during all periods of the year suggesting that a possible mechanism of viral overwintering was continued cycling of AHSV between small numbers of *C. imicola* and hosts (mostly unvaccinated donkeys and mules) during the winter months as suggested by Mellor [18]. In this regard it is particularly interesting to note that in Morocco the virus probably overwintered in 1989–1990 in the region of Larache (the last outbreaks in 1989, and the first in 1990, were in Larache province), and in 1990–1991 it may have overwintered in the region of Marrakech (in July 1991, when the virus reappeared, the first cases were in Marrakech province). Larache and Marrakech are both areas where adult *C. imicola* have been found at all times of the year.

This scenario, in which AHSV overwintered in certain areas of Morocco by cycling among a low number of adult *C. imicola* and equine hosts, prompts an important question: why were there no reported outbreaks of AHS in horses during the winter period? One possibility is that the apparent lack of cases in horses is entirely probabilistic: given the small number of adult *C. imicola* and, in most instances, a relatively small population of horses that remain unprotected by vaccination, the probability of a susceptible horse being bitten by an infected *C. imicola* is extremely small. A second possibility is that cases of AHS in horses do occur in winter, but at an incidence that is too low to be brought to the attention of the relevant authorities or for those authorities to act. This seems unlikely during an epizootic of a notifiable disease. In this regard, it is interesting to note that there was a report of a case of AHS in an equid in Marrakech province in December 1990 [25], 7 months before the epizootic officially commenced in that province in 1991 [1], although it has been suggested that this report was an error (M. El Harrak, pers. comm.).

In contrast to *C. imicola*, there is relatively little evidence to suggest that *C. obsoletus* and *C. pulicaris* contributed significantly to the epizootic of AHSV in Morocco. Both species have relatively large peaks in trap catch size in spring (compared to *C. imicola*) while outbreaks of AHS were limited to summer and autumn. Outbreaks of AHS were relatively rare in the area of Morocco in which *C. pulicaris* appears to be particularly abundant, and there were no outbreaks south of the Anti-Atlas in an area where *C. obsoletus* are abundant.

Acknowledgements

We thank the personnel at all of our sites for their help and cooperation with the vector surveillance, and A. Belamine, M. Ait Lachguer, B. Benzaouia, K. Seghrouchni, O. Amar and A. El Adlouni for the sorting of insect catches and identification of *Culicoides*. This work was financed by the EU project TS3-CT920151.

References

1. Anon (1992) La peste équine au Maroc. Epizooties de 1989, 1990 et 1991. Données epidemiologiques et strategie de lutte. Report of the Direction de l'Elevage. Ministère de l'Agriculture et de la Reforme Agraire, Rabat, Morocco
2. Bailly-Choumara H, Kremer M (1970) Deuxième contribution a l'étude des *Culicoides* du Maroc (Diptera, Ceratopogonidae). Cah O.R.S.T.O.M., Ser Entomol Med Parasitol 8: 383–391

3. Barnard BJH, Bengis R, Keet D, Dekker EH (1994) Epidemiology of African horse-sickness: duration of viraemia in zebra (*Equus burchelli*). Onderstepoort J Vet Res 61: 391–393
4. Baylis M, El Hasnaoui H, Bouayoune H, Touti J, Mellor PS (1997) The spatial and seasonal distribution of African horse sickness and its potential *Culicoides* vectors in Morocco. Med Vet Entomol 11: 203–212
5. Baylis M, Bouayoune H, Touti J, El Hasnaoui H (1998) Use of climatic data and satellite imagery to model the abundance of *Culicoides imicola* Kieffer, the vector of African horse sickness, in Morocco. Med Vet Entomol 12 (in press)
6. Boorman J (1974) *Culicoides* (Diptera: Ceratopogonidae) from Cyprus. Cah O.R.S.T.O.M., Ser Entomol Med Parasitol 12: 7–13
7. Boorman J (1986) Presence of bluetongue virus vectors on Rhodes. Vet Rec 118: 21
8. Boorman JPT, Wilkinson PJ (1983) Potential vectors of bluetongue in Lesbos, Greece. Vet Rec 113: 395–396
9. Capela R, Sousa C, Pena I, Caeiro V (1993) Preliminary note on the distribution and ecology of *Culicoides imicola* in Portugal. Med Vet Entomol 7: 23–26
10. Diaz-Laviada M, Arias M, Sanchez-Vizcaino JM (1995) New developments in African horsesickness diagnosis. In: The development of diagnostic and preventative methods by genetic engineering for African horsesickness and related orbiviruses. National Institute of Animal Health, Japan, pp 67–79
11. HMSO (1983) Tables of temperature, relative humidity, precipitation and sunshine for the world. Part 4. Africa, the Atlantic Ocean south of 35° N and the Indian Ocean. Her Majesty's Stationery Office, London
12. Hess WR (1988) African horse sickness. In: Monath TP (ed) The arboviruses: epidemiology and ecology, vol 2. CRC Press, Boca Raton, pp 1–18
13. Howarth FG (1985) Biosystematics of the *Culicoides* of Laos (Diptera: Ceratopogonidae). Int J Entomol 27: 1–96
14. Laaberki A (1969) Evolution d'une épizootie de Peste équine au Maroc. Bull Off Int Epiz 71: 921–936
15. Lhafi A, Tber A, Fikri A, Laghzaoui K (1992) African horse sickness in Morocco: the epizootics of 1989 and 1990. In: Walton TE, Osburn BI (eds) Bluetongue, African horse sickness, and related orbiviruses. CRC Press, Boca Raton, pp 205–216
16. Macfie JWS (1943) Ceratopogonidae (Diptera) from Egypt. Proc R Entomol Soc Lond A 12: 145–159
17. Meiswinkel R (1989) Afrotropical *Culicoides*: A redescription of *C. (Avaritia) imicola* Kieffer, 1913 (Diptera: Ceratopogonidae) with description of the closely allied *C. (A.) bolitinos* sp. nov. reared from the dung of the African buffalo, blue wildebeest and cattle in South Africa. Onderstepoort J Vet Res 56: 23–39
18. Mellor PS (1994) Epizootiology and vectors of African horse sickness virus. Comp Immunol Microbiol Infect Dis 17: 287–296
19. Mellor PS, Boned J, Hamblin C, Graham S (1990) Isolations of African horse sickness virus from vector insects made during the 1988 epizootic in Spain. Epidemiol Infect 105: 447–454
20. Mellor PS, Boorman J, Jennings M (1975) The multiplication of African horse-sickness virus in two species of *Culicoides* (Diptera, Ceratopogonidae). Arch Virol 47: 351–356
21. Mellor PS, Boorman JPT, Wilkinson PJ, Martinez-Gomez F (1983) Potential vectors of bluetongue and African horse sickness viruses in Spain. Vet Rec 112: 229–230
22. Mellor PS, Jennings DM, Wilkinson PJ, Boorman JPT (1985) *Culicoides imicola*: a bluetongue virus vector in Spain and Portugal. Vet Rec 116: 589–590
23. OIE (1988) Bull Off Int Epiz 100

24. OIE (1989) Bull Off Int Epiz 101
25. OIE (1991) Bull Off Int Epiz 103: 461
26. Sellers RF (1981) Bluetongue and related diseases. In: Gibbs EPJ (ed) Virus diseases of food animals – a world geography of epidemiology and control. Academic Press, London, pp 567–583 (Disease monographs, vol 2)
27. Szadziewski R (1984) Ceratopogonidae (Diptera) from Algeria. VI. *Culicoides* Latr. Polski Pis Entomol 54: 163–182
28. Theiler A (1921) African horse sickness (*Pestis equorum*). S Afr Dep Agric Sci, Government Printer, Pretoria (bulletin 19)
29. Wellby MP, Baylis M, Rawlings P, Mellor PS (1996) Effect of temperature on survival and rate of virogenesis of African horse sickness virus in *Culicoides variipennis sonorensis* (Diptera: Ceratopogonidae) and its significance in relation to the epidemiology of the disease. Bull Entomol Res 86: 715–720
30. Wirth WW, Hubert AA (1989) The *Culicoides* of South East Asia (Diptera: Ceratopogonidae). American Entomological Institute, Gainesville, Florida (Memoires of the American Entomological Institute, No 44)

Authors' address: Dr. M. Baylis, Institute for Animal Health, Ash Road, Pirbright, Surrey GU24 0NF, U.K.

Studies of the mortality rate of *Culicoides imicola* in Morocco

M. Baylis[1], **J. Touti**[2], **H. Bouayoune**[3], **L. Moudni**[4], **B. Taoufiq**[5], and **H. El Hasnaoui**[6]

[1] Institute for Animal Health, Pirbright, U.K.
Laboratoire Régional d'Analyses et de Recherches Vétérinaires d'Agadir[2], de Fès[3], d'Oujda[4], de Marrakech[5], de Casablanca[6], Morocco

Summary. Daily mortality rates of female *Culicoides imicola* were found for eight sites in Morocco in 1994 and for six sites in 1995. The mortality rates were found by operating Pirbright-type light traps for a number of consecutive nights in late summer or autumn and finding the parous rate assuming a feeding interval of 3 to 5 days. The mortality rates were calculated according to established methods. In Morocco the daily mortality rates were found to vary from about 5% per day (Arbaoua, 1994, 1995 and Sidi Moussa 1995) up to 20–25% per day (Berkane, Marrakech, Tangier). In general, estimates of daily mortality rate were consistent between the two years of study. Among sites, daily mortality rate was significantly correlated with the average night-time minimum wind speed but not mean or maximum night-time wind speeds, or with temperature, humidity or saturation deficit. The observed mortality rates suggest that at Arbaoua, were 1 000 flies to become infected with African horse sickness virus, at least 330 would live long enough to take 3 or more infective blood meals on hosts. At Berkane, the survival rate per 1 000 is less than 10. In general, the pattern observed for daily mortality rate, combined with the relative population sizes of *C. imicola* in Morocco, agree well with the observed distribution of African horse sickness in the country during the 1989–1991 epizootic.

Introduction

Vectorial capacity is a measure of the capability of a species of blood sucking insect to transmit parasites to the animals on which it feeds [8]. Many factors contribute to vectorial capacity, one of the most important being the daily mortality rate of the insect. There are two reasons. First, the daily mortality rate affects the population size of the insect and second, a proportion of the feeding population must live longer than the *extrinsic incubation period* of the parasite (i.e. the development time or maturation rate of the parasite within the vector).

For example, at 25°C the extrinsic incubation period of African horse sickness virus (AHSV) in the biting midge *Culicoides variipennis* is about 8–10 days [13]. If the insects feed one day after emergence, and then every three days, only females 10 or more days old may transmit AHSV to other hosts. The daily mortality rate will determine what fraction of the population live long enough for this to occur.

A widely used method for estimating the daily mortality rate of blood sucking insects is examination of the population age structure. This is possible because, often, the proportion of old individuals in a population is indicative of the mortality rate. Generally, a population dominated by young individuals has a high mortality rate (preventing many of these youngsters becoming old). A population with a relatively high proportion of old individuals has a lower mortality rate. Critical to this interpretation is the assumption that there is no age-specific immigration or emigration, which may alter the population age structure without mortality. For *Culicoides* such age-specific immigration or emigration is not known to occur.

How may the population age structure of *Culicoides* be found? For many species, including the field vector of AHSV, *C. imicola*, it is relatively simple to sort field-caught female adults into two age groups. First, *nulliparous* flies that have not yet taken a blood meal and have not oviposited; these flies have a transparent or white abdomen. Second, *parous* flies which have taken a blood meal and have laid one or more egg batches. Parous *C. imicola* deposit a red pigment in the abdomen, darkening the overall colour [7]. In general, nulliparous flies become parous within the first few days of adult life.

The method of estimating the daily mortality rate of mosquitoes and *Culicoides* from the population age structure has been investigated in depth in a series of papers by Birley [4, 5]. The principle is as follows. On any single day the population contains both parous and nulliparous females. Several days later, when this part of the population seek to feed again, all of these flies should now be parous. The actual number of parous flies caught at this time, as a fraction of the nulliparous plus parous catch the time before, gives the *parous rate*, an estimate of the survival rate over the intervening period. As an example, assuming that *C. imicola* females feed every three days, if on day 0 the catch comprised 200 nulliparous and 800 parous flies while on day 3 the catch contained 600 parous flies, we can estimate that 600 of 1000 potentially parous flies have survived, giving a survival rate of 0.6 (60%) and a mortality rate of 0.4 (40%). This is over 3 days; the daily survival rate is $\sqrt[3]{(0.6)} = 0.84$ (84% per day), giving a daily mortality rate of 0.16 or 16% per day. Note that *mortality rate* in this sense means *loss from the population*: it does not distinguish between death and dispersal as causes of this loss.

We have used the basic methodology devised by Birley to estimate the daily mortality rates of *C. imicola* at a number of sites in Morocco in the summers of 1994 and 1995, and we have then attempted to correlate these rates with climatic variables recorded at the same sites at the same time.

Materials and methods

Sites

Mortality rates may only be estimated where sufficient *C. imicola* are caught per day to allow reliable estimation of the parous rate. Of the 20 sites of the vector surveillance system operating in Morocco since 1993 [3], at only 8 were *C. imicola* deemed sufficiently abundant for mortality rate estimation. These 8 sites were Agadir, Arbaoua (near Larache), Berkane, Marrakech, Sablons (near Rabat), Sidi Moussa (near Khemisset), Tangier and Taroudant. All eight sites were examined in 1994; Agadir, Arbaoua, Berkane, Marrakech, Sablons and Sidi Moussa were re-examined in 1995. In each case, traps were run for a number of consecutive nights (from 18 to 58, average = 26), of which some nights may have been missed out. Details of sites and the durations of trapping are given in Table 1.

Collecting and sorting of insects

Female *C. imicola* were sorted into 5 groups: nulliparous/unfed; nulliparous/blood fed; parous/unfed; parous/blood fed; and gravid. Nulliparous and parous females were distinguished on the basis of red pigment in the abdomen, as described by Dyce [7]. Blood fed individuals were identified from the presence of red blood visible through the wall of the abdomen. Gravid females contained mature egg batches. Only the nulliparous/unfed and parous/unfed groups of females were used in the calculation of parous rates.

Table 1. Details of the sites used for *Culicoides imicola* mortality rate studies in Morocco in 1994 and 1995

Site	Province	Year	Start date[a]	End date[a]	Length of run[b]	Days missed[c]
Agadir	Agadir	1994	30/9/94	8/11/94	40	7
		1995	11/9/95	30/9/95	20	0
Arbaoua	Larache	1994	6/6/94	6/7/94	31	1
		1995	11/9/95	30/9/95	20	2
Berkane	Berkane	1994	20/9/94	10/10/94	21	0
		1995	11/9/95	2/10/95	22	0
Marrakech	Marrakech	1994	19/9/94	7/10/94	20	6
		1995	11/9/95	2/10/95	20	0
Sablons	Rabat	1994	1/10/94	21/10/94	21	1
		1995	16/9/95	10/10/95	25	1
Sidi Moussa	Khemisset	1994	11/10/94	6/12/94	58	3
		1995	11/9/95	2/10/95	22	0
Tangier	Tangier	1994	1/9/94	29/9/94	29	1
Taroudant	Taroudant	1994	22/9/94	9/10/94	18	0

[a] Start and end of insect collection for mortality rate calculation
[b] Number of days between start and end date
[c] Number of days between start and end date on which traps were not operated

Estimation of daily mortality rates

Following Birley and Rajagopalan [5], parous rates were calculated for the sequences of consecutive daily catches at each site on the assumption that *C. imicola* females feed every 1, 2, 3, 4, 5, 6 or 7 days. Those authors suggest that it is possible to find which of these feeding frequencies (and gonotrophic cycle lengths) is most correct by calculating the degree of correlation between the nulliparous + parous catch and the parous catch 1–7 days later; if *C. imicola* feed every 2 days correlations will be higher for feeding frequencies of 2, 4 and 6 days than 1, 3, 5 and 7 days. If they feed every 3 days the correlations for 3 and 6 days will be the highest. Here, we found this method to be ineffective: the degree of correlation was low and there was rarely any periodicity in goodness of fit. Best estimates of feeding frequency also varied from site to site. Therefore, on the basis that other studies have found *Culicoides* feeding frequencies to vary from 3–5 days [4, 6, 12, 14], daily mortality rates were calculated for 3, 4 and 5 day frequencies and then averaged. The estimates for 3, 4 and 5 days were generally quite similar: the greatest difference was only 12% with a mean of 7%, suggesting that the 3–5 day average is a reasonable estimate of daily mortality rate.

This method of calculating daily mortality rate makes numerous assumptions about the natural history of *C. imicola* and the sampling process. These include: female *C. imicola* require a single blood meal per gonotrophic cycle; all females have a gonotrophic cycle of the same duration; and survival is independent of age.

Climatic data

Climatic data were recorded hourly using automatic weather stations (Campbell Scientific datalogger) located less than 100 metres from the trap site. Data were processed to give daily maxima, means and minima of several variables that might, *a priori*, be expected to have an effect on adult mortality rates. These variables were air temperature, humidity, saturation deficit and wind speed. The daily maxima, means and minima were averaged for the periods that mortality rates were estimated at each site. *Culicoides imicola* is nocturnal and therefore only night-time (20.00–06.00 h) weather data were used in the calculation of climatic variables.

Unfortunately, hourly weather data for Berkane in 1995 were not recorded. The automatic weather station was also programmed to record daily maxima, averages and minima, but it was found that the records of daily minimum wind speed were almost all at the default zero level, suggesting a problem. Analysis of the data from the other sites revealed that the minimum wind speeds in 1995 were not significantly different to those in 1994 (ANOVA, $F_{1,4} = 1.32$, $P > 0.3$) and therefore the minimum wind speed at Berkane in 1995 was given the same value as in 1994. This problem did not apply to the maximum or average wind speeds at Berkane, or to the minima of other climatic variables.

Results

The daily mortality rates of *C. imicola* in 1994 and 1995 are shown in Fig. 1. The mortality rates varied from 3.9% per day (Sidi Moussa, 1995) to 25.6% per day (Tangier, 1994), with an average of 14.9% per day.

In general, the daily mortality rates at a given site were fairly consistent between 1994 and 1995. This is particularly clear at Arbaoua, Marrakech and Sablons where very similar estimates were obtained in both years. The largest difference between the two years was seen at Sidi Moussa where the estimates of daily mortality rate were approximately 15% in 1994 but under 5% in 1995. The

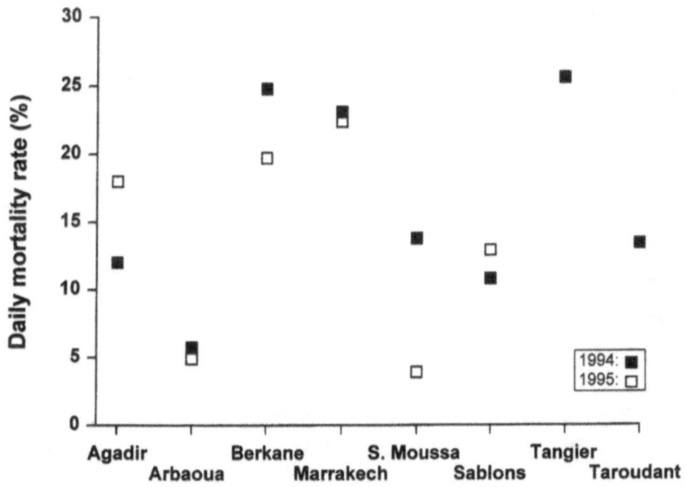

Fig. 1. The daily mortality rate of *Culicoides imicola* at 8 sites in Morocco in 1994 and 1995

effect of year on daily mortality rate was not significant by Analysis of Variance ($F_{1,5}$ = 0.23, P = 0.5). There were marked differences among the sites in the estimates of daily mortality rate. At Arbaoua the mortality rate in the two years averaged about 5% per day, while at Berkane and Marrakech they were about 22.5% per day. The effect of site on daily mortality rate was significant by Analysis of Variance ($F_{7,5}$ = 5.24, $P < 0.05$).

Correlation coefficients between daily mortality rate and several climatic variables are shown in Table 2. The only significant relationship was between the daily mortality rate and the minimum daily wind speed. This significant relationship is shown in Fig. 2. It is clear that the high daily mortality rates at Berkane and Tangier are related to the high wind speeds recorded at those sites, while the low daily mortality rates at Arbaoua are related to low wind speeds.

Table 2. Product moment correlation coefficients between the daily mortality rate of *Culicoides imicola* and several climatic variables in Morocco

	Climatic variable	Correlation coefficient
Humidity	daily maximum	–0.246
	daily average	0.264
	daily minimum	–0.216
Air temperature	daily maximum	–0.088
	daily average	–0.104
	daily minimum	0.064
Wind speed	daily maximum	0.278
	daily average	0.427
	daily minimum	0.604[a]
Saturation deficit	daily average	–0.066

[a] $p < 0.05$

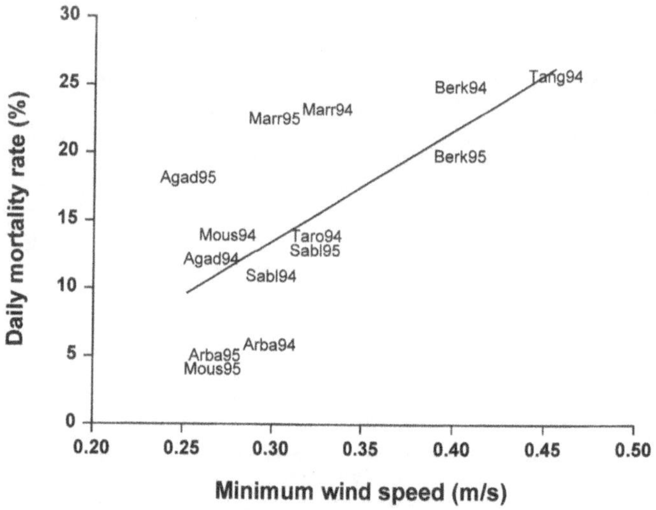

Fig. 2. The relationship between minimum night-time wind speed and the daily mortality rate of *Culicoides imicola* at 8 sites in Morocco in 1994 and 1995. *Agad* Agadir; *Arba* Arbaoua; *Berk* Berkane; *Marr* Marrakech; *Mous* Sidi Moussa; *Sabl* Sablons; *Tang* Tangier; *Taro* Taroudant

Discussion

In this paper we have shown that the daily mortality rate of *C. imicola* in Morocco may vary considerably from site to site. Such variation might arise from a number of sources. First, climatic or other factors (eg, insecticide use, abundance and quality of resting sites during the day) may affect the actual daily mortality rate. Second, several factors may lead to bias in the age structure of trapped *Culicoides*, which does not change the daily mortality rate but does change the parous rate in traps and hence the estimate of the daily mortality rate. Such factors include trap type, time of day, distance of trap from breeding site, and host access to breeding site [9]. In our study, both trap type and time of day were common to all sites and should not, therefore, have introduced any bias. However, the other two factors may have varied from site to site, and in some cases from year to year and might therefore have introduced some bias. Sampling bias would tend to lessen the chance of detecting an association between any climatic factors and mortality. Our observation that there is an association between daily mortality rate and minimum daily wind speed suggests that any effects of bias were relatively small.

The degree of variation in daily mortality rate between sites in Morocco (from 4 to 25%) is high. This may have profound implications for the transmission of AHSV by *C. imicola* during an epizootic. In Morocco in summer the daily mean temperature is around 25°C (Baylis, unpubl. data). It has been observed that at this temperature the extrinsic incubation period of AHSV in *C. variipennis* is about 8–10 days. If we then assume that during the peak time of AHSV transmission in Morocco the extrinsic incubation period of the virus in *C. imicola* is 9 days, and that the insects feed every 3–5 days, it is possible to calculate the expected proportion of insects that survive long enough to infect hosts when feeding. Such estimates are shown for 3 sites in Table 3.

Table 3 shows that at Arbaoua, where the daily mortality rate is very low, a very high proportion of *C. imicola* that take an AHSV-infecting blood meal are expected to live long to develop a full infection and possibly transmit the virus in subsequent feeds: indeed, approximately 33.7% of flies should live long enough to take 3 or more such feeds. At Sablons, where the daily mortality rate is approximately twice that of Arbaoua, this percentage is only 7.9%. At Berkane, with a daily mortality rate approximately 4 times greater than that of Arbaoua, the percentage is less than 1%. This example neatly demonstrates that the risk to livestock of becoming infected from the bite of an infected fly increases disproportionately with decrease in daily mortality rate.

This risk is a composite of a number of factors including both daily mortality rate and the population size of *C. imicola*. At Arbaoua the daily mortality rate is low, and the population size of *C. imicola* is very high [3], suggesting that the risk should be very high. Indeed, during the recent epizootic of AHS in Morocco the number of cases reported in Larache province was very high [1]. In contrast, at Marrakech *C. imicola* is again abundant but the daily mortality rate is very high, and this may explain why the number of reported cases of AHS in Marrakech province was relatively few [1]. This same factor, combined with the slightly lower population sizes of *C. imicola* found in the east of Morocco may have contributed to the lack of AHS cases reported in Berkane province [1].

As just described, the observed daily mortality rates and population sizes of *C. imicola* are in large part consistent with the geographical distribution of AHS cases in Morocco in 1989–1991. A notable exception is Tangier, where the daily mortality rate was found to be very high but the province suffered many cases of AHS [1]. We have shown that the most likely cause of the high daily mortality rate at Tangier is high wind speed at night. This wind speed may be a result of the location of the trap site near to the coast and the strait of Gibraltar, where wind speeds would be expected to be high, rather than a general phenomenon of the province. In other words, in the province of Tangier, away from the coast, the wind speeds are probably quite low allowing a low daily mortality rate of *C. imicola* and a high transmission rate of AHSV to susceptible animals.

It is unknown how higher wind speeds lead to a high mortality rate of *C. imicola*. One possibility is that wind movements contribute to desiccation of the insects; however, were this to be the case one would expect humidity and/or saturation deficit to also have been correlated with the observed daily mortality rates. A second, more likely possibility is that wind simply blows the insects away, leading to loss through dispersal. Such an effect of wind has been suggested previously based on the relationship between wind direction and the geographical spread of epizootics of bluetongue and AHS [10, 11]. Ours is the first study to show that an expected entomological consequence of this effect, evidence of more loss through dispersal in areas of higher wind speeds, is in fact found.

The significance of the minimum, as opposed to the mean or maximum night-time wind speeds as a factor affecting the daily mortality rates of *C. imicola* is not

Table 3. Estimates of the number of *Culicoides imicola* that survive long enough to take 1 or more, 2 or more and 3 or more infective blood meals on hosts, based on several estimates of feeding interval and daily mortality rate. The extrinsic incubation period of AHSV in *C. imicola* is taken to be 9 days. Values are the number of survivors out of 1000 initially infected with AHSV

Site	Daily mortality rate	Feeding interval = 3 days Number of infective feeds (first is taken on day 9)			Feeding interval = 4 days Number of infective feeds (first is taken on day 12)			Feeding interval = 5 days Number of infective feeds (first is taken on day 10)		
		1+	2+	3+[a]	1+	2+	3+[b]	1+	2+[a]	3+[b]
Arbaoua	5.3%	613	520	442	520	418	337	580	442	337
Sablons	11.9%	320	219	149	219	132	79	282	149	79
Berkane	22.2%	104	49	23	49	18	7	81	23	7

[a] These estimates are identical because both refer to day 15 after infection
[b] These estimates are identical because both refer to day 20 after infection

clear. It seems likely, perhaps, that at night the midges tend to be active at the time of the lowest wind speed and therefore that the impact of wind speed on mortality rate depends on how low the wind speeds actually are.

There is mounting evidence to suggest that wind speed may be a critical climatic factor in the ecology of *C. imicola* and, hence, the epidemiologies of the viruses it transmits. It has been shown that minimum wind speed, in combination with a satellite-derived correlate of soil moisture (and hence, probably, the frequency of larval breeding sites), is an important determinant of the distribution of *C. imicola* in Morocco [2]. The present study suggests that this effect arises from the interplay of wind speed with adult daily mortality rates through either death or, more likely, dispersal.

Acknowledgements

We are very grateful to the personnel at all of our sites for their help and cooperation with the light trapping and guarding of the weather stations, and A. Belamine, M. Ait Lachguer, B. Benzaouia, K. Seghrouchni, O. Amar and A. El Adlouni for the sorting of insect catches and identification of *Culicoides*. We thank Cynthia Lord and Philip Mellor for comments on the manuscript. This work was financed by the EU project TS3-CT920151.

References

1. Anonymous (1992) La peste équine au Maroc. Epizooties de 1989, 1990 et 1991. Données epidemiologiques et strategie de lutte. Report of the Direction de l'Elevage, Ministére de l'Agriculture et de la Reforme Agraire, Rabat, Morocco
2. Baylis M, Bouayoune H, Touti J, El Hasnaoui H (1998) Use of climatic data and satellite imagery to model the abundance of *Culicoides imicola* Kieffer, the vector of African horse sickness virus, in Morocco. Med Vet Entomol 12
3. Baylis M, El Hasnaoui H, Bouayoune H, Touti J, Mellor PS (1997) The spatial and seasonal distribution of African horse sickness and its potential *Culicoides* vectors in Morocco. Med Vet Entomol 11: 203–212
4. Birley MH, Boorman, JPT (1982) Estimating the survival and biting rates of haematophagous insects, with particular reference to the *Culicoides obsoletus* group (Diptera, Ceratopogonidae) in southern England. J Anim Ecol 51: 135–148
5. Birley MH, Rajagopalan PK (1981) Estimation of the survival and biting rates of *Culex quinquefasciatus* (Diptera: Culicidae). J Med Entomol 18: 181–186
6. Braverman Y, Linley JR, Marcus R, Frish K (1985) Seasonal survival and expectation of infective life of *Culicoides* spp. (Diptera: Ceratopogonidae) in Israel, with implications for bluetongue virus transmission and a comparison of the parous rate in *C. imicola* from Israel and Zimbabwe. J Med Entomol 22: 476–484
7. Dyce AL (1969) The recognition of nulliparous and parous *Culicoides* (Diptera: Ceratopogonidae) without dissection. J Aust Entomol Soc 8: 11–15
8. Dye C (1990) Epidemiological significance of vector-parasite interactions. Parasitology 101: 409–415
9. Mullens BA (1985) Sampling bias and the problem of age survivorship determination in *Culicoides*. In: Bluetongue and related orbiviruses. Alan R Liss, New York, pp 207–211 (Progress in Clinical and Biological Research, vol 178)
10. Sellers RF (1980) Weather, host and vector – their interplay in the spread of insect-borne animal virus diseases. J Hyg 85: 65–102

11. Sellers RF, Pedgley DE, Tucker MR (1978) Possible windborne spread of bluetongue to Portugal, June–July 1956. J Hyg 81: 189–196
12. Walker AR (1977) Adult lifespan and reproductive status of *Culicoides* species (Diptera: Ceratopogonidae) in Kenya, with reference to virus transmission. Bull Entomol Res 67: 205–215
13. Wellby M, Baylis M, Rawlings P, Mellor PS (1996) Effect of temperature on survival and rate of virogenesis of African horse sickness virus in *Culicoides variipennis sonorensis* (Diptera: Ceratopogonidae) and its significance in relation to the epidemiology of the disease. Bull Entomol Res 86: 715–720
14. Work TM, Mullens BA, Jessup DA (1991) Estimation of survival and gonotrophic cycle length of *Culicoides variipennis* (Diptera: Ceratopogonidae) in California. J Am Mosq Control Assoc 7: 242–249

Authors' address: Dr. M. Baylis, Institute for Animal Health, Ash Road, Pirbright, Surrey GU24 0NF, U.K.

Modelling the distribution and abundance of *Culicoides imicola* in Morocco and Iberia using climatic data and satellite imagery

M. Baylis and **P. Rawlings**

Institute for Animal Health, Pirbright, U.K.

Summary. Relative abundances of *Culicoides imicola* at 22 sites in Morocco were compared with climatic variables, altitude and the NDVI (Normalised Difference Vegetation Index, a satellite derived measure of photosynthetic activity) of the same sites. Abundances were negatively correlated with wind speed and positively correlated with the average and minimum NDVI ($NDVI_{min}$). There were no significant correlations with air or soil temperatures, relative humidity, saturation deficit, rainfall, altitude or mean annual maximum or range of NDVI. The best 2-variable model, which combined wind speed and $NDVI_{min}$ as predictors, explained over 50% of the variance in abundance. It is suggested that wind speed affects the abundance of *C. imicola* by causing adult mortality while $NDVI_{min}$ provides a measure of the availability of *C. imicola* breeding sites. Data from 27 sites in Iberia yielded broadly similar results to those found in Morocco although the great abundance of *C. imicola* at Milfontes (Portugal) could not be accounted for. Several northern sites where the $NDVI_{min}$ is sufficiently high to suggest the presence of *C. imicola* but where it appears to be absent indicate that there may be a northern limit to the distribution of *C. imicola* in Iberia that is independent of $NDVI_{min}$. The potential use of $NDVI_{min}$ to predict the distribution of outbreaks of African horse sickness was investigated using data from the 1989–1990 epizootic in northern Morocco. Within the cluster of outbreaks in Larache province is a corridor of very low $NDVI_{min}$ in which few or no outbreaks were reported.

Introduction

The 1987–1991 epizootic of African horse sickness (AHS) in Spain, Portugal and Morocco was unprecedented in its duration and prompted studies of the effects of climate, and climate change, on the seasonal and spatial distributions of the *Culicoides* vectors of AHS virus (AHSV) in the region. *Culicoides imicola* is the only proven field vector of AHSV and several lines of evidence have suggested that it was by far the most important vector during the Iberian and Moroccan epizootics [3, 20]. Climatic studies were undertaken using a network of automatic weather stations and associated light traps spread across the region (Spain, 16 sites; Portugal, 11 sites; Morocco, 22 sites). Preliminary analysis of the presence or absence of *C. imicola* at the sites in Iberia has been presented by Rawlings and coworkers (this volume). In Morocco, *C. imicola* was absent from only one study site although there was a very large range in abundance [3] and here, therefore, we attempt to identify climatic correlates of the relative abundances of the vector at the 22 Moroccan sites. The resulting climate-driven model is then tested for robustness by examining its fit to the entomological and climatic data obtained previously in Iberia.

Sensors on several earth-orbiting satellites record images that may act as surrogate climatic variables – that is, the images are correlated to a greater or lesser degree with certain climatic variables recorded on the earth's surface [11]. One such surrogate climatic variable is the Normalised Difference Vegetation Index (NDVI), which is a measure of photosynthetic activity [26] and is correlated with functions of moisture, such as soil moisture [10, 18], saturation deficit [22] and rainfall [22, 25]. Terrestrial arthropods tend to be profoundly affected by moisture levels [9] and, in recent years, NDVI has been used successfully to model the distributions of several arthropod disease-vectors: tsetse flies [21, 23, 24]; phlebotomid sandflies [7]; *Anopheles* [28, 31, 32]; *Aedes* [12]; and hard ticks [8, 19, 23]. In many cases NDVI has proved more effective than ground-based climatic variables at modelling arthropod distributions (e.g., [23]). Possible reasons are that the detailed recordings of weather stations may be a less appropriate measure of climate than the actual response of vegetation (although the latter may be influenced by additional factors, including soil type and slope), and that satellite imagery, with its coarse resolution, records the average over a large area while weather stations record at very specific point locations that may not be representative of the general area. Other benefits derived from the use of NDVI for the development of predictive models are that there is a high temporal frequency of the imagery (dekadal – 10 day – images are widely used), there is global coverage (data are available for the whole of the earth's surface so interpolation between weather station sites is not necessary) and that, for historical, less-detailed images at least, data are available either at low or no cost. For these reasons NDVI was investigated as a surrogate climatic variable for modelling the distribution of *C. imicola* in Morocco and Iberia.

Finally, we compare the NDVI of northern Morocco with the distribution of outbreaks of AHS in the region in 1989 and 1990.

Materials and methods

Trap sites, trapping procedure and estimation of C. imicola *abundance in Morocco*

Details of trap sites, the trapping procedure and the method of calculating the relative abundance of *C. imicola* are described by Bouayoune et al. (this volume). The relative abundances of *C. imicola* to be correlated here with climatic and remotely-sensed variables are given in Table 1.

Automatic weather stations in Morocco

Automatic weather stations (Campbell Scientific Limited, Shepshed, UK) were installed at each of the 22 sites in Morocco, usually within 20 m of the light trap. Each station comprised a BDR320 datalogger with 3 probes recording 4 weather variables: an A100R switching anemometer recording wind speed (± 0.1 m/s with stalling speed 0.25 m/s) at 2 m above ground level, an HMP35AC probe recording air temperature (± 0.4°C) and humidity (± 3%) at 1.5 m above ground level, and a 103 thermistor recording soil temperature (± 0.4°C) at a depth of 2–4 cm.

Climatic variables for Morocco

The weather variables (air temperature, soil temperature, relative humidity and wind speed) were read at 5 min intervals, and stored as hourly averages (1 h record per variable). Monthly maxima and minima, average daily maxima and minima, and means were derived

Table 1. Annually-averaged mean daily catches of *C. imicola* at 22 sites in Morocco, 16 sites in Spain and 11 sites in Portugal

Moroccan sites		Spanish sites		Portuguese sites	
Tangier	20.32	La Almoraima	5.99	Castro Marim	1.19
Arbaoua	72.27	Mijas	3.34	V. N. Milfontes	36.55
M'diq	3.46	Cañada	0	Barrancos	21.75
Chefchaouen	0.70	Santa Fé	0	Evora	0.27
Imzouren	0.07	Constantina	1.74	Muge	0.02
Berkane	5.27	Guadalupe	0	Castelo Branco	0.02
Sablons	44.27	Azuaga	3.22	Vagos	0
Sidi Moussa	23.64	Calzada de Calatrava	1.19	Carvalhos	0
Tissa	0.11	Sagrajas	0.56	Freixo de Espada	0.20
Guercif	0.24	Aguas Nuevas	0	Alijo	0
Oujda	0.37	Nambroca	0.05	Monçao	0
Jemaa Oulad Abbou	0	Ceclavin	1.87		
Adarouche	0.22	Talayuela	19.35		
Youssoufia	0.44	Talavera de la Reina	0.15		
Marrakech	40.18	Getafe	0		
Beni Mellal	1.86	La Mortera Palomar	0		
Agadir	3.31				
Taroudant	4.57				
Bouizakarn	0.35				
Goulimime	0.68				
Amerzgane	1.74				
Errachidia	2.04				

Raw catch data were log transformed, annually-averaged and then detransformed for presentation. Raw data were maximum daily catches of *C. imicola* in each 2-week period of the year (Morocco) or monthly-averaged mean daily catches of *C. imicola* (Iberia). Note that mean daily catches of *C. imicola* for sites in Morocco calculated according to the methods used in Iberia are, on average, only 30% of the values shown here

Table 2. Climatic variables used in statistical analysis

	Monthly maximum	Monthly average daily maximum	Monthly mean	Monthly average daily minimum	Monthly minimum
Annual maximum	MxMx	MxAvMx *Largest average daily maximum of year*	MxAv *Largest monthly mean of year*	MxAvMn *Largest average daily minimum of year*	MxMn
Annual average	AvMx *Average monthly maximum of year*	AvAvMx *Average daily maximum of year*	AvAv *Average of year*	AvAvMn *Average daily minimum of year*	AvMn *Average monthly minimum of year*
Annual minimum	MnMx	MnAvMx *Lowest average daily maximum of year*	MnAv *Lowest monthly mean of year*	MnAvMn *Lowest average daily minimum of year*	MnMn

An annual maximum, average and minimum were taken for the five monthly records. The variables in the corners were not used because there was no averaging in their derivation. Approximate descriptions of the 11 remaining variables are given in italics. Eleven climatic variables were obtained in this way for wind speed, air temperature, relative humidity and soil temperature from these hourly records (5 monthly records per variable). Fifteen annual records per variable were then derived from the monthly records by taking maxima, means and minima of the five monthly records (Table 2). Of the 15 annual records, 4 were not used in statistical analyses because their derivation did not include any averaging. Finally, the 11 remaining annual records per variable were averaged for 1994 and 1995, giving 11 climatic records per variable. These climatic records were used in statistical analyses.

Daily mean saturation deficits, a measure of the drying power of air based on both air temperature and relative humidity [9] were calculated by averaging the daily maximum and minimum deficits: the former was calculated from the daily maximum temperature and minimum relative humidity (highest saturation deficit); the latter from the daily minimum temperature and maximum relative humidity (lowest saturation deficit). A monthly average saturation deficit was calculated from the daily means. Annual maximum, mean and minimum saturation deficits were then derived from the monthly averages, and averaged across 1994 and 1995 as before.

Rainfall at the 22 sites was estimated in two ways. First, long-term (1925–1949) average rainfall for each site was obtained from the *National Atlas of Morocco*. Second, rainfall data in 1994 and 1995 were purchased from the Division de Météorologie, Casablanca. Such data were not available for all sites, however, and where data were available, rain collection sites were usually several kilometres from our light traps.

Satellite images of Morocco

NDVIs with a pixel resolution of about 7.6 km^2 were obtained from the FAO-ARTEMIS whole-Africa 10-day maximum value composite data set, available on CD-ROM. Raw images were long-term means (1981–1991) for 36 periods of the year, three approximately 10-day periods per month. Using the software programme IDA 4.1 the three images per month were combined to generate one image per month wherein each pixel took the value of the largest NDVI from the three base images. This maximum value composite further reduced the effects of cloud cover on the recorded NDVIs. In this manner 12 monthly NDVI images of Morocco (the long-term average NDVI for January, February etc) were created. These 12 images were then combined to generate 4 annual images: the average annual maximum NDVI ($NDVI_{max}$, equivalent to the annual maximum NDVI, averaged across the years 1981–1991), the average annual mean NDVI ($NDVI_{avg}$), the average annual minimum NDVI ($NDVI_{min}$, equivalent to the annual minimum NDVI, averaged across the years 1981–1991) and the average annual range in NDVI ($NDVI_{rng}$). Fitting a digitised map of the coastline of Morocco to these images suggested that, for Morocco, the registration of the images to a coordinate system was slightly incorrect (in other words, the coastline evident on the satellite images was slightly displaced from the digitised coastline); images were re-registered by changing the centre of each image from 1°N, 17°E to 1.06°N, 16.97°E. NDVIs of the 22 sites in Morocco were obtained from each of these newly registered images by taking the maximum record of the 3 x 3 array of pixels centred over the precise coordinates of each site.

Insect, climatic and NDVI data for Iberia

Monthly-averaged mean daily catches of *C. imicola* from May 1993 to April 1995 are presented by Rawlings et al. [10] for 10 sites in Spain and 7 in Portugal. At the remaining sites (6 in Spain, 3 in Portugal) *C. imicola* appears to be absent. For easier comparison with Moroccan data, the monthly catch data of Rawlings et al. [10] for 1993–1994, after an initial $\log_{10}(n+1)$ transformation for normalisation, were converted to annual averages for each site. At several sites data are missing for some months, however, and therefore, prior to taking annual averages, missing values were estimated by interpolation. The method of interpolation was to calculate the proportion of the annual catch obtained in each month from the 5 sites in which all 12 months' data are available (1993–1994, Milfontes, Barrancos; 1994–1995, Almoraima, Mijas, Constantina). These proportions were then used to estimate monthly catches for the remaining sites where necessary. Annual averages, shown in Table 1, were then derived from a combination of observed and estimated monthly averages.

Climatic monitoring in Iberia was undertaken using the same equipment as that used in Morocco; in Spain, however, the anemometers (for wind speed) were set at only 1 m or less above ground level, compared to nearer 2 m in Portugal and Morocco.

Data for an 11th Portuguese site, Monçao (Rawlings et al., this volume) were included in the analyses.

NDVIs of sites in Iberia were obtained using a similar procedure to that used for Morocco. Monthly images were derived from NASA's Pathfinder AVHRR Land dataset, however, and were kindly supplied in IDA 4.1 format by the TALA Research Group, Oxford University.

Results

Development of model using Moroccan data

Means and ranges of the 54 independent variables across the 22 sites in Morocco, and their degree of correlation with the annually-averaged mean daily catch of *C. imicola* at the same sites, are shown in Table 3. Ten of the 11 wind speed variables were significantly and negatively correlated with the mean daily catch. The most significant of the relationships (that between the mean daily catch of *C. imicola* and $\text{Windspeed}_{\text{MxAvMx}}$) is shown in Fig. 1A. At higher wind speeds mean catches were all low, while at lower wind speeds both high and low catches of *C. imicola* were obtained. Following linear regression, the site of Bouizakarn was identified as an outlier on the basis of its large standardized residual (> 2).

There were no significant correlations between the annually-averaged mean daily catch of *C. imicola* and any of the air temperature, relative humidity, soil temperature, saturation deficit, rain or altitude variables. There were also no significant correlations with either NDVI_{max} or NDVI_{rng}. There were positive correlations with both NDVI_{avg} and NDVI_{min}. The relationship between the mean daily catch of *C. imicola* and NDVI_{min} is shown in Fig. 1B. Following linear regression both Chefchaouen and Marrakech were identified as outliers on the basis of large standardized residuals (> 2). An image of the NDVI_{min} of Morocco is shown in Fig. 2A.

Two-variable models of the annually-averaged mean daily catch of *C. imicola*, comprising one of the 11 wind speed variables and one of the four NDVI variables, were examined with Best Subsets Regression using the statistical package Minitab. The most significant two-variable model combined $\text{Windspeed}_{\text{MnAvMn}}$ and NDVI_{min}. The use of two variables considerably improved the amount of variance explained by the predictor variables: the adjusted r^2 was 53.2% for the two variable model, compared to 38.6% ($\text{Windspeed}_{\text{MxAvMx}}$) and 24.2% ($\text{NDVI}_{\text{min}}$) for the one variable models. In the two variable model each variable had a significant effect ($\text{Windspeed}_{\text{MnAvMn}}$, $t = -3.66$, P = 0.002; NDVI_{min}, $t = 4.15$, P = 0.001). The fitted values of the two-variable model (i.e. predicted

Table 3. Descriptive statistics of 54 climatic variables and vegetation indices obtained from 22 sites in Morocco, and correlation coefficients (r) with the annually-averaged mean daily catch of *Culicoides imicola* at the same sites. Values shown are the mean and range from among the 22 sites

	Wind speed (m/s)[a]			Air temperature (°C)		
	mean	range	r	mean	range	r
AvMx	4.69	2.29–7.22	–0.591**	31.8	26.5–35.9	0.053
MxAvMx	3.57	1.60–5.82	–0.644**	34.0	27.3–39.6	–0.153
AvAvMx	2.68	1.28–4.34	–0.596**	25.5	22.0–28.7	–0.071
MnAvMx	1.93	0.96–3.29	–0.542**	17.5	15.7–20.9	0.009
MxAv	1.79	0.87–3.10	–0.562**	26.8	21.5–32.1	–0.154
AvAv	1.33	0.71–2.26	–0.558**	18.7	15.0–20.4	0.121
MnAv	0.92	0.48–1.63	–0.480*	11.0	7.5–12.9	0.127
MxAvMn	0.66	0.32–1.31	–0.483*	20.1	13.5–26.0	–0.056
AvAvMn	0.46	0.29–0.76	–0.483*	12.8	7.3–15.1	0.251
MnAvMn	0.34	0.25–0.54	–0.439*	5.3	0.4–9.2	0.218
AvMn	0.28	0.25–0.39	–0.377	8.2	3.1–10.8	0.281
	relative humidity (%)			soil temperature (°C)		
AvMx	96.6	74.8–100	0.099	31.1	22.2–38.7	–0.090
MxAvMx	95.1	77.9–99.0	0.189	38.2	25.9–50.0	–0.112
AvAvMx	84.6	52.0–96.5	0.260	26.9	20.0–32.7	–0.091
MnAvMx	72.5	28.4–91.7	0.292	14.9	9.5–19.4	–0.090
MxAv	79.6	55.4–90.4	0.192	31.2	24.1–37.6	–0.070
AvAv	63.5	34.8–80.0	0.289	21.3	17.8–25.3	0.179
MnAv	50.3	17.1–71.6	0.310	11.4	5.5–14.2	0.169
MxAvMn	55.9	31.4–73.9	0.199	26.0	16.4–30.6	0.158
AvAvMn	39.1	17.3–58.9	0.300	17.2	10.2–20.9	0.389
MnAvMn	27.6	8.9–51.6	0.289	8.3	1.6–11.1	0.369
AvMn	17.4	6.4–33.6	0.315	13.8	6.7–17.3	0.400
Saturation deficit (mbar)						
Annual min.	5.3	2.7–8.1	–0.216			
Annual mean	13.2	6.5–21.4	–0.248			
Annual max.	24.9	10.4–43.7	–0.243			
Total annual rainfall (mm)						
Long-term mean	254	32–692	0.378			
Met-station	372	78–887	0.102			
Altitude (m)	353	0–1275	–0.297			
NDVImax	0.38	0.16–0.57	0.320			
NDVIavg	0.26	0.12–0.45	0.445*			
NDVImin	0.18	0.08–0.35	0.528*			
NDVIrng	0.20	0.06–0.40	–0.009			

[a] The stalling speed of the wind speed gauge is 0.25 m/s
* $P < 0.05$; ** $P < 0.01$

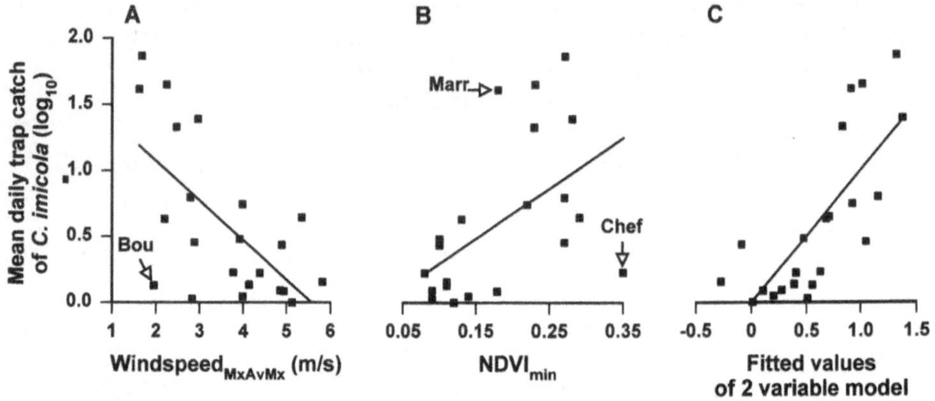

Fig. 1. Relationships between certain climatic variables (x-axis) and the log-transformed annually-averaged mean daily trap catch of *Culicoides imicola* (y-axis) at 22 sites in Morocco. Lines are regression lines, with sites found to be outliers (standardized residual > 2) labelled by name. **A** Independent variable = Ws_{MxAvMx}, the average daily maximum wind speed of the windiest month; *Bou* Bouizakarn. **B** Independent variable = $NDVI_{min}$, the mean annual minimum NDVI; *Marr* Marrakech, *Chef* Chefchaouen. **C** Independent variable = the fitted values from a regression of Ws_{MnAvMn} and $NDVI_{min}$ on the annually-averaged mean daily catch of *C. imicola* (reproduced with permission from [2])

catches of *C. imicola* based on $Windspeed_{MnAvMn}$ and $NDVI_{min}$) are compared with the observed catches of *C. imicola* in Fig. 1C. No sites were identified as outliers (in terms of standardized residuals > 2) in this two-variable model.

Three-variable models of the annually-averaged mean daily catch of *C. imicola*, combining one of the 11 wind speed variables, $NDVI_{min}$, and one other climatic variable, were also investigated with Best Subsets Regression. Only one 3-variable model had a substantially greater adjusted r^2 value than the best 2-variable model (60.1% compared to 53.2%); this model combined $Windspeed_{MnAvMn}$, $NDVI_{min}$ and $Airtemp_{MxAvMn}$. In the 3-variable model $Windspeed_{MnAvMn}$ ($t = -4.5$, $P < 0.001$) and $NDVI_{min}$ ($t = 4.6$, $P < 0.001$) were both highly significant predictor variables; $Airtemp_{MxAvMn}$ ($t = -2.1$, $P = 0.053$) approached but did not reach significance.

Since the addition of a third variable did not greatly increase the fit of the model to the observations, and the third variable itself did not reach significance, models incorporating 4 or more variables were not investigated. It is concluded therefore that a 2-variable model, combining $Windspeed_{MnAvMn}$ (the average daily minimum wind speed of the least-windy month of the year) and $NDVI_{min}$ (the average annual minimum NDVI) is the most appropriate for modelling the abundance of *C. imicola* in Morocco. It is important to note, however, that the best of the wind speed variables differs between the one-and two-variable models of *C. imicola* abundance, raising the question of which variable(s) is most applicable for predictive modelling. The wind speed variables were generally strongly correlated with each other ($r = 0.424$ to 0.966) and the better performance of one variable over another is probably more a result of statistical chance than any

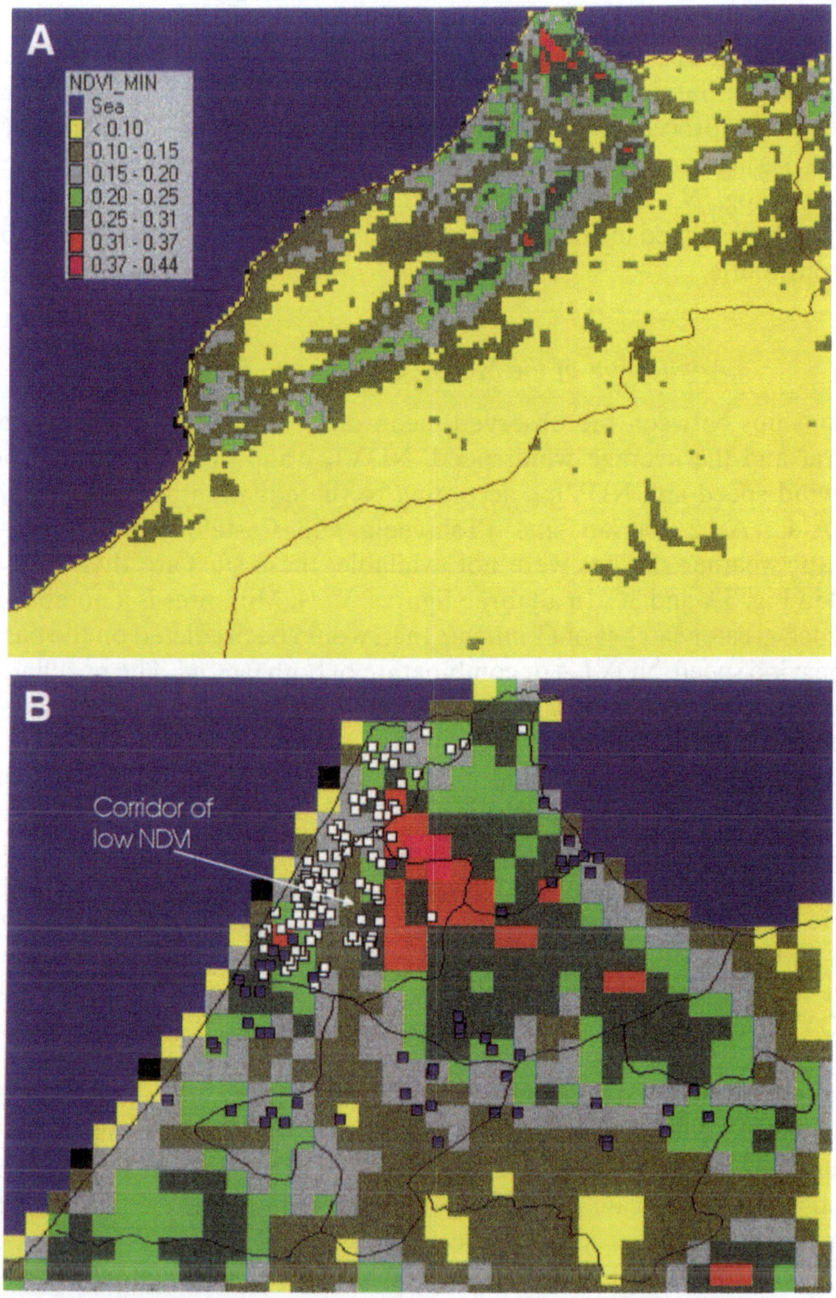

Fig. 2. A Satellite image of Morocco showing the mean annual minimum Normalised Difference Vegetation Index ($NDVI_{min}$). NDVI provides an instantaneous measure of the level of photosynthetic activity and is correlated with moisture levels such as soil moisture, rainfall etc. $NDVI_{min}$ is the annual minimum NDVI (i.e. the NDVI at the driest time of year, which is late summer or autumn in Morocco), averaged across years. Pixel size is approximately 8 x 8 km. **B** $NDVI_{min}$ of northern Morocco, showing the locations of reported outbreaks of African horse sickness in 1989 (white squares) and 1990 (blue squares). Thin black lines are provincial boundaries. Clearly visible is a corridor of low $NDVI_{min}$, in which there were very few disease outbreaks

biological effect. The average wind speed at a given site is probably sufficient and is an easier variable to obtain. A two-variable model of the abundance of *C. imicola* in Morocco, combining $NDVI_{min}$ and the average wind speed (Windspeed$_{AvAv}$) is highly significant, with an adjusted r^2 of 47.1%; the best-fit regression equation is: $\log_{10} N = 0.886 + 3.330(NDVI_{min}) - 0.664(Windspeed_{AvAv})$, where N is the annually-averaged mean daily catch of *C. imicola*; this is the model applied to data from Iberia.

Application of the Moroccan model to Iberian data

Relationships between the observed mean daily catches of *C. imicola* at sites in Iberia and the average wind speed, $NDVI_{min}$ and predicted catches (based upon wind speed and $NDVI_{min}$ according to the equation above) are shown in Fig. 3A–C. At 2 Iberian sites (Talayuela and Castelo Branco) data from automatic weather stations were not available; these sites are therefore not included in Fig. 3A and 3C. In all three figures, V. N. Milfontes is a notable outlier, with much greater catches of *C. imicola* than would be predicted on the basis of its average wind speed, $NDVI_{min}$ or combination of both factors. The peculiarly great abundance of *C. imicola* at V. N. Milfontes is not explicable on the basis of the Moroccan model.

If V. N. Milfontes is accepted as an outlier, the relationship between wind speed and mean daily catch in Iberia has a broadly similar shape to that found in Morocco (see Fig. 1A) although it is not significant. In contrast, as in Morocco, there is a significant, positive relationship between $NDVI_{min}$ and mean daily catch in Iberia (Fig. 3B, solid line; $F_{1,25} = 9.41$, $P < 0.01$, $r^2 = 27.3\%$). A particularly strong outlier in this relationship is Talayuela, where *C. imicola* appears to be highly abundant despite a relatively low $NDVI_{min}$. It is interesting to note, however, that the $NDVI_{min}$ of Talayuela is equal to that of Marrakech (0.18) in Morocco where the great abundance of *C. imicola* appears to be explicable on the basis of low wind speed. Unfortunately the mean wind speed at Talayuela is not known.

If V. N. Milfontes is accepted as an outlier, the relationship between the predicted mean daily catch of *C. imicola* in Iberia (based on the Moroccan relationship between catch, wind speed and $NDVI_{min}$) and the observed catch is positive and approaches significance (Fig. 3C, solid line; $F_{1,22} = 3.66$, $P < 0.07$, adjusted $r^2 = 14.3\%$). Of particular interest in Fig. 3C are several sites where, on the basis of wind speed and $NDVI_{min}$, *C. imicola* is predicted to be abundant but where the midge is actually rare or absent. These sites, shown as open squares in Fig. 3, are Alijo, Carvalhos, Freixo de Espada, Muge and Monçao (in Portugal) and La Mortera Palomar (in Spain). The Portuguese sites are 5 of the 6 most northern sites in Portugal; La Mortera Palomar is the most northern of the Spanish sites (see Fig. 1 of Rawlings et al., this volume). At the sixth northern Portuguese site, Vagos, *C. imicola* is absent but this is not unexpected from the model on the basis of its high wind speeds. If these northern sites (and V. N. Milfontes) are considered to be outliers, there is a strong, positive relationship between the predicted and observed trap catches in Iberia (Fig. 3C, dashed line;

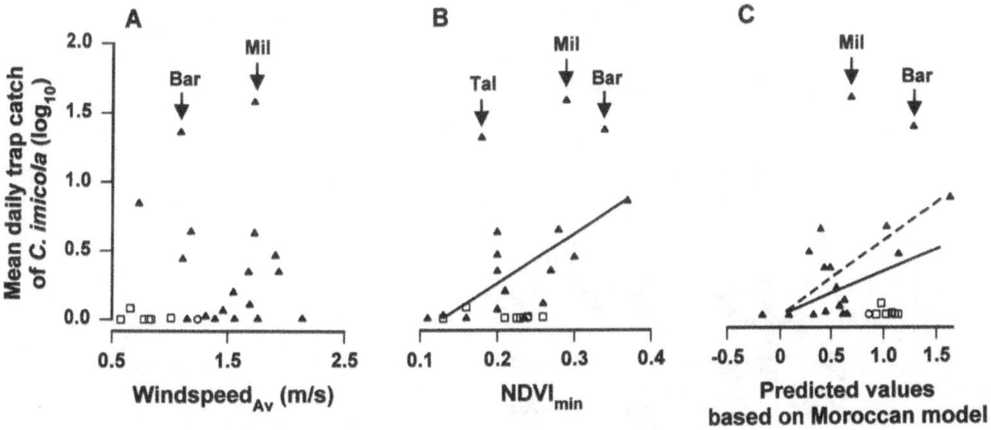

Fig. 3. Relationships between certain climatic variables (x-axis) and the log-transformed annually-averaged mean daily trap catch of *Culicoides imicola* (y-axis) at 25–27 sites in Iberia. **A** Independent variable = $Windspeed_{Av}$, the average daily wind speed; *Bar* Barrancos, *Mil* V. N. Milfontes. **B** Independent variable = $NDVI_{min}$, the mean annual minimum NDVI; *Tal* Talayuela. The best fit regression line is shown. **C** Independent variable = the predicted values for Iberia based on a 2 variable model derived from Moroccan data (see text); *solid line* best fit regression line, with V. N. Milfontes excluded; *dashed line* best fit regression line with northern sites and V. N. Milfontes excluded. *Open squares* northern sites; *open circle* Guadalupe

$F_{1,15}$ = 11.55, P < 0.005, adjusted r^2 = 39.7%). *Culicoides imicola* is absent at Guadalupe, the most eastern of the Spanish sites (shown as an open circle in Fig. 3), although the model predicts it to be abundant.

$NDVI_{min}$ and the distribution of outbreaks of AHS in northern Morocco in 1989 and 1990

As described elsewhere [3], records of the 1989–1991 epizootic of AHS in Morocco, held at the Direction d'Elevage, Ministère de l'Agriculture et de la Reforme Agraire, Rabat, were examined by M. B. to obtain the names and localities of all villages in Morocco where outbreaks of AHS were reported during the 3 years. The latitudes and longitudes of these villages were obtained from 1:50,000 Morocco survey maps; subsequently, these coordinates were plotted upon a digitised outline of Morocco (see Fig. 5 of Bouayoune et al., this volume). In 1989, a total of 512 cases of AHS were reported from the 3 most northern provinces of Morocco. These provinces are the nearest to Spain, from where the virus is presumed to have been introduced. In 1990 the epizootic spread south and east, and 555 cases of AHS were reported from 6 provinces.

In Fig. 2B the locations of the reported outbreaks of AHS in northern Morocco in 1989 (white squares) and 1990 (blue squares) are overlaid onto an image of the $NDVI_{min}$ of the same region. In Larache province there is a corridor of low $NDVI_{min}$ within which very few outbreaks of AHS were reported. The

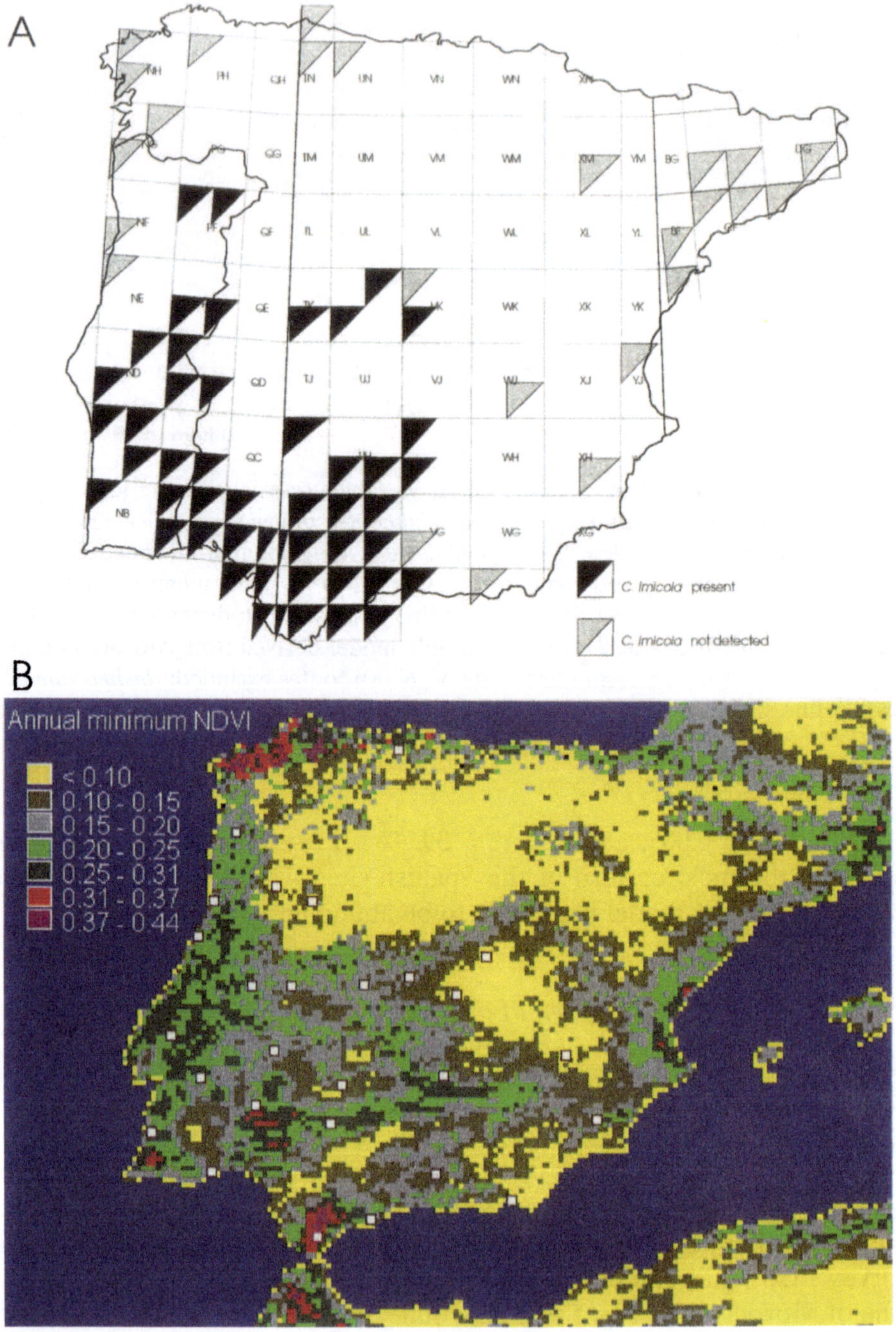

Fig. 4. A Grid map of the distribution of *Culicoides imicola* in Iberia by 50 x 50 km UTM (reproduced with permission from [20]). **B** Satellite image of Iberia showing the mean annual minimum NDVI. White squares are monitoring sites used in the present study

corridor is surrounded to the west and the east by areas of higher $NDVI_{min}$ in which there were numerous reported outbreaks.

Discussion

There is considerable diversity in the abundance of *C. imicola* at the 22 sites studied in Morocco [3]. The analyses presented here suggest that over half of this variation may be explainable by two climatic variables: wind speed and the remotely-sensed vegetation index, $NDVI_{min}$.

Finding a significant effect of wind speed on the mean daily catch of *C. imicola* was unexpected. It is well known that high wind speeds exert a negative impact on flying insects [27], including *Culicoides* [29] but in most if not all previous studies the negative impact exerted by wind speed is reported to be on the flight activity (activity rate) of the insects. We suggest that, in addition to any effect on flight activity, wind speed also affects the actual abundance of *C. imicola.* This is based on two observations: first, there were significant, negative correlations between wind speed and the abundance of *C. imicola* among the 22 sites in Morocco, (even though abundance was calculated in a manner designed to minimise any confounding effects of weather on activity rate or trap efficiency [3]); second, an entirely separate study found a significant correlation between wind speed and the mortality rates of adult *C. imicola* (Baylis et al., this volume).

A serious concern over the use of wind speeds in predictive modelling is the extent to which the wind speed measured by an automatic weather station at a site is affected by local topology (in terms of trees, buildings, local hills). It seems likely that the topology of a farm can drastically affect the wind speed - and hence *C. imicola* abundance – within it and therefore predictive modelling based on wind speeds from synoptic stations may not prove worthwhile. This remains to be confirmed.

The relationship between $NDVI_{min}$ and *C. imicola* abundance in Morocco operates, presumably, via certain climatic factor(s) that affect the midge and of which $NDVI_{min}$ is a remotely-sensed surrogate climatic variable. It has been shown elsewhere that in North Africa $NDVI_{min}$ is strongly correlated with total annual rainfall [2] and those authors argue that the relationship between rainfall, $NDVI_{min}$ and the abundance of *C. imicola* may operate via soil moisture levels. They suggest that soil moisture, which is a function of rainfall (and certain other factors such as soil porosity and slope), determines the extent or suitability of an area for breeding sites of *C. imicola,* while simultaneously affecting vegetation growth and NDVI.

The most suitable breeding sites of *C. imicola* are areas of wet, organically-enriched soil or mud [4, 5, 13, 29, 30] that are usually covered with grass [14, 17]. Smaller numbers may be found at the edges of puddles or pools [5]. Although breeding sites for *C. imicola* must be moist, there have been no clear demonstrations that the level of moisture affects the suitability of the breeding site. It has recently been suggested, however, that the Port Elizabeth area of South Africa is *C. imicola*-free as a result of sandy soils leading to low soil moisture and poor nutrient content [16], while sloping terrain (which generally has low water content) is known to be unfavourable for *C. imicola* [15].

While soil moisture appears to be an important factor in breeding sites for *C. imicola*, it also affects the NDVI. In Botswana a six-year study of NDVI, soil moisture and rainfall led to the conclusion that "NDVI is controlled by soil moisture in the concurrent month" [10]. A ground-truthing study in India found a linear relationship between NDVI and soil moisture in the root zone [18].

Given that *C. imicola* breeds in wet soil, and that soil moisture affects NDVI, it is perhaps not surprising that we have identified a relationship between NDVI and *C. imicola* abundance in Morocco. The specific relationship with the annual minimum NDVI (as opposed to the annual maximum, mean or range in NDVI) may result from the concurrence in peak numbers of adult *C. imicola* – and hence demand for breeding sites – with the most barren time of year. Soil that is sufficiently moist to act as a breeding site in late summer or autumn is also soil capable of supporting some vegetation at that time of year. This relationship is manifested in the correlation of *C. imicola* abundance in Morocco with $NDVI_{min}$.

In Fig. 2A there are two areas in Morocco with noticeably high $NDVI_{min}$: the north-northwest, and a long strip running in a southwest to northeast direction across the country. In 1989 and 1990 cases of AHS in Morocco were limited to the north-northwestern region of Morocco [1] and high catches of *C. imicola* have been obtained consistently in this area [3]. The long strip running southwest to northeast, which lies just to the west of the Atlas mountains, has a high $NDVI_{min}$ presumably because it is watered by runoff from the hills; this area has not been surveyed intensively for *C. imicola* but it is interesting to note that of the 177 cases of AHS that were reported from 20 provinces in 1991 [1], 108 (61%) were in the 8 provinces (Azilal, Beni Mellal, Boulmane, Fès, Ifrane, Khenifra, Marrakech, Taza) that overlie this strip. Conversely, in the west of Morocco there is a large area of low $NDVI_{min}$. Very low numbers or no *C. imicola* were reported from two sites situated in this area [3] and AHS was not reported from there in 1989–1991, even though it has the greatest density of equids in Morocco [1]. In general, therefore, it appears that there is considerable consistency between the $NDVI_{min}$ of Morocco, the catches of *C. imicola* reported by [3] and cases of AHS reported by [1]. This pattern is further strengthened by the observation that in Larache province of northern Morocco there is a corridor of low $NDVI_{min}$ within which there were very few reported cases of AHS in 1989 or 1990. It is possible (but not yet demonstrated) that within this corridor *C. imicola* are relatively rare because of low soil moisture and as a consequence outbreaks of AHS were very few.

The relationship between $NDVI_{min}$ and the abundance of *C. imicola* extends beyond Morocco into Iberia. We report here a significant correlation between $NDVI_{min}$ and the mean daily catch of *C. imicola* at 27 sites in Iberia. In Iberia *C. imicola* has so far been reported only from the southwestern region of the peninsula (Fig. 4A) and all cases of AHS during the 1987–1990 epizootic occurred in this region [20]. Figure 4B shows that this area of the peninsula is consistently of high $NDVI_{min}$.

The most important benefit derived from developing a model is, arguably, the insight gained when it fails to work. Here, the application of the 2-variable Moroccan model to Iberian data resulted in some success but there were several sites that did not fit the expected pattern. It is particularly interesting to consider these sites in detail. First, V. N. Milfontes in Portugal yielded much larger catches of *C. imicola* than are expected on the basis of its $NDVI_{min}$ and wind speed. A separate study of the Iberian data (Rawlings et al., this volume) also found V. N. Milfontes to be an exception to trends observed at the other sites and further study of this site, especially in terms of larval breeding sites, is required. A second exceptional site is Guadalupe in southeastern Spain. The Moroccan model predicts *C. imicola* to be fairly abundant at this site, yet the species appears to be absent. As suggested elsewhere, the geographical spread of *C. imicola* across Iberia may not be complete (Rawlings et al., this volume) and it is possible that the east of Spain, where large areas such as Valencia and the Ebro valley are of high $NDVI_{min}$ (Fig. 4B), is yet to be invaded by the species. Finally, there are several sites in central and northern Portugal (Alijo, Carvalhos, Freixo de Espada, Muge and Monçao), and the far north of Spain, at which the Moroccan model predicts *C. imicola* to be abundant yet it is very rare or absent. Since *C. imicola* is present at 2 of these sites (Freixo de Espada, Muge), albeit at very low density, and 1 *C. imicola* has previously been reported from Alijo [6], this pattern cannot be attributed to incomplete geographical spread. Instead, it seems likely that at central to northern latitudes in Iberia the Moroccan model is no longer applicable; instead, a third climatic variable, limiting the distribution of *C. imicola*, may become of overriding importance. A strong candidate variable here is temperature; it is intuitively obvious that at some point low temperatures must impose a limit to the distribution of *C. imicola*, regardless of the $NDVI_{min}$ and wind speed, and an air temperature variable approached significance as a predictor of the abundance of *C. imicola* in Morocco after the effects of $NDVI_{min}$ and wind speed had been accounted for. Rawlings et al. (this volume) also suggest that temperature is the most important climatic variable determining the presence or absence of *C. imicola* at sites in Iberia. The data presented here, therefore, provide preliminary evidence for a temperature-determined northern limit to the distribution of *C. imicola* in Iberia.

Acknowledgements

We thank our many colleagues in Iberia and Morocco, especially Hassan Bouayoune, Ruben Capela, Hassan El Hasnaoui, Lola Ortega, Isabel Pena, Maria José Pro and Jamal Touti, and all workers who helped in the collection of light trap or climatic data. Special thanks to David Rogers for encouragement and assistance with the satellite imagery and supply of the Pathfinder NDVI data, Fred Snijders for providing the ARTEMIS NDVI dataset, Eric Pfirman for IDA 4.1 and Philip Mellor for comments on the manuscript. This work was financed by the EU projects TS3-CT920151 and 8001-CT91-0211.

References

1. Anonymous (1992) La peste équine au Maroc. Epizooties de 1989, 1990 et 1991. Données epidemiologiques et strategie de lutte. Report of the Direction de l'Elevage, Ministère de l'Agriculture et de la Reforme Agraire, Rabat, Morocco
2. Baylis M, Bouayoune H, Touti J, El Hasnaoui H, Mellor PS (1998) Use of climatic data and satellite imagery to model the abundance of *Culicoides imicola* Kieffer, the vector of African horse sickness virus, in Morocco. Med Vet Entomol 12 (in press)
3. Baylis M, El Hasnaoui H, Bouayoune H, Touti J, Mellor PS (1997) The spatial and seasonal distribution of African horse sickness and its potential *Culicoides* vectors in Morocco. Med Vet Entomol 11: 203–212
4. Braverman Y (1978) Characteristics of *Culicoides* (Diptera, Ceratopogonidae) breeding places near Salisbury, Rhodesia. Ecol Entomol 3: 163–170
5. Braverman Y, Galun R, Ziv M (1974) Breeding sites of some *Culicoides* species (Diptera, Ceratopogonidae) in Israel. Mosq News 34: 303–308
6. Capela R, Sousa C, Pena I, Caeiro V (1993) Preliminary note on the distribution and ecology of *Culicoides imicola* in Portugal. Med Vet Entomol 7: 23–26
7. Cross ER, Newcomb WW, Tucker CJ (1996) Use of weather data and remote sensing to predict the geographic and seasonal distribution of *Phlebotomus papatasi* in southwest Asia. Am J Trop Med Hyg 54: 530–536
8. Daniel M, Kolár J (1990) Using satellite data to forecast the occurrence of the common tick *Ixodes ricinus* (L.). J Hyg Epidemiol Microbiol Immunol 34: 243–252
9. Edney EG (1957) The Water Relations of Terrestrial Arthropods. Cambridge University Press, Cambridge
10. Farrar TJ, Nicholson SE, Lare AR (1994) The influence of soil type on the relationships between NDVI, rainfall, and soil moisture in semiarid Botswana. II. NDVI response to soil moisture. Remote Sensing Environ 50: 121–133
11. Hay SI, Tucker CJ, Rogers DJ, Packer MJ (1996) Remotely sensed surrogates of meteorological data for the study of the distribution and abundance of arthropod vectors of disease. Ann Trop Med Parasitol 90: 1–19
12. Linthicum KJ, Bailey CL, Davies G, Tucker CJ (1987) Detection of Rift Valley fever viral activity in Kenya by satellite remote sensing imagery. Science 235: 1656–1659
13. Lubega R, Khamala CPM (1976) Larval habitats of common *Culicoides* Latreille (Diptera, Ceratopogonidae) in Kenya. Bull Entomol Res 66: 421–425
14. Meiswinkel R (1989) Afrotropical *Culicoides*: a redescription of *C. (Avaritia) imicola* Kieffer, 1913 (Diptera: Ceratopogonidae) with description of the closely allied *C. (A.) bolitinos* sp. nov. reared from the dung of the African buffalo, blue wildebeest and cattle in South Africa. Onderstepoort J Vet Res 56: 23–39
15. Meiswinkel R (1995) Afrotropical *Culicoides*: biosystematics of the Imicola group, subgenus *Avaritia* (Diptera: Ceratopogonidae). With special reference to the epidemiology of African horse sickness. M Sc thesis, University of Pretoria, Pretoria
16. Meiswinkel R (1997) Discovery of a *Culicoides imicola*-free zone in South Africa: preliminary notes and potential significance. Onderstepoort J Vet Res 64: 81–86
17. Mellor PS, Pitzolis G (1979) Observations on breeding sites and light-trap collections of *Culicoides* during an outbreak of bluetongue in Cyprus. Bull Entomol Res 69: 229–234
18. Narasimha Rao PV, Venkataratnam L, KrishnaRao PV, Ramana KV (1993) Relation between root zone soil moisture and normalized difference vegetation index of vegetated fields. Int J Remote Sensing 14: 441–449
19. Randolph SE (1993) Climate, satellite imagery and the seasonal abundance of the tick *Rhipicephalus appendiculatus* in southern Africa: a new perspective. Med Vet Entomol 7: 243–258

20. Rawlings P, Pro M-J, Pena I, Ortega M-D, Capela R (1997) Spatial and seasonal distribution of *Culicoides imicola* in Iberia in relation to the transmission of African horse sickness virus. Med Vet Entomol 11: 49–57
21. Rogers DJ, Hay SI, Packer MJ (1996) Predicting the distribution of tsetse flies in West Africa using temporal Fourier processed meteorological satellite data. Ann Trop Med Parasitol 90: 225–241
22. Rogers DJ, Randolph SE (1991) Mortality rates and population density of tsetse flies correlated with satellite imagery. Nature 351: 739–741
23. Rogers DJ, Randolph SE (1993) Distribution of tsetse and ticks in Africa: past, present and future. Parasitol Today 9: 266–271
24. Rogers DJ, Williams BG (1993) Monitoring trypanosomiasis in space and time. Parasitology 106: s77–s92
25. Schultz PA, Halpert M (1993) Global correlation of temperature, NDVI and precipitation. Adv Space Res 13: 277–280
26. Sellers PJ (1985) Canopy reflectance, photosynthesis and transpiration. Int J Remote Sensing 6: 1355–1372
27. Service MW (1978) The effect of weather on mosquito biology. In: Weather and Parasitic Animal Disease. World Meteorological Organisation, Geneva. Technical Note No. 159, pp 151–167
28. Thomson MC, Connor SJ, Milligan PJM, Flasse SP (1996) The ecology of malaria – as seen from earth-observation satellites. Ann Trop Med Parasitol 90: 243–264
29. Walker AR (1977) Seasonal fluctuations of *Culicoides* species (Diptera: Ceratopogonidae) in Kenya. Bull Entomol Res 67: 217–233
30. Walker AR, Davies FG (1971) A preliminary survey of the epidemiology of bluetongue in Kenya. J Hyg Camb 69: 47–60
31. Wood BL, Beck LR, Washino RK, Hibbard KA, Salute JS (1992) Estimating high mosquito-producing fields using spectral and spatial data. Int J Remote Sensing 13: 2813–2826
32. Wood BL, Beck LR, Washino RK, Palchick SM, Sebesta PD (1991) Spectral and spatial characterization of rice field mosquito habitat. Int J Remote Sensing 12: 621–626

Authors' address: Dr. M. Baylis, Institute for Animal Health, Ash Road, Pirbright, Surrey GU24 0NF, U.K.

Effect of temperature on African horse sickness virus infection in *Culicoides*

P. S. Mellor, P. Rawlings, M. Baylis, and **M. P. Wellby**

Institute for Animal Health, Pirbright Laboratory, Pirbright, U.K.

Summary. This paper shows that both the infection rate and the rate of virogenesis of African horse sickness virus (AHSV) within vector *Culicoides* are temperature dependent. As temperature is reduced from permissive levels the lifespan of the vector itself is extended but the rate of virogenesis decreases and infection rate falls dramatically so that at 10°C virtually all midges are free from virus by 13 days post infection (dpi). When vectors that had been kept at this temperature for 35 days were moved to a permissive temperature for 3 days; however, the apparent zero infection rate increased to 15.5%. It therefore appears that at low temperature (≤ 15°C) AHSV does not replicate but virus may persist in some vectors at a level below that detectable by traditional assay systems and when the temperature later rises to permissive levels virus replication is able to commence. On the basis of this information an overwintering mechanism for AHSV is suggested. The temperature at which the immature stages of *Culicoides* are reared may also influence infection with AHSV. A 5–10°C rise in larval developmental temperature resulted in an increase in oral infection rate of a normally non-vector species of *Culicoides*, from < 1% to > 10%. A mechanism is suggested.

Introduction

African horse sickness virus (AHSV) is a dsRNA virus within the family *Reoviridae.* To date 9 internationally recognised serotypes have been identified. The virus causes a non-contagious, infectious, arthropod-borne disease of equines and in susceptible populations of horses its effects can be devastating, with mortality rates frequently exceeding 80 or 90%. The virus is therefore rightly considered as the cause of one of the most lethal horse diseases which accordingly has been allocated OIE List "A" status.

All 9 serotypes of AHSV are enzootic in parts of sub-Saharan Africa but periodically epizootics occur beyond this region [14]. The most recent of these was an epizootic of AHSV serotype 4 in the western Mediterranean area. This epizootic persisted from July 1987 until the end of 1991 and at various times during this period was recorded in Spain, Portugal and Morocco [14].

The only confirmed field vector of AHSV is the biting midge *Culicoides imicola* [14]. This species is widespread in sub-Saharan Africa, North Africa, the Middle East and parts of southern Asia [11]. In 1982 *C. imicola* was first discovered to also occur in Europe, in southern Spain [17]. Subsequently it has been found as far north as Madrid in that country [19] and the River Douro at 41° 17' N in northern Portugal [6].

The northern limits of the known distribution of *C. imicola* in Europe are characterised by a large summer-winter temperature differential. Mean monthly temperatures vary from a summer high of about 26°C to a winter low of 5°C, although daily maxima and minima can vary from 40°C to –8°C. Such temperature patterns are likely to have profound effects upon the epidemiology of arboviral diseases such as AHS, impacting particularly upon aspects of the biology of vector and of the virus, both of which are poikilothermic, and upon interactions between the two [1]. Nevertheless the net or overall effect of temperature upon the dynamics of arbovirus transmission can be difficult to predict and is likely to vary with each virus-vector combination. Generally speaking, however, at higher temperatures the rate of virogenesis within a vector is usually faster and the vector may blood-feed more frequently, these two factors will tend to enhance the probability of transmission; conversely, the lifespan of the vector may be shortened, which tends to lessen transmission potential. As temperature decreases virogenesis usually slows and, at some point specific for each species of arbovirus, will cease altogether; at lower temperatures, however, the lifespan of the vector may be extended. The likelihood of virus transmission by arthropod vectors is therefore a function of the interaction of these two opposing sets of trends. Additionally there may be other, so far ill-defined but more subtle effects of temperature upon the ability of arthropod vectors to transmit arboviruses, for example upon the threshold levels or efficacy levels of the various barriers to virus replication and dissemination which exist within a variable proportion of most vector species [13].

Since we were unaware of any published data dealing with the effects of temperature on AHSV replication in and transmission by, potential vector species of *Culicoides*, these observations prompted the present investigation into the subject.

Materials and methods

Insects and virus

Culicoides imicola, the major vector of AHSV in Africa and Europe, has never been colonised and so was not readily available for this work. The American midge, *C. variipennis*, which is a confirmed laboratory vector of AHSV [5, 15] and which has been colonised in our laboratory for a number of years, was therefore used as a model. From this colony a strain of midges with a high level of oral susceptibility to AHSV was bred (designated PIR-s-3) by rearing solely from the progeny of proven susceptible females for 3 generations. Midges from this colony were then used for the present work without further selection. The PIR-s-3 colony has been tested 6 times over 20 generations for oral susceptibility to AHSV and has a mean susceptibility rate of 71% (range 53–81%) when incubated at 25°C and assayed for virus at 7 days post infection (dpi) by virus isolation in tissue culture.

Culicoides nubeculosus colonised from wild-caught midges caught in southern England [12] were also used in this study. This species has a standard oral susceptibility to AHSV of < 1% when maintained at 25°C and fed upon virus at titres of $10^{7.0}$ $TCID_{50}$ per ml or less ([16], P.S. Mellor, unpubl. obs.).

The virus used was AHSV serotype 9 originally obtained from the Onderstepoort Veterinary Institute, South Africa. This virus had been passaged initially by intracerebral inoculation of sucking mice, followed by 2–5 passages in baby hamster kidney (BHK-21) cells. The virus was stored as a tissue culture supernatant at 4°C or at –70°C. Fresh virus was prepared when the titre of the stored virus fell to < $10^{7.0}$ $TCID_{50}$ per ml.

When 2–4 days old, *Culicoides* of both species were fed an infecting blood meal of defibrinated horse blood mixed with an equal volume of virus suspension (final titre $10^{6.7-7.0}$ $TCID_{50}$ per ml) using established methods [15], except that on some occasions a stretched parafilm 'M' (American National Can) membrane was used in place of the 1-day-old chick skin. All midges were allowed to feed for 20–30 minutes to ensure full engorgement. Female *C. variipennis* and *C. nubeculosus* generally imbibe about 0.2 µl of blood [J.P.T. Boorman – unpubl. obs.] suggesting that when feeding upon a blood-virus suspension of $10^{6.7-7.0}$ $TCID_{50}$ per ml each midge would probably ingest between 10^3 and 10^4 $TCID_{50}$ of virus; this is the day 0 value.

Effect of temperature on survival rates and AHSV replication in Culicoides

In a series of experiments approximately 3,000 *C. variipennis* were fed upon AHSV-9 suspended in horse blood and the survival rates, infection rates and rates of virogenesis measured. After the offer of a blood meal, midges were anaesthetized with carbon dioxide and engorged females separated from the unfed individuals. The engorged midges were then divided into 4 groups and held in waxed cardboard pill boxes with fine mesh tops. Sucrose (10%) was provided as an energy source. The midges were maintained at 80–90% RH in constant temperature incubators or insectaries set at 10 ± 0.5, 15 ± 0.5, 20 ± 1.0 or 25 ± 1.0°C. After 35 days surviving midges from the 10°C incubator were moved to the 25°C incubator.

At intervals of 1 or more days samples of live midges (5–50 in number) were removed from each of the 4 groups and stored at –70°C until required for virus assay. Midges were assayed either individually (to assess infection rate) or in pools (to determine titre of virus) using established methods [15, 22]. The titre of virus per midge was calculated using the method of Spearman-Karber as quoted by [7]. When necessary the presence or absence of virus was confirmed using the AHSV antigen detection ELISA [9].

Estimates of the survival rates of *C. variipennis* at each of the 4 temperatures were made at intervals of 1–3 days by deducting the number of individuals dying naturally from the number of survivors recorded at the previous observation period. When measuring survival rates a 5th group of uninfected, blood-fed *C. variipennis* was also maintained at the additional temperature of 28°C.

Effect of larval rearing temperature on the oral susceptibility of C. nubeculosus *to AHSV*

Freshly laid eggs of *C. nubeculosus* were placed in 3 separate incubators maintained at 25 ± 1.0, 30 ± 1.0 or 35 ± 1.0°C and were reared through to adults using established methods

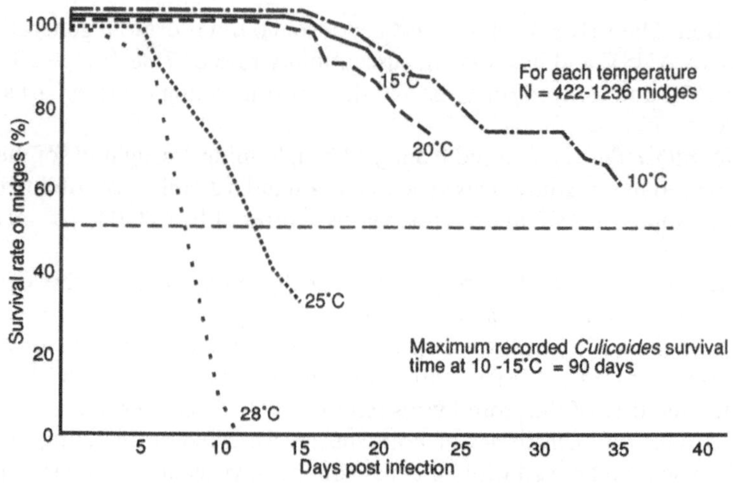

Fig. 1. *Culicoides variipennis* survival rates at 10, 15, 20, 25 and 28°C

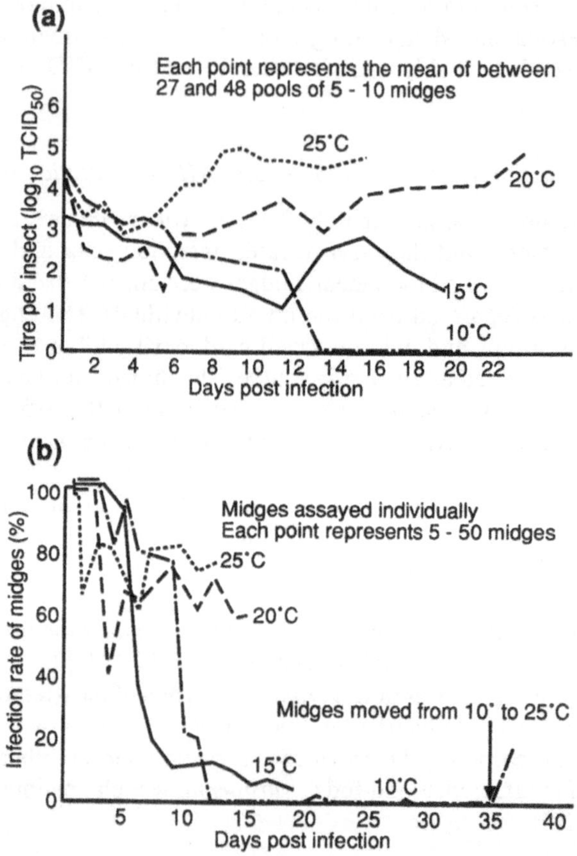

Fig. 2. a Replication of African horse sickness virus serotype 9 in *Culicoides variipennis* maintained at different temperatures (titre of virus in infecting bloodmeal $10^{6.7-7.0}$ $TCID_{50}$ per ml. **b** Oral infection rates of African horse sickness virus serotype 9 in *Culicoides variipennis* maintained at different temperatures (titre of virus in infecting bloodmeal $10^{6.7-7.0}$ $TCID_{50}$ per ml)

[3]. Adult female midges were then offered an AHSV-infected blood meal and engorged midges were maintained at 25°C for periods of 7–10 days before being assayed individually for AHSV (as described earlier) in order to determine the infection rate.

Results

Effect of temperature on survival rates and AHSV replication in Culicoides

Figure 1 shows that the survival rates at 10, 15 and 20°C are very similar to each other with around 90% of midges still alive after 14 days incubation at these temperatures. At 25°C, however, survival was reduced to 40% within the same time period, and at 28°C all midges were dead within 11 days.

Figure 2 confirms that infection rate and the rate of virogenesis of AHSV in the vector are also related to temperature. At 25°C the greatest mean virus titre per midge was reached by 8–9 dpi and the infection rate stabilised between 70 and 80%. At 20°C virogenesis was significantly slower and peak virus titre per midge was not reached until 23 dpi, although by this time virus replication had reached a level similar to the maximum reached at 25°C. The infection rate at 20°C was also reduced to 50–70%. At 15°C there was an overall decline in virus titre per midge with time, although between 13 and 15 dpi a few pools of midges (3) were found to contain $10^{3.0-4.0}$ $TCID_{50}$ of virus per midge, suggesting that virogenesis may still be possible at this temperature. Furthermore the infection rate at 15°C decreased dramatically to less than 15% by 9 dpi. At 10°C there was no detectable virogenesis at all and by 13 dpi virtually all midges tested were free from virus, except for 2 individuals which contained trace amounts only. However when 58 surviving midges that had been kept at 10°C for 35 days were moved to 25°C for 3 days the infection rate increased from apparently zero to 15.5% (9/58) (Fig. 2).

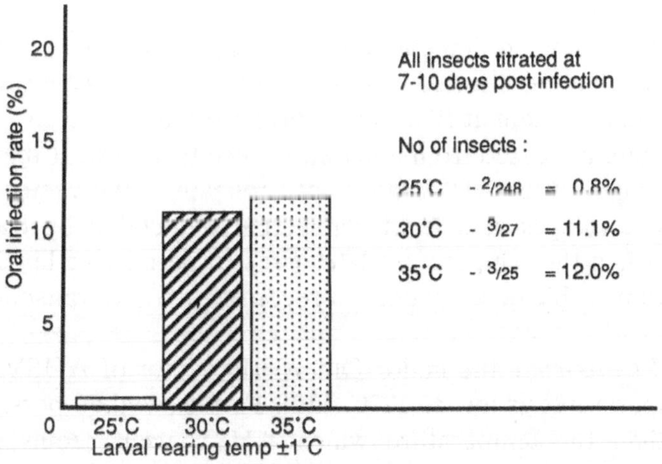

Fig. 3. Oral susceptibility to African horse sickness virus serotype 9 of *Culicoides nubeculosus* when reared from egg to pupa at different temperatures and kept as adults at 25°C (titre of virus in infecting bloodmeal $10^{5.8-6.0}$ $TCID_{50}$ per ml)

Effect of larval rearing temperature on the oral susceptibility of C. nubeculosus *to AHSV*

Figure 3 shows that the standard oral susceptibility rate of adult female *C. nubecalosus* for AHSV-9 is approximately 1% when the virus titre of the infecting blood meal is between $10^{5.8-6.0}$ $TCID_{50}$ per ml and when the larval rearing temperature is 25°C. When larval rearing temperature is increased to 30 or 35°C, however, the oral infection rate of adults produced from these larvae is increased to > 10%. The virus titre in these orally infected adults, when measured after 7–10 days incubation at 25°C, ranged from $10^{3.5-5.0}$ $TCID_{50}$ per midge.

Discussion

AHSV and vector species of Culicoides

The results presented in this paper show that at 25°C AHSV infects a high proportion of orally susceptible *Culicoides* and that the virus replicates rapidly to high titre in these individuals. Furthermore, earlier studies have shown that at 25°C transmission becomes possible by at least some infected individuals as early as 10–13 dpi ([14], P.S. Mellor – unpubl. obs.). At 20°C virus replication is slower and transmission has not been recorded earlier than 18 dpi (P.S. Mellor – unpubl. obs.). At 15°C virus replication appears to be minimal so that the titre per infected midge is generally low, while the infection rate falls dramatically and transmission has not been recorded at this temperature. In this context it may be of significance to note that recent work on the related bluetongue virus has reported the presence of a dissemination barrier in *C. variipennis* [8]. When low titres of virus are released from gut cells into the haemocoel this barrier appears to prevent dissemination of virus to the salivary glands and hence stops transmission. Such a barrier to AHSV dissemination would also tend to reduce the likelihood of transmission of this virus, particularly at low temperatures when virus titres in the haemocoel are likely to be minimal.

At 10°C there was no detectable virogenesis at all in *C. variipennis* and by 13 dpi virtually all midges tested were free from virus. However, when surviving midges that had been kept at 10°C for 35 days were moved to 25°C, within 3 days the infection rate increased from apparently zero to 15.5%. It therefore appears that at low temperatures AHSV does not replicate in the vector but virus may persist in some midges at a level below that detectable by traditional assay systems and when the temperature later rises to permissive levels (i.e. > 15°C) virus replication is able to commence or recommence and transmission may then be possible.

Adults of *C. imicola*, the major Old World vector of AHSV, can be active down to temperatures as low as 12°C ([20], R. Meiswinkel, pers. comm.), some 3°C lower than the temperature which AHSV itself seems to require (in *C. variipennis* at least) and they can survive in an inactive state at even lower temperatures. This suggests that *C. imicola* may be capable of transmitting AHSV only over the warmer, more southerly parts of its European range and perhaps not at all or only during very restricted times of the year in the cooler parts of its distribution (i.e., summer or early autumn). It also suggests that should *C. imicola* continue its northerly expansion in range through Europe, its ability to support virus replication and to transmit will progressively decrease with increasing latitude, unless of course this is accompanied by some form of climate change. This is clearly not the whole story, however. The results presented in this paper also suggest that AHSV can persist at undetectable levels in vector insects kept at low temperatures and that once the temperature rises to > 15°C (e.g., in spring) virus replication will be able to recommence and transmission may then be possible. Since the lifespan of individual adult *Culicoides* is dramatically extended at low temperatures, to as much as 90 days in some cases [4] this set of circumstances could constitute an effective overwintering mechanism for the virus in the absence of vertebrate involvement and also in the absence of detectable virus in the vectors. This mechanism may be summarised as follows: (1) AHSV seems to require temperatures ≥ 15°C in order to replicate in and be transmitted by, vector *Culicoides*. (2) Adults of the major AHSV vector, *C. imicola*, are active down to temperatures as low as 12°C, and can survive even cooler temperatures in an inactive state. (3) Since the virus requires a higher minimum temperature than does the vector this suggests that replication and transmission may be possible over only part of the range of *C. imicola* or during restricted times of the year. (4) During cold periods, however, virus can survive for extended periods at undetectable levels in adult vectors whose own lifespan is extended at these temperatures, to as long as 90 days, in some cases. (5) Once temperatures rise to a suitable level (≥ 15°C) virus replication in the vectors commences and transmission may then be possible. (6) This set of circumstances provides an effective overwintering mechanism for AHSV in the absence of vertebrate involvement and also in the absence of detectable virus in the vectors.

AHSV and "non-vector" species of Culicoides

The northern European midge *C. nubeculosus* has a standard oral susceptibility rate for AHSV in the order of 1% or less when the virus titre in the infecting blood meal is between $10^{5.8-6.0}$ $TCID_{50}$ per ml and when using insects colonised at 25°C. Furthermore, unlike *C. variipennis*, this rate of oral susceptibility cannot be enhanced by selective breeding from susceptible parents and 4 generations of such selection have failed to change it significantly (P. Rawlings, unpubl. obs.). Nevertheless, when the temperature at which *C. nubeculosus* immature stages are reared was raised to 30–35°C not only did the developmental time from egg to adult decrease dramatically to produce significantly smaller adults but the oral susceptibility rate increased to > 10% and virus replicated to levels suggesting that transmission would be possible [10, 21]. Similar results have also been obtained when using *C. nubeculosus* and bluetongue virus (P. Rawlings – unpubl. obs.).

Since selective breeding had no effect upon the *C. nubeculosus* oral susceptibility rate, this is clearly not a genetically controlled, heritable trait such as is exhibited by *C. variipennis* and which is probably due to an increased susceptibility of the gut cells to virus infection. In the case of *C. nubeculosus* it is likely that the increase in environmental development temperature not only gives rise to smaller adults but to more "fragile" individuals which have an increased incidence of the "leaky gut" phenomenon [2]. In such individuals virus is able to pass directly from the ingested blood meal in the gut lumen, into the haemocoel without the necessity of first infecting and replicating in the gut cells. Once in the haemocoel it is well documented that many arboviruses will replicate and then may be transmitted, even by normally non-vector insects [13]. Such a sequence of events may be envisaged as a hybrid mechanism whereby infection is initiated by a mechanical event (i.e., passage of virus from the lumen of the midgut through the gut wall and into the haemocoel, without replication in the gut cells) but transmission is the result of a series of subsequent biological events (i.e., virus infection of the salivary gland cells, replication of virus in the salivary gland cells and release of progeny virus particles into the salivary ducts).

Should this work with *C. nubeculosus* be confirmed, it will be a significant finding since it suggests that under appropriate environmental conditions (i.e., increasing temperature) many species of *Culicoides* not normally considered to be vectors may be able to transmit AHSV. Indeed the isolations of AHSV-4 made from mixed pools of *C. obsoletus* and *C. pulicaris* in Spain during 1988 [18], two *Culicoides* species which are not normally considered to be AHSV vectors, may be an example of the "leaky gut" phenomenon in operation; potentiated by the unusually warm conditions prevailing at the time. Increasing temperature due to climate change will tend to increase the likelihood of such unusual isolations.

References

1. Ando M, Carcavallo R, Epstein PR, Haines A, Jendritzky G, Kalkstein LS, McMichael AJ, Odongo RA, Patz I, Piver WT, Slooff R (1996) Effects on biological disease agents. In: McMichael AJ, Haines A, Slooff R, Kovats S (eds) Climate change and human health. WHO, Geneva, pp 71–105
2. Boorman J (1960) Observations on the amount of virus present in the haemolymph of *Aedes aegypti* infected with Uganda S yellow fever and Semliki forest viruses. Trans R Soc Trop Med Hyg 54: 362–365
3. Boorman J (1974) The maintenance of laboratory colonies of *Culicoides variipennis* (Coq.), *Culicoides nubeculosus* (Mg.) and *Culicoides riethi* (Kieff.) Diptera, Ceratopogonidae. Bull Entomol Res 64: 371–377
4. Boorman J (1990) A review of *Culicoides* subgenus *Avaritia* species (Insecta, Diptera, Ceratopogonidae), vectors of viruses of sheep, cattle and horses, with particular reference to *Culicoides imicola* in Europe and the Mediterranean region. A report prepared for the Overseas Development Administration of the British Government, Natural History Museum, London
5. Boorman J, Mellor PS, Penn M, Jennings M (1975) The growth of African horse sickness virus in embryonating hens eggs and the transmission of virus by *Culicoides variipennis*. Arch Virol 47: 343–349
6. Capela R, Sousa C, Pena I, Caeiro V (1993) Preliminary note on the distribution and ecology of *Culicoides imicola* in Portugal. Med Vet Entomol 7: 23–26
7. Finney DJ (1994) Statistical method in biological assay, 2nd ed. Griffin, London

8. Fu H (1996) Mechanisms controlling infection of *Culicoides* biting midges with bluetongue virus. PhD thesis, University of Hertfordshire
9. Hamblin C, Mertens PPC, Mellor PS, Burroughs JN, Crowther JR (1991) A serogroup specific enzyme-linked immunosorbant assay for the detection and identification of African horse sickness viruses. J Virol Methods 31: 285–292
10. Jennings DM, Mellor PS (1988) The vector potential of British *Culicoides* species for bluetongue virus. Vet Microbiol 17: 1–10
11. Meiswinkel R (1989) Afrotropical *Culicoides*: a redescription of *C. (Avaritia) imicola* Kieffer, 1913 (Diptera: Ceratopogonidae) with description of the closely allied *C. (A) bolitinos* sp.nov. reared from the dung of the African buffalo, blue wildebeest and cattle in South Africa. Onderstepoort J Vet Res 56: 23–39
12. Mellor PS (1971) A membrane feeding technique for the infection of *Culicoides nubeculosus* and *Culicoides variipennis* with *Onchocerca cervicalis*. Trans R Soc Trop Med Hyg 65: 199–201
13. Mellor PS (1990) The replication of bluetongue virus in *Culicoides* vectors. In: Roy P, Gorman BM (eds) Bluetongue viruses. Curr Top Microbiol Immunol 162: 143–161
14. Mellor PS (1993) African horse sickness virus; transmission and epidemiology. Vet Res 24: 199–212
15. Mellor PS, Boorman J, Loke R (1974) The multiplication of main drain virus in two species of *Culicoides*. Arch Ges Virusforsch 46: 105–110
16. Mellor PS, Boorman J, Jennings M (1975) The multiplication of African horse sickness virus in two species of *Culicoides*. Arch Virol 47: 351–356
17. Mellor PS, Boorman JPT, Wilkinson PJ, Martinez-Gomez F (1983) Potential vectors of bluetongue and African horse sickness viruses in Spain. Vet Rec 112: 229–230
18. Mellor, PS, Boned J, Hamblin C, Graham S (1990) Isolations of African horse sickness virus from vector insects made during the 1988 epizootic in Spain. Epidemiol Infect 105: 447–454
19. Rawlings P, Pro M-J, Pena I, Ortega M-D, Capela R (1997) Spatial and seasonal distribution of *Culicoides imicola* in Iberia in relation to the transmission of African horse sickness virus. Med Vet Entomol 11: 49–57
20. Sellers RF, Mellor PS (1993) Temperature and the persistence of viruses in *Culicoides* during adverse conditions. Rev Sci Tech OIE 12: 733–755
21. Weaver SC, Scherer WF, Cupp EW, Castello DA (1984) Barriers to dissemination of Venezuelan encephalitis virus in the middle American enzootic vector mosquito *Culex (Melanoconion) taeniopus*. Am J Trop Med Hyg 33: 953–960
22. Wellby MP, Baylis M, Rawlings P, Mellor PS (1996) Effect of temperature on survival and rate of virogenesis of African horse sickness virus in *Culicoides variipennis sonorensis* (Diptera: Ceratopogonidae) and its significance in relation to the epidemiology of the disease. Bull Entomol Res 86: 715–720

Authors' address: Dr. P. S. Mellor, IAH-Pirbright, Ash Road, Pirbright, Woking, Surrey, GU24 0NF, U.K.

Duration of repellency of various synthetic and plant-derived preparations for *Culicoides imicola*, the vector of African horse sickness virus

Y. Braverman and **A. Chizov-Ginzburg**

Kimron Veterinary Institute, Bet Dagan, Israel

Summary. Objectives of the study were threefold: to find a safer and longer lasting repellent of the biting midge *Culicoides imicola* than di-ethyl toluamide (DEET), to examine whether the current recommendations in Israel for application of repellents during an outbreak of *C. imicola*-borne pathogens are justified; and to examine whether plant-derived preparations that have no known detrimental side effects are potential replacements of synthetic repellents. Of the seven repellents tested, those inferior to DEET were: oregano and Herbipet which showed a slight non-significant repellency for 2 h and 1 h respectively and Stomoxin which showed significant ($P < 0.05$) repellency for only 1 h. As the active ingredient of Stomoxin is permethrin, this suggests that recommendations to spray animals with this insecticide to prevent the spread of *C. imicola*-borne pathogens are not useful. Tri-Tec14™ showed significant ($P < 0.05$) repellency with respect to controls for 2 h only, but performed similarly to, or slightly better than DEET. The repellents clearly superior to DEET were: the plant-derived material Ag1000 that repelled significantly ($P < 0.05$) with respect to controls for up to 4 h following a similar pattern to but somewhat more strongly than DEET, and pyrethroid-T which exerted significant ($P < 0.05$) repellency for 9h. Pyrethroid-T proved to be the best repellent tested and if sprayed nightly it might provide protection from *C. imicola* borne pathogens.

Introduction

An outbreak of African horse sickness (AHS), caused by African horse sickness virus (AHSV) (*Reoviridae, Orbivirus*) serotype 9, occurred in Israel (Palestine) in 1944 and since then the disease has not been recorded in the country [6]. The major old world vector of AHSV, the biting midge *Culicoides imicola*, is widely distributed in Israel [3], however, and future outbreaks must be anticipated. In the event of an outbreak of *C. imicola* borne viral disease, current recommendations (Circular of the Veterinary Services) in Israel include the application of repellents to livestock. Several controlled studies on the application of repellents on humans and farm animals have monitored the duration of their effects on *Culicoides* spp (e.g. [9, 16, 18]), but there have been no published studies on *C. imicola*. Most studies only compared the repellency of the various compounds with that of DEET (e.g. [11, 15]), without monitoring the exact duration of the repellency. Repellents have the advantage of being effective against a wide range of arthropod disease vectors and are considered to be a strategic means of disease control [22], whereas vaccines are specific to each disease and only a few vaccines are effective against arthropod-borne diseases [7]. The general purpose of the present study was to select, from among the various available synthetic and plant-derived preparations, adequate repellents that in a proper application could be used to protect animals against African horse sickness and other *Culicoides*-borne diseases. The more specific purposes were: (a) to find a repellent that could provide protection from the bites of *C. imicola* for the entire night or, as a minimum objective, to find a safer and longer lasting repellent than DEET; (b) to test whether the current recommendations in Israel for repellent application during an outbreak of *C. imicola*-borne pathogens (e.g. bluetongue virus) are correct; and (c) to test environmentally friendly plant-derived preparations which have no known detrimental side-effects, as potential replacements for the synthetic repellents.

Materials and methods

Locality and situation

All studies were conducted at the Bet Dagan (32° 00' N 34° 49' E), Kimron Veterinary Institute horse stable, which houses seven horses. During many years of suction light trapping, catches at this site consisted almost entirely of *C. imicola*. *Culicoides imicola* have repeatedly been found to breed in the damp horse manure under the water troughs of the horses (Y. Braverman, unpubl. data).

Materials

The following preparations were tested for repellency: (1) a plant-derived extract code-named Ag1000; (2) Oregano 15% (Analyt Ltd, Hadera, Israel); (3) Herbipet (batch no. 18711), a flea repellent derived from a 4% mixture of plant extracts from the Labiatae, comprising the oils of sage (*Salvia officinalis*), rosemary (*Rosmarinus officinalis*) and oregano (*Oreganum vulgare*) (Analyt Ltd, Hadera, Israel); (4)Tri-Tec14, a fly repellent with cypermethrin and pyrethrins as major ingredients (Farnham, Phoenix, Arizona, USA); (5) Pyrethroid-T, a type II pyrethroid that contains the α-cyano, 3-phenoxybenzyl moiety [8]; (6) DEET, 15% (N,N-diethyl-1,3-methylbenzamide) provided by Sano-Brunus Plants Ltd., Hod Ha'Sharon, Israel; (7) Stomoxin, a preparation containing 20% permethrin (Cooper, McDougall & Robertson, Erembodegen, Belgium).

Procedure

As Venter et al. [21] reported 10–70% success in membrane feeding of wild-caught *C. imicola* and our experience in feeding of *C. imicola* on farm animals had also shown variable success, an alternative in vitro method, using repellent-impregnated bed nets, as recommended by WHO [17], was chosen. The horses in the stable served as the main source of attraction for *C. imicola* and this was aided by mercury vapour lamps (HPW 125 W-TS, Belgium) of the DuToit [4] suction trap (Fig. 1). Preparations under test were impregnated onto rounded pieces of polyester bed net of diameter 66 cm (area 0.34 m^2), with about 30 holes per cm^2 (196 per square inch). Each piece of bed net was weighed, immersed in the solution of the test preparation for 30 min, dried for another 30 min, and then fixed using a rubber band to the 53 cm-diameter base of a suction light trap. The mesh size of the netting permitted free entrance of *C. imicola* to the trap. Each net was used for only one experimental preparation and was washed following two or three applications, depending on how much dirt had accumulated. For each experimental preparation, new pieces of nets were used. The amount of preparation absorbed was calculated per m^2 of net (Table 1), according to the

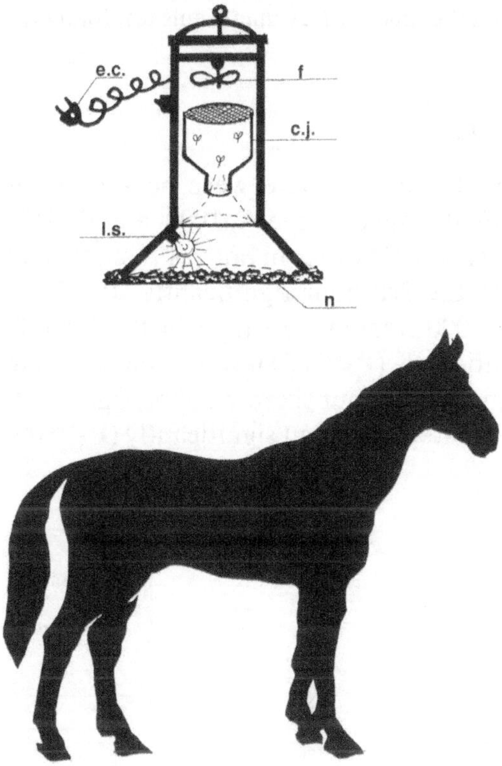

Fig. 1. Impregnated bed net fitted onto the entrance of the trap: *ec* electrical cord, *f* fan, *cj* collecting jar, *ls* light source, *n* netting

Table 1. Weights of nets impregnated with experimental repellents

Weights (g/m^2) of active ingredients on impregnated nets				
Experimental repellent	Dates of testing	Experimental compound	DEET	Ag1000
Ag1000	Jun. 27–Jul. 12	0.003	0.009	–
Oregano	Aug. 21–Sep. 2	0.004	0.009	–
Herbipet	Sep. 8–Sep. 19	0.0004	0.01	–
Tri-Tec14	Sep. 25–Oct. 9	0.008	0.01	0.003
Stomoxin	Oct. 23–Nov. 4	0.022	0.005	0.002
Pyrethroid-T	Nov. 10–Nov. 21	0.01	0.008	0.004

WHO method [17]. Experiments began on 8 July 1996, when suction light traps fitted with bed nets started to operate nightly from 18:30 to 06:30, and continued until 3 December 1996, when suction light traps fitted with bed nets were operated from 17:00 to 05:00. The suction light trap catches were sorted hourly and counted.

Experimental design and statistical analysis

Each experiment consisted of four bed net treatments: (1) untreated, (2) 15% DEET, (3) solvent or Ag1000 and (4) the test repellent. The locations of the four suction light traps fitted with bed nets of the various treatments were randomized for four nights in a Latin square design. Analysis of variance tests were performed with the General Linear Model (GLM) of Statistical Analysis System [14]. Where the GLM found a significant ($P < 0.05$) effect of treatment, means were compared using the Student-Newman Keuls test for multiple comparisons.

Results

In the first 2 h of the Ag1000 experiment (Fig. 2), there were no significant differences between the numbers of *C. imicola* caught in the Ag1000- and DEET-treated traps, but there were significantly ($P < 0.05$) fewer in both of them than in the untreated or solvent-treated traps. In the 3rd hour significantly ($P < 0.05$) fewer *C. imicola* were caught in the Ag1000-treated trap than in the DEET-treated trap, and both of them caught significantly ($P < 0.05$) fewer *C. imicola* than the untreated and solvent-treated nets. In the 4th hour there were no significant differences between DEET- and Ag1000-treated traps, but significantly ($P < 0.05$)

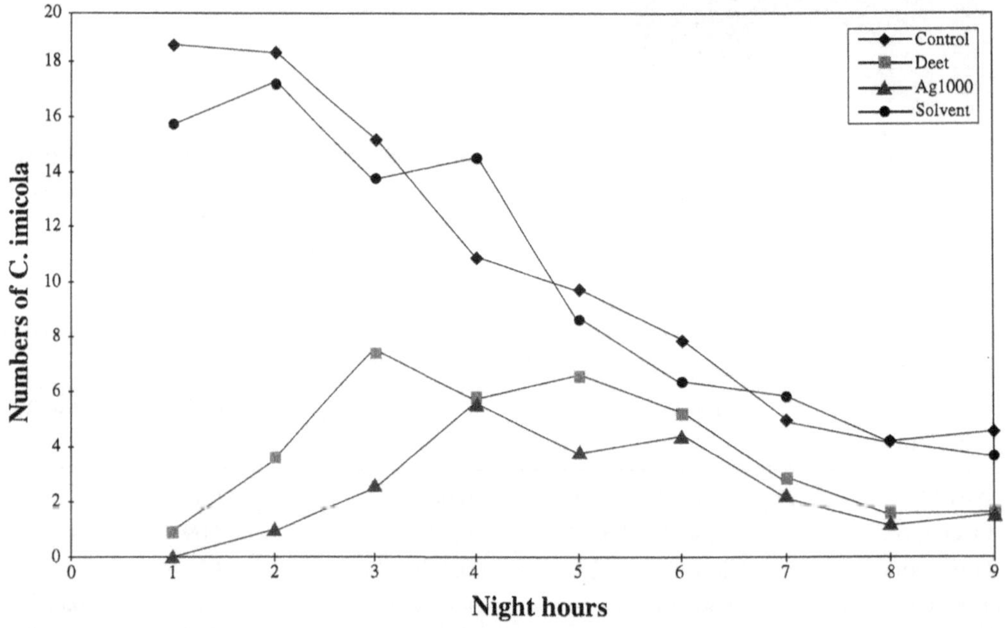

Fig. 2. Mean catches of *Culicoides imicola* during 9 h at night in suction light traps covered with netting of 4 treatments: untreated, DEET-treated, Ag1000-treated and solvent-treated

more *C. imicola* were caught in the solvent-treated trap than in the other treatments. There was no effect of site on the catches of *C. imicola* but there was a significant ($P < 0.02$) effect of day.

The oregano experiment (Fig. 3) showed significant ($P < 0.0003$) differences between treatments. For the first 4 h the DEET-treated trap caught significantly ($P < 0.05$) fewer *C. imicola* than the untreated, oregano- and solvent-treated traps; for the first 2 h the oregano-treated trap caught fewer *C. imicola* than the untreated or solvent-treated traps but the difference was not significant. There were no significant effects of site on the catches of *C. imicola* but there was a significant ($P < 0.001$) effect of day.

In the first hour of the Herbipet experiment (Fig. 4) the DEET-treated trap caught significantly ($P < 0.02$) fewer *C. imicola* than the other treatments; the Herbipet-treated trap caught fewer *C. imicola* than the untreated and solvent-treated traps but this was not significant. There were non-significant differences from the 2nd hour onwards. Figure 4 shows that during the first 4 h the number of *C. imicola* caught in the solvent-treated trap was greater than that caught in the untreated trap. The solvent, which was water containing 0.03% emulsifiers (polyoxyethylene sorbitan, monostearate sorbitan C) possibly had some attractancy for *C. imicola.* There were no significant effects of day or site.

The Tri-Tec14 experiment (Fig. 5) showed significant ($P < 0.01$) differences between treatments. In the first 2 h the untreated trap caught significantly ($P < 0.05$) more *C. imicola* than the DEET-, Tri-Tec14- and Ag1000-treated traps. Thereafter there were no significant differences between treatments apart from in the 5th hour when the untreated trap caught significantly ($P < 0.05$) more *C. imicola* than the Tri-Tec14- and Ag1000-treated traps. There were no significant effects of day or site.

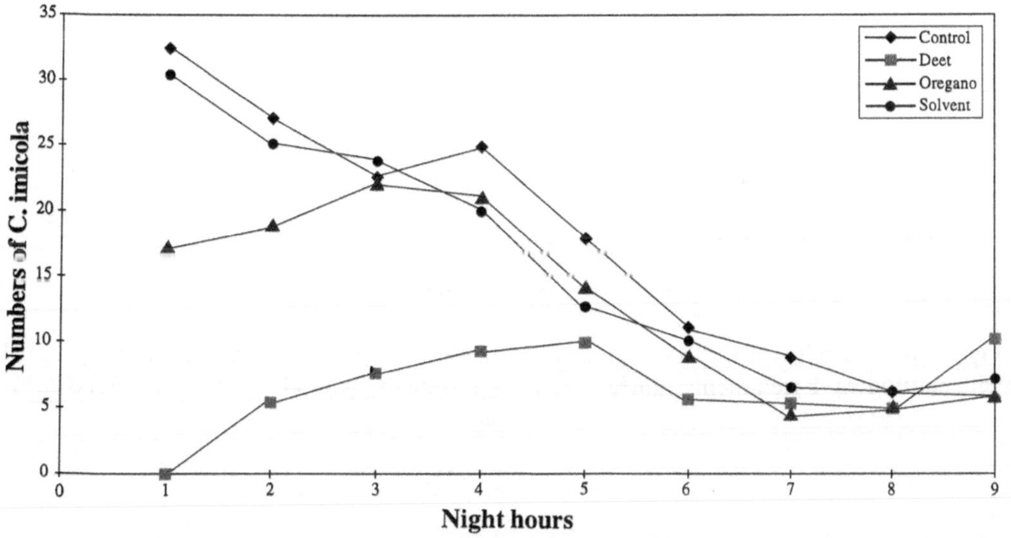

Fig. 3. Mean catches of *Culicoides imicola* during 9 h at night in suction light traps covered with netting of 4 treatments: untreated, DEET-treated, oregano-treated and solvent treated

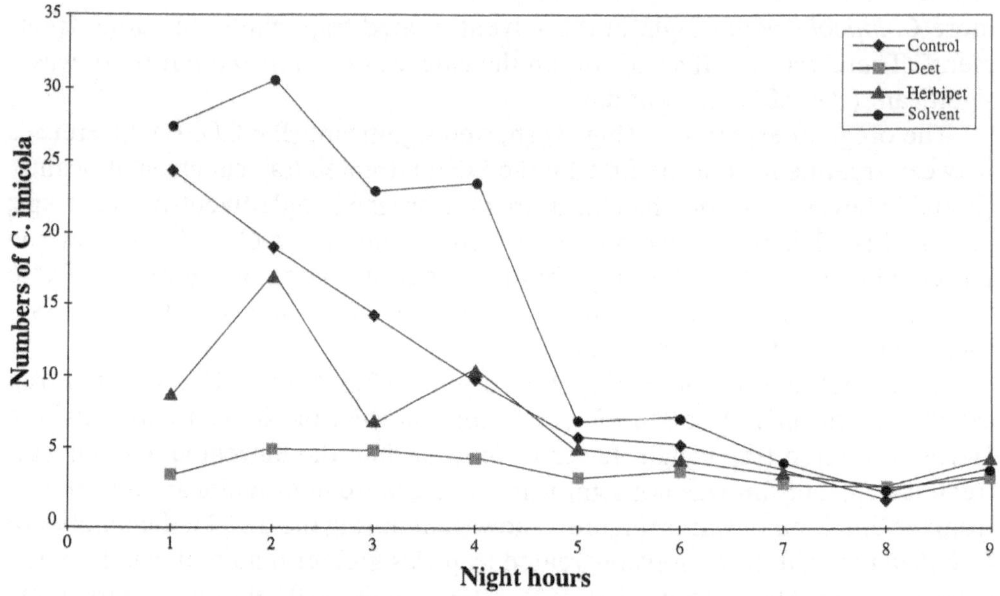

Fig. 4. Mean catches of *Culicoides imicola* during 9 h at night in suction light traps covered with netting of 4 treatments: untreated, DEET-treated, herbipet-treated and solvent treated

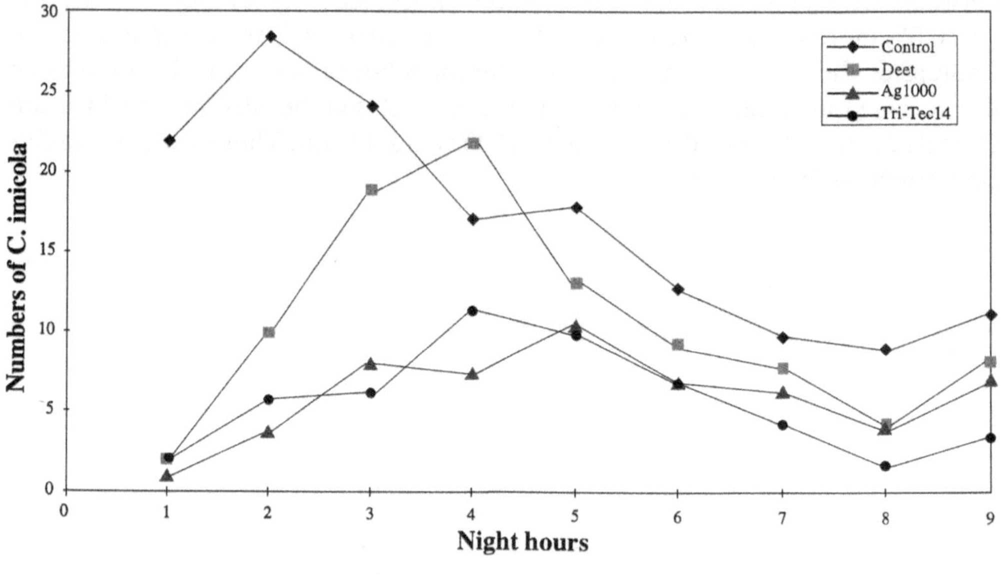

Fig. 5. Mean catches of *Culicoides imicola* during 9 h at night in suction light traps covered with netting of 4 treatments: untreated, DEET-treated, Ag1000-treated and Tri-Tec14-treated

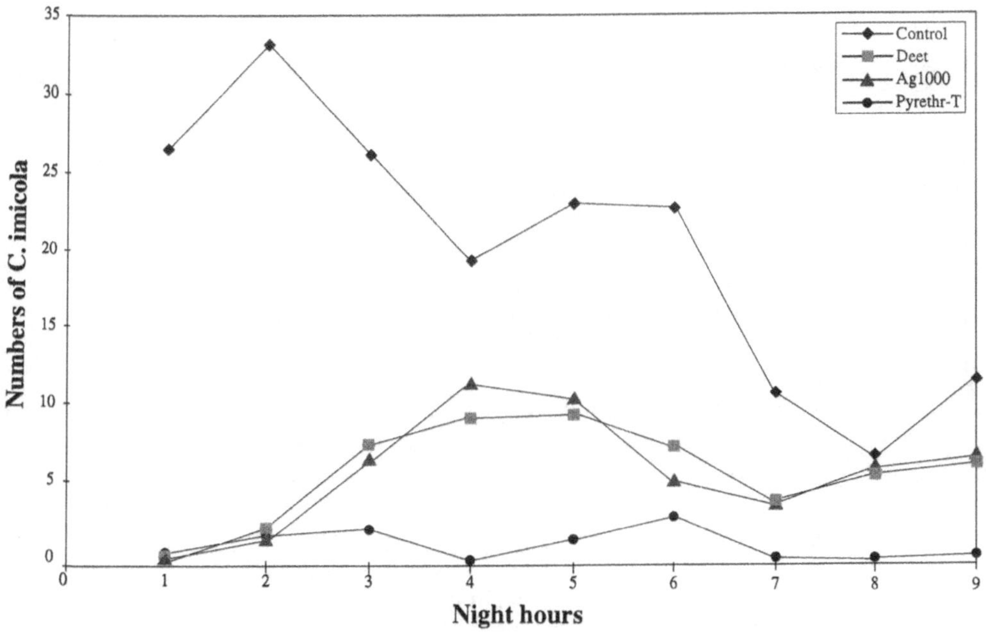

Fig. 6. Mean catches of *Culicoides imicola* during 9 h at night in suction light traps covered with netting of 4 treatments: untreated, DEET-treated, Ag1000-treated and Pyrethroid-T-treated

The pyrethroid-T experiment (Fig. 6) showed highly significant ($P < 0.0001$) differences between treatments. The untreated trap caught significantly ($P < 0.05$) more *C. imicola* than the DEET-, Ag 1000- or pyrethroid-T-treated traps for the first 7 h. The pyrethroid-T-treated trap appeared to catch fewer *C. imicola* than all other treatments from the 3rd to the 9th hour (i.e., there appears to have been repellency throughout the night); in the 4th and the 8th hours these differences were significant ($P < 0.05$). There were significant effects of both day ($P < 0.002$) and site ($P < 0.007$).

In the 1st hour of the Stomoxin experiment (Fig. 7), the Stomoxin-treated trap caught significantly ($P < 0.05$) fewer *C. imicola* than the untreated trap. Fewer *C. imicola* were caught in the DEET- and Ag1000-treated traps, although these differences were not significant. From the 2nd hour onwards there was no significant difference among the four treatments. There were no significant effects of day or site on the catches of *C. imicola*.

Discussion

The practical significance of this study stems from the fact that it was a field study and results can be immediately applied. Laboratory olfactometer studies provide results more quickly in comparing various repellents, but do not show the duration and repellency under real conditions. As the project was a field study which started in July and ended in December, the repellents were subjected to varying

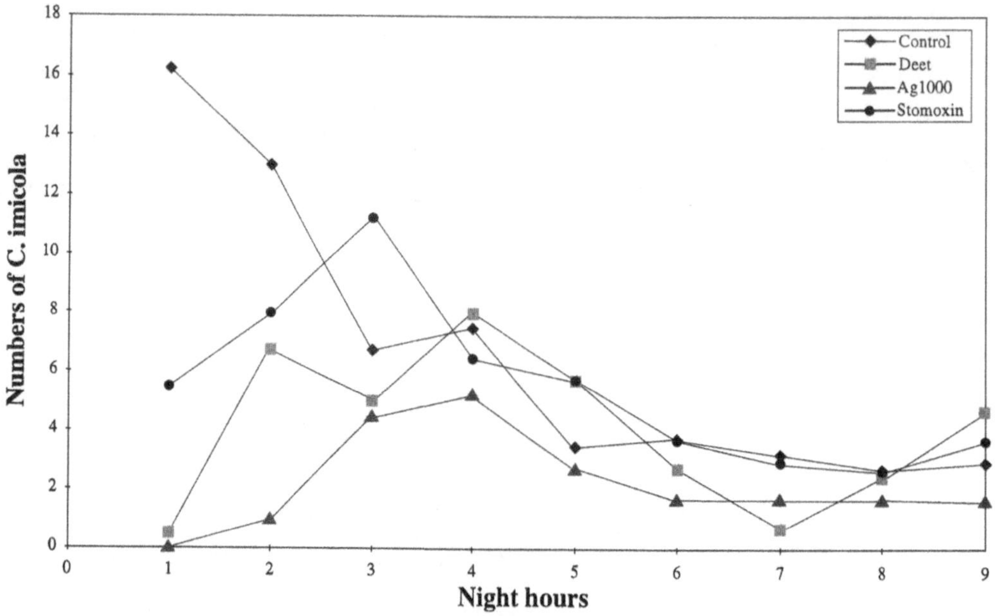

Fig. 7. Mean catches of *Culicoides imicola* during 9 h at night in suction light traps covered with netting of 4 treatments: untreated, DEET-treated, Ag1000-treated and Stomoxin-treated

climatic conditions. The night temperatures dropped from July to December and the relative humidity rose during this period. Thus, the release rates of the repellents most probably changed during this period also. The numbers of *C. imicola* were also increasing and peaked in October–November [3], as is usual in Israel. As large numbers of *C. imicola* prevail mainly in the first 4 h of the night any repellent that lasts this long is useful in reducing the bites of *C. imicola*, but does not ensure complete protection from *Culicoides*-borne pathogens. Our previous experience (unpublished data) with repellent-treated horses affected by *Culicoides* hypersensitivity, where the causative agent was *C. imicola*, matched the present results obtained with repellent-impregnated bed nets. The repellency pattern of the plant-derived Ag1000 was better than but similar to that of DEET; the effective repellency of each of them lasted about 4 h. A similar duration of repellency was also found for *Stomoxys calcitrans* L., when a 50% concentration of DEET was used to protect cattle [2]. Certain tabanids were repelled for 3 h when 75% DEET was applied to horses [2]. Sjogren [19] tested commercial formulations in which DEET was the primary active ingredient against the black gnat *Leptoconops kerteszi* attacking humans, but did not find any repellency effect. Only a new formulation, 5734, which contained DEET and other isomers, was able to repel *L. kerteszi* for 2 h. In a recent study [13] where 12.5% DEET was used against *L. americanus* biting humans, effective protection was achieved for 2 to 4 h. As repeated use of DEET might cause systemic toxicity [1] encephalopathy [5] and cardiovascular effects [10] in humans and side effects in humans and animals [2, 12] a presumed safe plant product such as Ag1000 has an obvious advantage. The trap with the oregano-treated net caught non-significantly fewer *C. imicola* in the first 2 h than those with the solvent-treated and control nets, but it was significantly ($P < 0.05$) inferior to DEET for the first 4 h and, therefore, it is not useful as a repellent. Herbipet was inferior to DEET and exerted nonsignificant repellency for 2 h. The Herbipet solvent curve showed nonsignificant attractancy for *C. imicola* to the solvent-treated net in the first 4 h. Tri-Tec14 showed significant ($P < 0.05$) repellency for only 2 h, but nonsignificant repellency was shown until the 5th hour. Tri-Tec14 performed as a repellent similarly to DEET and is not recommended for use as a repellent. Stomoxin, in which the active ingredient is permethrin, repelled *C. imicola* significantly ($P < 0.05$) for only 1 h; the weakness of the repellency effect of this compound is known [17, 20]. Of the seven tested repellents, the only one that showed repellency throughout the night was pyrethroid-T. Furthermore, pyrethroid-T showed significantly ($P < 0.05$) stronger repellency than DEET and Ag1000 in the 4th and 8th hours. Our study suggests that this compound might be the most effective of those examined here in helping to limit or prevent the spread of African horse sickness and other *Culicoides*-borne animal diseases.

Acknowledgements

We thank Mrs T. Lavy of the Kimron Veterinary Institute for her assistance in calculating the weights of the impregnated repellents and Dr. A. Gnizi and Mrs M. Zarchi of the Volcani Center for their help in the statistical analysis.

References

1. Ambrose AM (1959) Pharmacologic and toxicologic studies on N,N diethyltoluamide. Toxicology 1: 97–115
2. Blume RR, Roberts RH, Eschle JL, Matter JJ (1971) Tests of aerosols of deet for protection of livestock from biting flies. J Econ Entomol 64: 1193–1196
3. Braverman Y, Galun R (1973) The occurrence of *Culicoides* in Israel with reference to the incidence of bluetongue. Refuah Vet 30: 121–127
4. DuToit RM (1944) The transmission of bluetongue and horse-sickness by *Culicoides*. Onderstepoort J Vet Sci Anim Ind 19: 7–16
5. Edwards DL, Johnson CE (1987) Insect repellent induced toxic encephalopathy in a child. Clin Pharm 6: 496–498
6. Goor S (1952) Infectious diseases of equines in Palestine (1921–1948). Refuah Vet 9: 36–38
7. Gupta RJ, Rutledge LC (1994) Role of repellents in vector control and disease prevention. Am J Trop Med Hyg 50: 82–86
8. Hansen SR, Stemme KA, Villar D, Buck WB (1994) Pyrethrins and pyrethroids in dogs and cats. Compedium 16: 707–712
9. Kitaoka S, Morii T, Kosuge M (1965) Field experiments on the repellents to chicken-biting midges. Jpn J Sanit Zool 16: 244–248
10. Leach GJ, Russell RD, Houpt JT (1988) Some cardiovascular effects of the insect repellent N,N-diethyl-m-toluamide (DEET). J Toxicol Environ Health 25: 217–225
11. Magnon GJ, Robert LL, Kline DL, Roberts LW (1991) Repellency of two deet formulations and Avon Skin-So-Soft against biting midges (Diptera: Ceratopogonidae) in Honduras. J Am Mosq Control Assoc 7: 80–82

12. Miller JD (1992) Anaphylaxis associated with insect repellent. N Engl J Med 307: 1341–1342
13. Perich MJ, Strickman D, Wirtz RA, Stockwell SA, Click JI, Burge R, Hunt G, Lawyer PG (1995) Field evaluation of four repellents against *Leptoconops americanus* (Diptera: Ceratopogonidae) biting midges. J Med Entomol 32: 306–309
14. SAS Institute (1982) SAS users' guide: statistics. SAS Institute, Cary, NC, USA
15. Schreck CE, Kline DL (1981) Repellency determinations of four commercial products against six species of ceratopogonid biting midges. Mosq News 41: 7–10
16. Schreck CE, Kline DL (1983) Area protection by use of repellent-treated netting against *Culicoides* biting midges. Mosq News 43: 338–342
17. Schreck CE, Self LS (1985) Treating mosquito nets for better protection from bites and mosquito-borne disease. WHO/VBC/85.914. World Health Organization
18. Schreck CE, Smith N, Govern TP (1979) Repellency of selected compounds against two species of biting midges (Diptera: Ceratopogonidae: *Culicoides*). J Med Entomol 16: 524–527
19. Sjogren RD (1971) An effective repellent for Leptoconops kerteszi Kieffer (Diptera: Ceratopogonidae). Mosq News 31: 115–116
20. Tomlin C (1994) The pesticide manual: 10th ed. British Crop Protection Council Publications, Farnham, pp 783–785
21. Venter GJ, Hill E, Pajor ITP, Nevill EM (1991) The use of a membrane feeding technique to determine the infection rate of *Culicoides imicola* (Diptera Ceratopogonidae) for 2 bluetongue virus serotypes in South Africa. Onderstepoort J Vet Res 58: 5–9
22. Webster WR, Gard GP, St George TD, Kirkland PD (1992) The Australian bluetongue control strategy. In: 2nd International Symposium on Bluetongue, African Horse Sickness and Related Orbiviruses, Paris 17–21 June 1991. CRC Press, Boca Raton, pp 843–850

Authors' address: Dr. Y. Braverman, Kimron Veterinary Institute, P.O.B. 12, Bet Dagan 50250, Israel.

Molecular biology

New generation of African horse sickness virus vaccines based on structural and molecular studies of the virus particles

P. Roy and **G. Sutton**

Department of Biochemistry, University of Oxford and NERC Institute of Virology and Environmental Microbiology, Oxford, U.K.

Summary. African horse sickness virus (AHSV) is a member of the genus *Orbivirus*, which also includes bluetongue virus (BTV) and epizootic haemorrhagic disease (EHDV) virus. These orbiviruses have similar morphological and biochemical properties, with distinctive pathobiological properties and host ranges. Sequencing studies of the capsid proteins have revealed evolutionary relationships between these viruses. Biochemical studies of the viruses together with the expression of individual proteins and protein complexes have resulted in the development of new generation vaccines. Baculovirus expressed AHSV VP2 provides protection against death caused by AHSV challenge. Similarly, BTV VP2 alone elicits protective neutralising antibodies against BTV in sheep, which is enhanced in the presence of VP5. Recent developments in biotechnology (multiple gene expression baculovirus systems) have made it possible to synthesise orbivirus particles that biochemically and immunologically mimic authentic virions but lack the genetic material. Particle doses as low as 10 μg elicit responses that are sufficient to protect sheep 15 months post vaccination, against virulent virus challenge. Moreover, knowledge of the three dimensional structure of these particles enables us to engineer them to deliver multiple foreign peptide components representing other viral epitopes (e.g. foot and mouth disease virus and influenza virus) in order to elicit protective immunity.

Introduction

With the advent of new technologies and a greater understanding of the molecular and structural biology of viruses, vaccine development has taken new directions that avoid the difficulties and the risks associated with live virus, or preparation of killed virus vaccine. Attention has been focussed on identifying the relevant proteins and sequences involved in protective immune responses to viral infection and on systems to produce and present these proteins, or protein sequences, to elicit the required protection.

Over the last few years our laboratory has been involved in studying the biochemical and molecular features of *Culicoides* transmitted orbiviruses, including African horse sickness virus (AHSV), to allow the rational design of suitable vaccines. Orbiviruses have similar morphological and biochemical characteristics, with distinctive biological properties, pathologies, and host ranges. Orbivirus particles contain seven structural proteins, organized in four concentric layers [31, 32]. The innermost components of the virus particle are composed of three minor proteins (VP1, VP4 and VP6) together with the ten segments of the double stranded (ds) RNA genome (see Fig. 1). These components are surrounded by a thin layer of a single major protein, VP3, 120 copies of which are arranged to form the subcore [14, 38]. This subcore is encapsidated by 260 trimers of another core protein, VP7, to form the core particle. The core has a knobbly surface appearance, with icosahedral symmetry [14, 31]. The outer capsid is composed of the two other major proteins, VP2 and VP5. When virus particles are viewed by electron microscopy and negative staining, the morphology of their outer surface is not clearly defined [40]. The topography of these orbivirus proteins and their structures, functions and biochemical characteristics have mainly been derived from studies on the prototype orbivirus, bluetongue virus (BTV), that infects sheep and cattle. This report will discuss our understanding of these viruses based on the BTV model. The genetic relationship and the structural features that are shared by these viruses are discussed, together with a brief account of new generation vaccines that have been developed in recent years.

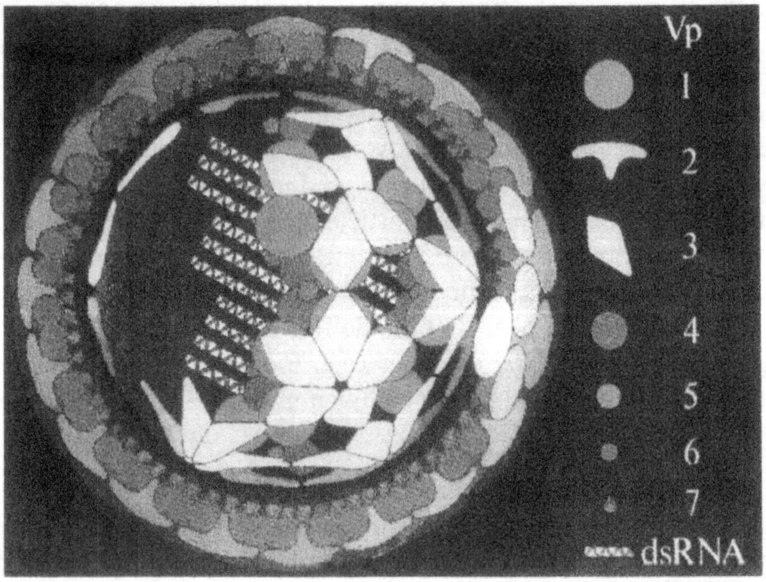

Fig. 1. Schematic diagram of the bluetongue virus particle based on data obtained from baculovirus expressed proteins and analysis by cryo-electron microscopy. The diagram shows VP2 and VP5 in the outer capsid surrounding the VP7 trimers on the core surface. VP3 forms the scaffold for VP7 trimers and encloses the three minor proteins, VP1, VP4 and VP6, as well as the ten double-stranded RNA segments

The evolutionary relationships among three Orbiviruses, as demonstrated by comparisons of their major capsid protein sequences

The dsRNAs of BTV 10 were isolated from infected cell extracts, individual RNA segments separated and full length cDNA clones representing each of the ten RNA segments were obtained (Fig. 2). The complete sequence of the BTV 10 genome is 19,218 base pairs (bp). Table 1 summarises the sizes of individual segments, as well as the predicted amino acid composition, location and sizes of all of the primary gene products of BTV 10 [11].

AHSV, BTV and EHDV induce seasonal diseases in both domestic and wild ruminants, and in certain regions of the world two or three of these viruses are sympatric, hence there is opportunity for them to infect the same vectors and host species. This raises the question of whether the viruses have evolved from a common ancestor. To obtain such information, the sequences of the four major capsid proteins (namely VP2, VP3, VP5 and VP7) of EHDV 1 and AHSV 4 have been deduced from cDNA analyses, and compared with that of BTV 10 (Table 2).

The predicted amino acid sequence comparisons revealed that, of the four capsid proteins the innermost protein, VP3, is the most conserved, and the outermost protein, VP2, is the most variable. Some 57–58% of the aligned BTV 10 and EHDV 1 VP3 amino acids are identical with those of AHSV 4. This compares to an identity of 70% between the BTV and EHDV VP3 sequences. For the VP7 proteins 64% of the aligned amino acids are identical between BTV 10 and EHDV 1, while they share 44-46% amino acid residues when aligned with AHSV 4 [19–21, 33]. As expected, the outermost capsid protein VP2, of the three viruses, shares only 19–24% identical amino acids (see Table 2). VP2 is implicated in virus neutralisation [17, 27]. The other outer capsid protein VP5, is more conserved among the three viruses than VP2 [16, 30, 41]. On alignment, the VP5 of AHSV 4 has some 43–45% identical amino acids with that of BTV 10 and EHDV 1. Between BTV and EHDV, 62% of the aligned sequences are identical [20, 21]. Clearly, the three viruses share strong phylogenetic relationships, and the differences in the VP2 structures reflect the immune pressure of the host (Fig. 3).

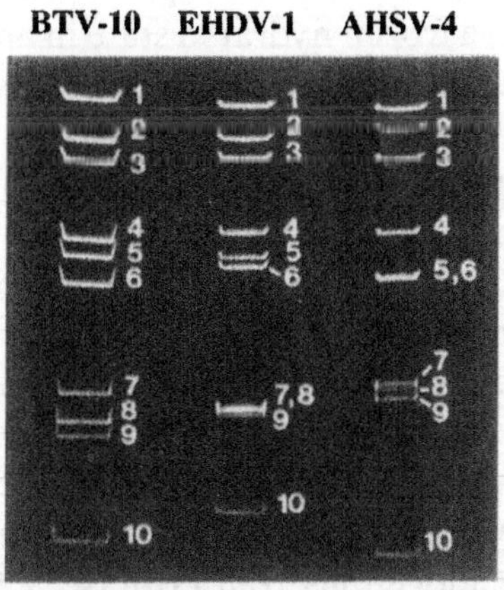

Fig. 2. SDS-PAGE profile of the RNA segments of BTV 10, EHDV 1 and AHSV 4

Table 1. BTV genes and gene products

Genome segment	No. of bp	Encoded protein(s)	Predicted mass (Da)	Location	Function
1	3954	VP1	149,588	Inner core	RNA polymerase
2	2926	VP2	111,112	Outer shell	Type specific structural protein
3	2772	VP3	103,344	Core	Structural protein, forms scaffold for VP7 trimers
4	2011	VP4	76,433	Inner core	Capping enzyme, guanylyltransferase
5	1639	VP5	59,163	Outer shell	Structural protein
6	1769	NS1 tubules	64,445	Non structural	Unknown, may be involved in morphogenesis
7	1156	VP7	38,548	Core surface	Group specific structural protein
8	1123	NS2 VIBs	40,999	Non structural	Binds mRNA
9	1046	VP6	35,730	Inner core	Helicase, ATPase, Binds ss & dsRNA
10	822	NS3	25,572	Non structural	Glycoproteins, aid virus release
		NS3A	24,020	Non structural	Glycoproteins, aid virus release

Three-dimensional structures of BTV core and virion

To determine the three dimensional structure of cores (nucleocapsids) and virions, we used electron cryo microscopy of unstained, unfixed particles and computer image reconstruction methods. The major advantage of electron cryo microscopy over conventional electron microscope techniques is that the biological molecules are observed in a frozen hydrated state in amorphous ice, which closely resembles the native aqueous state. The problems associated with heavy metal stains, fixatives and dehydration are thus avoided.

Electron cryo micrographs of cores derived from BTV revealed that they have a diameter of 69 nm and icosahedral symmetry with a triangulation number of 13 [31] (Fig. 4). The core structure is divided into two concentric layers of protein enclosing the RNA and minor proteins. The surface layer of the core is made up of clusters of VP7 trimers in triangular protrusions and which are organized into pentameric and hexameric units (rings) with channels between them (Fig. 4). There are a total of 780 VP7 molecules per particle, 132 channels and 260 knobs. The aqueous channels, which are about 70Å deep and 80Å wide at the surface, maybe involved in the endogenous transcriptase activity of the particles. The second major core protein VP3, forms a scaffold composed of 120 molecules in

Table 2. Conservation nature of three gnat transmitted orbiviruses

		Core				Outer capsid			
		VP3		VP7		VP5		VP2	
		BTV 10	EHDV 1	BTV 10	EHDV 1	BTV 10	EHDV 1	BTV 10	EHDV 1
EHDV 1	identity	79	–	64	–	62	–	23	–
	similarity	90	–	81	–	77	–	49	–
AHSV 4	identity	58	57	44	46	45	42	17	20
	similarity	76	75	66	67	63	64	44	44

total, which has a relatively smooth outer surface for attachment of the VP7 trimers [31, 38]. The VP7 and VP3 capsid shells enclosed the inner core, which is composed of the three minor proteins (VP1, VP4 and VP6) and the genomic dsRNAs. How these are organized with respect to each other, or to VP3 and VP7, has not been completely determined [38].

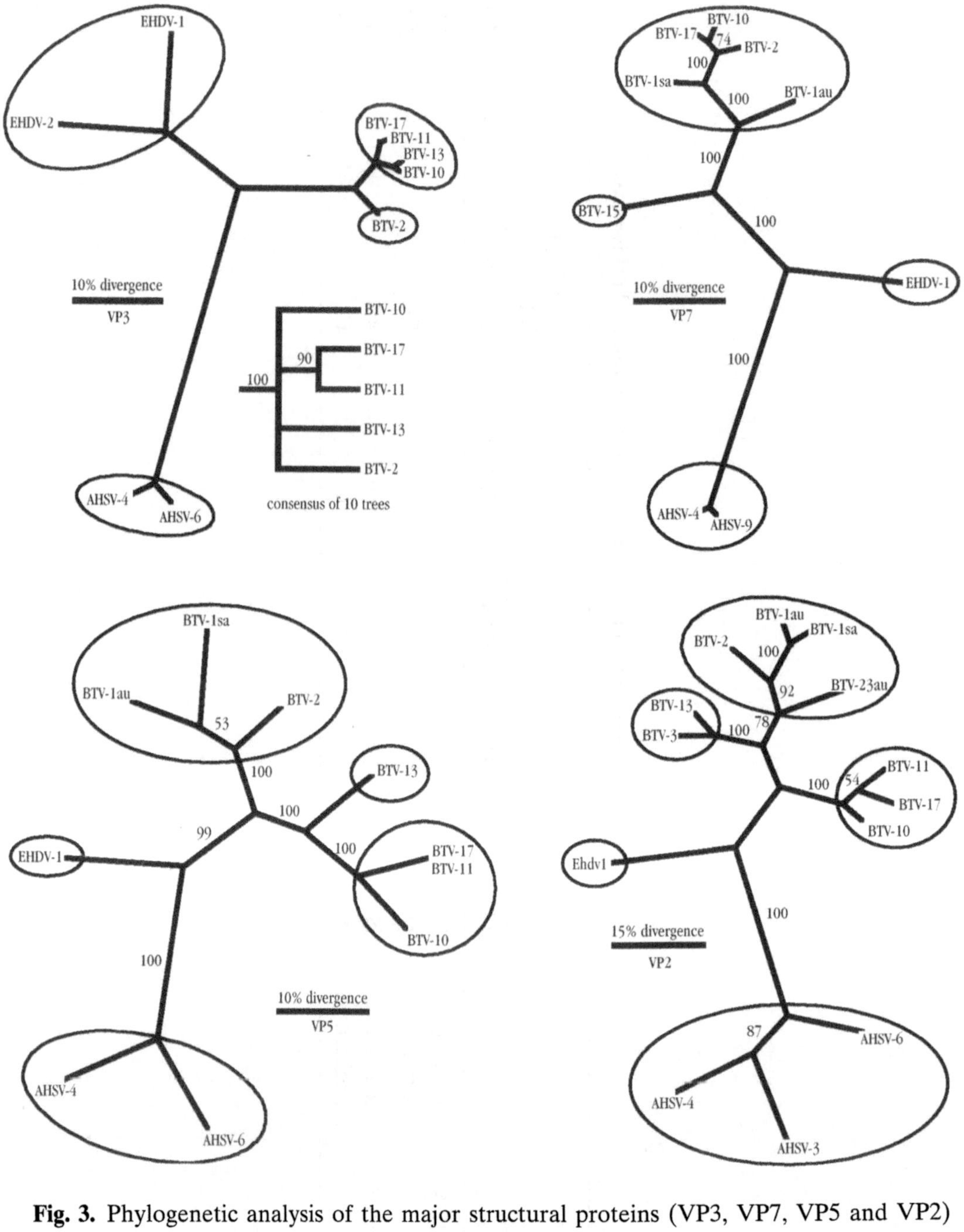

Fig. 3. Phylogenetic analysis of the major structural proteins (VP3, VP7, VP5 and VP2) from several orbiviruses

The three-dimensional reconstruction of BTV double capsid particles using image analysis of electron cryo-micrographs has revealed that the outer capsid of the virion has a well ordered morphology [14]. In contrast, when the complete BTV particle is viewed by negative staining methods, the outer capsid has a fuzzy appearance, with no evidence of clearly defined morphology [40]. The outer capsid has icosahedral symmetry and the two proteins VP2 and VP5 have distinctive shapes, one sail shaped, the other globular and almost spherical [14] (Fig. 5). The globular VP5 proteins, estimated to be 120 in number (although biochemical data suggest a total of 360 copies of VP5, indicating that they may be arranged as

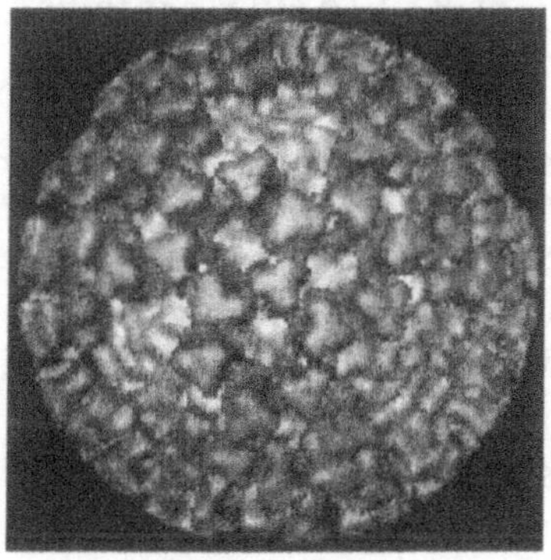

Fig. 4. Cryo-EM images of the two layers of CLP mimic that seen with the cores showing VP7 trimers and VP3

Fig. 5. Representation of the surface of the reconstructed icosahedral BTV virus particle obtained from cryo-electron microscopy. The globular regions and sail shaped spikes are arranged in a triskelion motif

trimers [38]), site neatly in the channels formed by each of the six membered rings of VP7 trimers. The sail shaped spikes, which project 4 nm beyond the globular proteins, are located above 180 of the VP7 trimers and form 60 triskelion type motifs that cover nearly all the VP7 molecules. These spikes are thought to be composed of VP2, the viral haemagglutinating protein, which also contains virus neutralising epitopes. The two proteins appear to form a continuous layer around the core except for holes on the five-fold axes.

Baculovirus expression vectors and production of BTV/AHSV proteins

Recent advances in gene manipulation have made it possible to express foreign genes in heterologous systems. The productivity and flexibility of insect baculovirus expression vectors and the ability of the baculovirus genome to incorporate (and express) large amounts of foreign DNA have permitted this system to be used for the expression of not only single genes, but also for the simultaneous expression of dual and multiple genes [6]. To accomplish this, several expression vectors have been developed, based on the resident promoters of the nuclear polyhedrosis virus of *Autographa californica* (AcNPV). One vector which has found particular favour is the strong promoter of a so called 'very late' polyhedrin gene (*ph*). The promoter of the *ph* gene has been exploited to express, one, two or multiple foreign gene products, either by replacing the *ph* gene or by duplicating the promoter, together with appropriate transcription termination signals [6]. A second very late gene, *p* 10, which is responsible for the synthesis of a non structural protein, has also been used successfully to express foreign proteins, either by replacing the resident *p* 10 gene or by duplicating the *p*10 promoter and inserting it elsewhere in the AcNPV genome, e.g., in proximity to the *ph* gene site [29].

Using these vectors, we have expressed all ten genes of BTV, synthesized BTV proteins and analysed the structure-function relationship of each gene and gene product. Each of the recombinant expressed BTV 10 proteins was characterized biochemically and immunologically. They all appeared to be identical to the corresponding authentic protein [4]. The functions of some of these gene products in virus replication and assembly processes have been subsequently investigated. The preliminary data obtained from these studies are summarized in Table 1. Subsequently, we have expressed a number of AHSV 4 and AHSV 6 genes and compared their structures and functions with those of BTV. As expected, the data confirmed that the component proteins are relatively highly conserved between these two orbiviruses.

Although the BTV VP7 molecules contain a high percentage of hydrophobic residues, BTV 10 VP7 synthesised by recombinant baculoviruses is soluble and, under the appropriate conditions, forms crystals of VP7 trimers [1]. This property has allowed the three dimensional structure of VP7 to be determined to a resolution of 2.6 Å [2, 12, 13]. Similarly, the upper domain of the AHSV 4 VP7 has been determined [2]. From these data it has been shown that the VP7 monomer is composed of two distinct domains (Fig. 6). The upper domain is formed by the central one third of the polypeptide chain (aa 121–249), folded into an anti parallel β-sandwich. The lower domain contains the N-terminal region composed of 5 α helices (aa 1–120) and four C terminal α helices (aa 250–349) that together form a bundle of nine helices with loops in between (Fig. 6). The two domains are twisted anticlockwise in the trimer so that the top domain of one monomer rests on the lower domain of the 3-fold related subunit.

In contrast to BTV, the AHSV VP7 forms flat hexagonal crystals in the cytoplasm of infected cells [7] as well as large disc shaped crystals when expressed by a recombinant baculovirus [8]. Due to this intrinsic insolubility, purification of a soluble form of the complete AHSV 4 VP7 has proved difficult and only the top domain has been crystallized [2]. The X ray structure of this top domain is very similar to that of the upper domains of BTV VP7 trimers (Fig. 7), as could be predicted from the high degree of sequence similarity between the BTV and AHSV VP7 proteins (overall: 67% similarity, 46% identity). BTV VP7 expressed by baculovirus has recently been crystallized in two forms exhibiting different packing of the molecule in the two crystals [2]. However, in both crystal forms VP7 exists as a trimer, and each trimer consists of two domains, upper and lower,

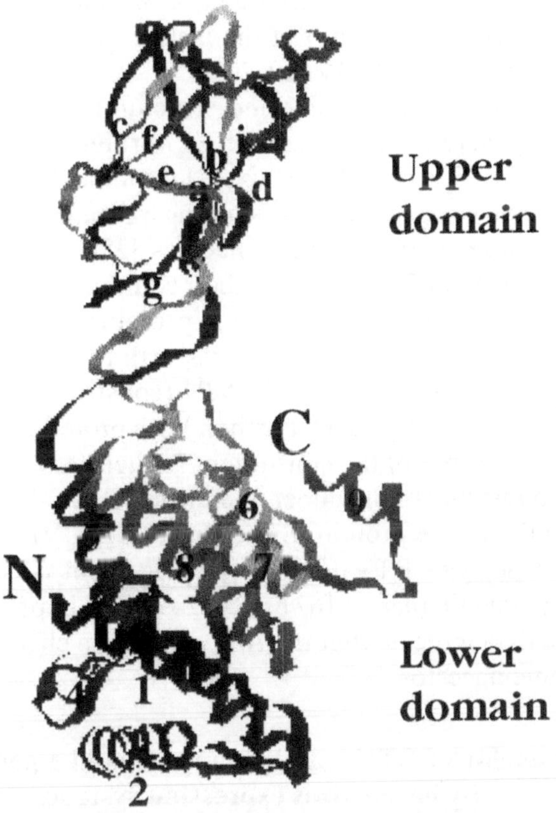

Fig. 6. A sketch of the structure of a VP7 monomer highlighting the secondary structural elements

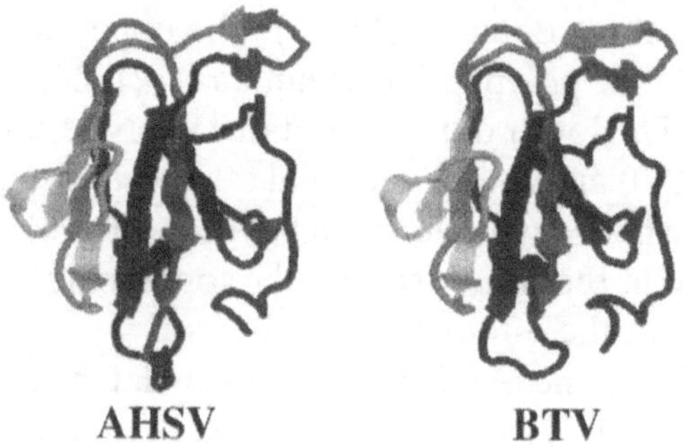

Fig. 7. Comparison of the top domains of Orbivirus VP7 showing the structural similarity between BTV and AHSV

giving an appearance of a tripod like structure. The X ray crystallographic analysis confirmed the data obtained by reconstruction studies.

The conservative nature of VP7 allows domains switching between the BTV and AHSV proteins

As discussed above, at the amino acid (aa) sequence level, BTV 10 VP7 protein is highly conserved among BTV serotypes (> 95% sequence identity, depending on the geographic origin (topotype) of the isolate) and is closely related to the VP7 proteins of AHSV and EHDV [19, 20].

Based on the available molecular and structural information, we have exchanged specific regions and domains of the VP7 proteins of BTV and AHSV and analysed the products. This has allowed us to determine if the chimeric molecules have equivalent molecular features (including protein solubility, folding and trimerisation) to those of the native molecules [28]. The effects of replacing specific aa of the AHSV VP7 protein with their counterparts from BTV VP7 were also investigated. The chimeric and mutant proteins were expressed from the derived recombinant baculoviruses. Chimeras with the central upper domain of BTV 10 VP7 replaced by that of AHSV 4 VP7 (construct BAB) formed trimers, as did the converse construct (ABA). Further, both proteins exhibited the expected conformational epitopes of the constituent sequences. Using BAB VP7 it was demonstrated that residues of the upper domain of AHSV 4 VP7 contribute to the observed insolubility of the protein. By contrast, ABA VP7 protein had similar solubility to the wild type BTV-10 VP7. Replacement of selected amino acid residues in the top domain (e.g., A167 by R; F209 by T) improved the solubility of BAB VP7. These data indicate that the two viruses are closely related and likely to origin from same ancestor.

Assembly of BTV core-like and virus-like particles by baculovirus expression systems

The flexibility of baculovirus expression vectors and the capacity of the baculovirus genome to accommodate large amounts of foreign DNA has allowed us to exploit the system for the simultaneous expression of multiple BTV genes in a single insect cell. 3D studies indicated that it might be possible to obtain a stable structure composed of the scaffolding protein VP3 and the core surface protein VP7, if they were synthesised and could be assembled in the correct conformation. Therefore we took advantage of the dual and multigene baculovirus vectors system to simultaneously co express either VP3 and VP7, or VP3, VP7, VP2 and VP5 in *Spodoptera frugiperda* cells. In both cases these experiments resulted in the production of synthetic particles, which although they were empty and lacked the three minor proteins and the viral RNA segments, resembled either native cores (core like particles: CLP) or intact virions (virus like particles: VLP). These particles have allowed us not only to answer some fundamental questions regarding viral assembly but also to develop candidate vaccines.

Our initial effort was to assemble the two major core proteins, VP3 and VP7. For this purpose, a dual baculovirus expression vector consisting of duplicated polyhedron promoter of AcNPV, with downstream transcription terminator sequences, was utilized to express the coding sequences of the genome segment 3 (VP3) and genome segment 7 (VP7) genes of BTV [9]. Recombinant baculoviruses synthesising both proteins were isolated and CLP were indeed produced and were distributed throughout the infected insect cells. Purified CLP were similar in size and appearance to cores prepared from BTV (see Fig. 8). VP3 and VP7 were the only protein components identified in the expressed particles and the molar ratios of these two proteins were similar to those of VP3 and VP7 derived from infectious BTV. The CLP appeared to lack nucleic acids when analysed by phenol chloroform extraction and alcohol precipitation.

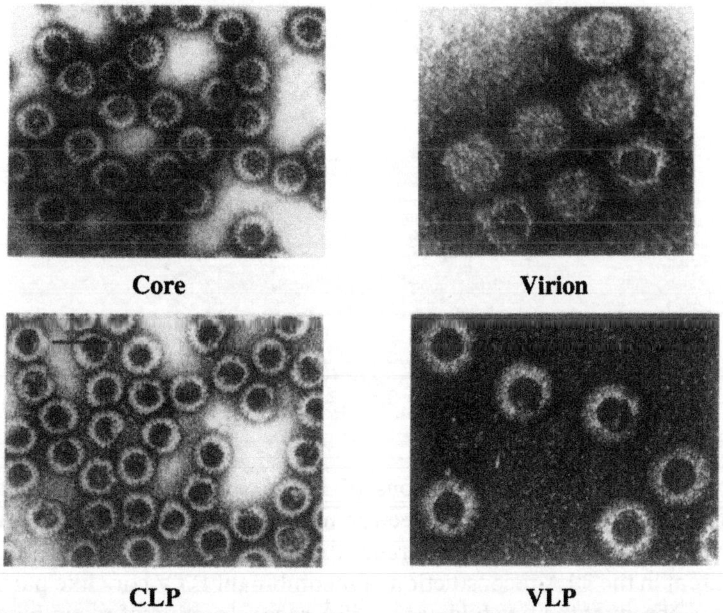

Fig. 8. Electron micrographs of authentic BTV derived cores and viral particles, and baculovirus expressed core-like and virus-like particles purified by differential centrifugation

We have subsequently constructed baculovirus multigene vectors to co-synthesise up to four BTV proteins in the same cell [4, 5]. In addition to the polyhedrin promoter, copies of the *p*10 promoter of AcNPV have been utilized to facilitate the high level co-expression of several proteins in each infected cell. For optimum VLP synthesis, a quadruple gene expression vector was used to synthesize the BTV VP2, VP3, VP4 and VP7 proteins. The expressed proteins assembled into virtually homogenous double capsid particles (see Fig. 8). Co-infections with single or dual gene expression vectors resulted in production of VLP containing variable amounts of the outer capsid proteins in different experiments [10]. The formation of complete VLP in the absence of the non structural proteins implies that these proteins are not necessary for the assembly of these double capsid particles or for CLP. VLP express high levels of hemagglutination activity, similar to that of BTV virions. Further, antibodies raised to the expressed particles gave high titers of neutralising activity against the homologous BTV serotype [10].

To determine whether CLP and VLP mimic the morphology of BTV derived core and virus particles, electron cryo-microscopy and image analysis was used to examine unstained and unfixed virus derived core particles, CLP and VLP,

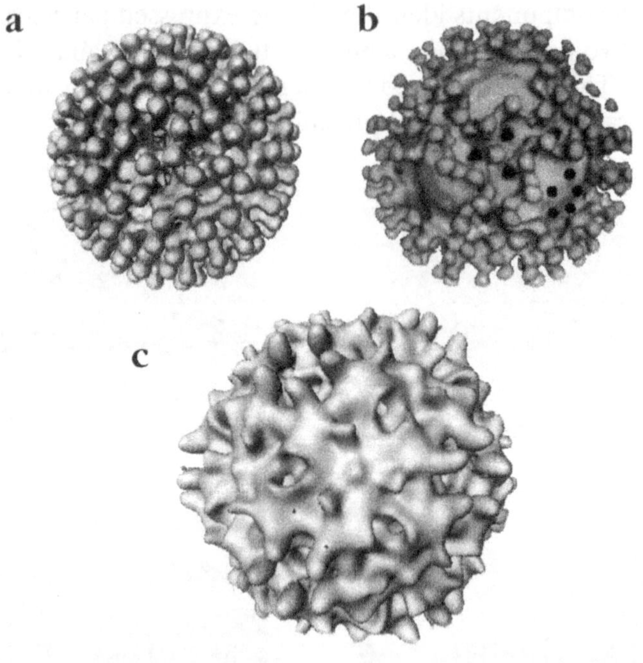

Fig. 9. Three-dimensional reconstructions of BTV particles and recombinant particles obtained using image analysis of cryo-electron micrographs. **a** 45 Å reconstruction of BTV cores showing the knobbly surface appearance due to VP7. **b** The trimeric nature of the VP7 is more apparent in the 30 Å reconstution of recombinant BTV core like particles although the trimers are absent at the five fold axes. **c** 55Å reconstruction of recombinant BTV virus like particles shows essentially the same features as the authentic virus (see Fig. 5). All views are down a three-fold axis

thereby avoiding heavy metal stains, fixture and dehydration [14, 15, 31]. These images have not only revealed an icosahedral structure of CLP, which is similar to that of authentic cores but have also yielded information not available from previous analyses of the structure of the native core particle [31]. Firstly this is because CLP are empty and therefore the contrast between the inner shell and the interior is greater. Secondly, this method provides information on the order of assembly of the core particle through analysis of incomplete CLP, i.e., CLP that lack a full complement of VP7. However, the VP7 spikes, which are seen more clearly, consist of an inner and outer domain and their presumed trimeric nature, suggested by authentic core particle reconstruction, is supported by the triangular shape of the spikes on the CLP (see Fig. 9).

The underlying subcore of the CLP is apparently composed of closely bonded VP3 molecules [15]. From the atomic structure of native cores, the subcore layer is composed of 12 icasahedrally related decamers of VP3, each centered at one of the vertices of the icosahedron [38]. Pores in this shell are essentially spaces between VP3 molecules.

The fact that VP7 trimers can interact with the VP3 subcore to form CLP was further exploited for identifying the regions of VP7 necessary for the formation of such particles. A number of site specific mutants and deletion or extension mutants of VP7 were generated and expressed in the presence of VP3. The data indicate that the intact carboxyl terminus of the VP7 molecule is essential for particle formation and that the amino terminus can be modified by the addition of some foreign sequences without compromising the ability to form the CLP structures. However, the CLP that were formed were physically less stable (e.g., under conditions of high salt concentration or long term storage) than those formed with unmodified VP7 and VP3 [3, 22].

Three dimensional reconstruction of VLP at 55Å resolution has also been undertaken. At this low resolution the structure of the VLP (860 Å in diameter) is comparable to that of authentic virions [14] and exhibits essentially the same basic features as well as a full complement of the four proteins (see Fig. 9).

Assembly of seven structural proteins

To determine whether the remaining three viral proteins, VP1, VP4 and VP6 can also be incorporated into CLP and VLP, similar experiments were performed involving co-infection of *S. frugiperda* cells with various combinations of recombinant baculoviruses. The largest BTV protein, VP1, is the putative viral polymerase. To discover whether the VP1 protein can be encapsidated within CLP, insect cell cultures were co-infected with two recombinant baculoviruses, the dual recombinant expressing VP3 and VP7 and a second single recombinant virus expressing VP1. The assembled CLP contained complexes of the three proteins. Further analysis showed that CLP consist of a subcore of VP3, which serves both as a frame work for the trimers of VP7 as well as interacting with VP1 [24]. Similar co-expression experiments showed VP4 can also be incorporated into CLP [23]. The construction of particles or complexes containing various combinations of

BTV protein should help elucidate the role of each protein in the morphogenic process of BTV.

To find out whether CLP without the minor protein components have the ability to interact with viral RNA species, various assay systems were developed that used purified CLP and single stranded RNA species synthesized in vitro. The accumulated data clearly indicate that the RNA binding affinity of CLP resides on VP3 and not on VP7 [25]. This type of approach will be helpful in exploring the encapsidation mechanism of viral RNA.

Assembly of heterologous VLP using different BTV serotypes

Some 24 serotypes of BTV are recognized (BTV 1, BTV 2, etc.). The outercapsid proteins (VP2 and VP5) exhibit the least conservation between BTV serotypes, in contrast to all the other virus coded proteins [37]. VP2 is the main serotype specific antigen. Antisera raised to the baculovirus expressed VP2 of individual BTV serotypes neutralize the homologous virus, and depending on the antigen, cross neutralize to some extent certain other BTV serotypes [18, 39]. These data indicate that some BTV serotypes are more closely related to each other than to the other serotypes.

The structural compatabilities of VP2 and VP5 between different BTV serotypes have been investigated in VLP formation. Heterologous VLP have been sought by co-infection of insect cells with appropriate recombinant baculoviruses [24]. Assembled particles were purified and analysed by electron microscopy and SDS PAGE to confirm their authenticity. The presence of VP2 and VP5 in VLP was demonstrated by hemagglutination and Western immunoblotting, respectively. Despite the high level of sequence variation amongst the different serotypes, the VP2 and VP5 proteins of six different BTV serotypes formed VLP with VP3 and VP7 from a heterologous source [26]. Some combinations (e.g., the VP2 of BTV 2 and the VP5 of BTV 10) did not form VLP. The data indicate that VP2 may require the VP5 of the same or a closely related serotype (e.g., BTV 10 or BTV 17), but may not form VLP efficiently with the VP5 proteins of a more distant serotype.

Assembly of chimeric particles between different Orbiviruses

Sequence comparisons of the four major capsid proteins, VP2, VP3, VP5 and VP7, of BTV 10 with those of EHDV 1 and AHSV 4 have revealed that the three viruses have generally comparable primary sequences for their capsid components [19]. Of the four major capsid proteins, the innermost, VP3, is the most conserved and the outermost, VP2, the most variable. The two major core proteins of BTV 10 and EHDV 1 share a high percentage of identical residues (79% for VP3, and 64% for VP7). However, when compared to AHSV 4, the VP3 and VP7 proteins of BTV and EHDV show less similarity (44-46%). It is likely that the most highly conserved sequences represent structurally important regions.

Intertypic CLP and VLP formation has been investigated using combinations of the cloned genes and baculovirus expression vectors. Preliminary studies have demonstrated that both the VP3 and VP7 proteins of BTV 10 and EHDV 1 are exchangeable, although the outercapsid proteins are not. The latter is not unexpected, since the VP2 proteins of these three orbiviruses share only 19-24% identical amino acids, although VP5 is more conserved, scoring up to 64% identity between BTV and EHDV and 43-45% between BTV or EHDV and AHSV.

Interestingly, minor proteins of BTV can be encapsidated into chimeric CLP that are formed by the VP3 of EHDV and the VP7 of BTV. Thus it appears that the functional regions of EHDV and BTV VP3 proteins involved in interacting with each of the minor proteins are conserved [23]. The ability of the VP3 of BTV (or EHDV) to interact with the VP7 of AHSV and the formation of CLP appears to be restricted. Occasionally a few particles have been identified, although they appear to be highly labile and sensitive to most physical treatment (unpubl. obs.).

Immunization of sheep with VLP affords long lasting protective immunity against BTV challenge

VLP synthesized by recombinant baculoviruses are biologically and immunologically similar to the virion. VLP exhibit high levels of haemagglutination activity, similar to those of authentic BTV. Antibodies raised to the expressed particles contain high titers of neutralizing activity against the homologous BTV serotype [10]. To analyse the protective effects and duration of VLP vaccination, a number of successful vaccination trials have been performed using BTV susceptible, 1 year old Marino sheep (BTV free) in insect proof isolation [35]. Here, we describe one such trial in which VLP (10 μg or 50 μg per sheep) representing BTV 10 and BTV 17 were employed [33].

Each group of eight sheep was immunized subcutaneously with VLP in saline containing the indicated amount of protein (Table 3) and suspended in 50% Montanide Incomplete Seppic Adjuvant (ISA-50, Seppic, Paris). Each animal received 2 ml of the mixturc. For control experiments, one group of sheep received only saline. The vaccinated animals were boosted with the same amounts of protein on day 21.

The neutralizing antibody titers of the vaccinated sheep were determined at weekly intervals over a sixty week period after the booster. Both types of VLP elicited (to various levels) antibodies that neutralized the homologous virus. In almost all cases these neutralizing titers remained high throughout the 60 week period. The neutralizing antibody titers for the animals that 50 μg doses of VLP were not significantly higher than those that received the 10 μg doses. Sheep vaccinated with the mixture of the two types of VLP induced antibodies that when tested by plaque reduction assays, neutralized both types of virus, as well as some related heterologous viruses (e.g., BTV 4). As expected, the six control sheep that were inoculated with saline remained sero-negative.

Table 3. Neutralizing antibodies of sheep at various intervals after vaccination with bluetongue virus like particles and their protective responses following challenge (after 14 months) with homologous and heterologous virulent bluetongue viruses

Sheep no.	Antibody to	Week 3	Week 5	Week 11	Week 15	Week 19	Week 23	Week 27	Week 31	Week 35	Week 39	Week 43	Week 47	Week 51	Week 56	Challenge virus	CRI*
1	BTV 10	24	32	24	32	32	32	32	32	32	24	16	16	16	16	BTV 10	0
2	BTV 10	32	64	32	32	24	32	32	32	NT	NT	NT	NT	NT	NT	NT	NT
3	BTV 17	64	128	128	128	64	64	64	64	64	32	32	32	32	32	BTV 17	0
4	BTV 17	64	96	128	128	128	128	128	128	NT	NT	NT	NT	NT	NT	NT	NT
5	BTV 10	96	256	64	32	32	32	32	32	32	32	32	32	32	32	BTV 10	0
6	BTV 10	48	128	64	64	64	64	64	64	64	64	48	48	48	48	BTV 10	0
7	BTV 17	32	64	64	64	64	64	64	64	64	64	64	64	64	64	BTV 17	0
	BTV 4	<4	12	12	4	<4	<4	<4	<4	<4	<4	<4	<4	<4	<4		
8	BTV 17	4	4	4	4	4	4	4	4	<4	<4	<4	<4	<4	<4	BTV 17	0
9	BTV 10	<4	4	4	4	4	4	8	8	8	8	4	4	4	4	BTV 4	7.1
	BTV 17	4	6	8	8	8	8	4	4	4	>4	>4	>4	>4	<4		
10	BTV 10	16	16	16	16	16	16	16	8	8	8	8	8	8	8	BTV 4	0
	BTV 17	16	16	12	8	4	4	>4	>4	>4	>4	>4	>4	>4	<4		
11	BTV 10	4	4	16	ND	16	16	16	16	16	16	16	16	16	16	BTV 11	3.1
	BTV 17	<4	<4	8	8	8	8	12	8	8	8	8	8	8	8		
12	BTV 10	16	24	32	32	32	32	64	32	32	32	32	32	32	16	BTV 11	3.9
	BTV 17	32	32	24	32	32	32	32	32	32	32	32	32	16	16		
13	BTV 10	48	256	64	64	64	64	64	64	64	64	32	32	32	32	BTV 11	3.0
	BTV 17	128	256	256	256	128	128	128	128	128	64	64	64	64	64		
	BTV 4	8	12	4	<4	<4	<4	<4	<4	<4	<4	<4	<4	<4	<4		
14	BTV 10	32	32	16	16	16	12	16	12	12	12	12	12	12	12	BTV 4	0.5
	BTV 17	64	64	32	32	32	32	32	32	32	32	32	32	32	32		
	BTV 4	4	4	4	<4	<4	<4	<4	<4	<4	<4	<4	<4	<4	<4		

(continued)

Table 3 *(continued)*

Sheep no.	Antibody to	Week 3	Week 5	Week 11	Week 15	Week 19	Week 23	Week 27	Week 31	Week 35	Week 39	Week 43	Week 47	Week 51	Week 56	Challenge virus	CRI*
15	BTV 10	32	256	64	64	32	32	32	32	32	32	32	32	32	16	BTV 4	0
	BTV 17	128	256	64	64	32	32	32	32	32	32	32	32	16	16		
	BTV 4	16	32	12	<4	8	<4	<4	<4	<4	<4	<4	<4	<4	<4		
16	BTV 10	64	64	32	32	32	32	32	32	32	32	32	32	32	32	BTV 4	0
	BTV 17	64	128	128	128	128	128	128	64	64	64	64	64	64	32		
17	BTV 10	32	32	32	32	24	32	32	16	16	16	16	16	16	16	BTV 11	0
	BTV 17	96	192	128	128	64	64	64	64	64	32	32	32	32	32		
18	BTV 10	24	32	24	16	16	16	16	16	12	16	16	8	8	8	BTV 11	2.3
	BTV 17	24	32	32	32	32	32	32	32	32	8	8	8	8	8		
	BTV 4	16	24	16	<4	<4	<4	<4	<4	<4	<4	<4	<4	<4	<4		
19	BTV 10	<4	<4	<4	<4	<4	<4	<4	<4	<4	<4	<4	<4	<4	<4	BTV 10	8.0
	BTV 17	<4	<4	<4	<4	<4	<4	<4	<4	<4	<4	<4	<4	<4	<4		
20	BTV 10	<4	<4	<4	<4	<4	<4	<4	<4	<4	<4	<4	<4	<4	<4	BTV 10	7.1
	BTV 17	<4	<4	<4	<4	<4	<4	<4	<4	<4	<4	<4	<4	<4	<4		
21	BTV 10	<4	<4	<4	<4	<4	<4	<4	<4	<4	<4	<4	<4	<4	<4	BTV 17	1.6
	BTV 17	<4	<4	<4	<4	<4	<4	<4	<4	<4	<4	<4	<4	<4	<4		
22	BTV 10	<4	<4	<4	<4	<4	<4	<4	<4	<4	<4	<4	<4	<4	<4	BTV 17	2.7
	BTV 17	<4	<4	<4	<4	<4	<4	<4	<4	<4	<4	<4	<4	<4	<4		
23	BTV 10	<4	<4	<4	<4	<4	<4	<4	<4	<4	<4	<4	<4	<4	<4	BTV 4	3.7
	BTV 17	<4	<4	<4	<4	<4	<4	<4	<4	<4	<4	<4	<4	<4	<4		
24	BTV 10	<4	<4	<4	<4	<4	<4	<4	<4	<4	<4	<4	<4	<4	<4	BTV 11	3.2
	BTV 17	<4	<4	<4	<4	<4	<4	<4	<4	<4	<4	<4	<4	<4	<4		

NT Not tested due to loss of identification; sheep 1, 2 received 2 doses of 10 μg BTV 10 VLPs; sheep 3, 4 received 2 doses of 1 μg BTV 17 VLPs; sheep 5, 6 received 2 doses of 50 μg BTV 10 VLPs; sheep 7, 8 received 2 doses of 50 μg BTV 17 VLPs; sheep 9–13 received 2 doses of mixtures BTV 10 (10 μg) and BTV 17 VLPs (10 μg); sheep 14–18 received 2 doses of mixtures of BTV 10 (50 μg) and BTV 17 VLPs (50 μg); sheep 19–24 received no VLPs The indicated weeks represent the weeks after the second VLP inoculation. Challenge virus was administered following the week 56 serum sample.

* *CRI* Critical reaction index

All of the sheep were challenged 14 months after the booster vaccination by the subcutaneous injection of virulent BTV. The animals that were challenged with the homologous viruses (BTV 10, BTV 17) were completely protected and showed no clinical reactions, even those that received 10 μg doses of VLP. No viraemia was detected in any of these animals after challenge. Some animals, with 50 μg VLP, were also protected to low level when challenged with heterologous virus. By comparison, the control animals developed high or moderate signs of disease e.g. BTV 10, clinical reaction index (CRI) 7.1–8.0; BTV 17, CRI 1.6–2.7 and produced viraemias (data not shown).

In summary, the data showed that long lasting protection against homologous BTV challenge was provided by vaccination with VLP. Some preliminary evidence was obtained for cross-protection, which depended on the challenge virus and the amounts of antigen used for vaccination [33].

The data clearly demonstrate that VLP are highly immunogenic even at low doses. The sheep given only 10 μg of VLP actually received 1 μg of VP2 (10–20% of the VLP mass). This compares well with the much higher amounts required for VP2 alone, or VP2 and VP5 mixtures in protection studies [36]. There are several possible explanations. First, when assembled into VLP the conformational presentations of the relevant epitopes on VP2 probably mimic those on the authentic virus. Second, both VP2 and VP5 are present. Third, the VP3 and VP7 may provide a necessary scaffold for VP2 and VP5 antigen presentation. Fourth, any of the four BTV proteins might have a direct role in eliciting cell mediated immunity induced by the BTV VLP.

It can also be anticipated from the results obtained that this technology has much to offer for development of vaccines for both veterinary disease and human diseases. Moreover, there is every reason to believe that it should be possible to make vaccine chimeras representing different BTV serotypes (e.g., involving the expression of several BTV VP2 genes).

Core-like particles without outer capsid proteins can mitigate the viral infection

The question of whether CLP containing the two conserved proteins VP3 and VP7 would provide a measure of homologous and heterologous BTV protection by cell mediated mechanism was investigated. For initial studies, two groups of five sheep each were used, one group of five sheep inoculated with 50 μg BTV 10 CLP in ISA50, and a second group of five sheep boosted on day 21, and were challenged with BTV 10 two weeks later.

All post challenged sheep developed viraemias and neutralizing antibodies (Table 4 [34]), however, with the exception of fever, the vaccinated sheep developed only slight clinical reactions whereas controls showed characteristic mouth and feet lesions in addition to fever. The average CRI of the vaccinated sheep was 3.5 whereas that of the control sheep was 9.0.

In summary, partial protection against BTV challenge was afforded by CLP vaccination. Further experiments are planned to study this phenomenon further.

Table 4. Vaccination of sheep with core-like particles

Sheep no.	Neutralizing antibody titers	Clinical reaction index (CRI)	Viraemia
	CLP		
1	320	1.5	+
2	160	5.6	+
3	320	3.8	+
4	160	2.7	+
5	160	4.0	+
	Controls		
6	320	11.4	+
7	320	9.5	+
8	320	12.3	+
9	320	6.8	+
10	160	5.1	+

The immunized and control sheep developed viraemias and high neutralizing antibody titers. However, clinical signs of bluetongue were reduced for the vaccinated animals (CRI 1.5–5.6, average 3.5) compared to the controls (CRI 5.1–12.3, average 9.0). Only the controls showed the characteristic mouth and feet lesions

Protection of horses against virulent African horse sickness virus challenge, by recombinant baculovirus expressed VP2

African horse sickness virus serotype 4 (AHSV 4) outer capsid protein VP2, or VP2 and VP5, prepared from single or dual recombinant baculovirus expression vectors grown in Sf 9 insect cells were administered in different amounts to horses and the neutralizing antibody responses measured (see Table 5). Six months later, the control and vaccinated horses were challenged with virulent AHSV 4 and monitored post challenge. The results indicate that two inoculations of extracts containing the outer capsid proteins VP2 and VP5, or VP2 alone, in doses of 5 μg of VP2 or more per horse were sufficient to elicit protection against AHS (see Tables 5 and 6). The recombinant VP2 protein is potentially a candidate vaccine for AHS in horses.

CLP and VLP as vehicles for foreign immunogens

Synthesis of morphological entities in eukaryotic cells using an expression system opens up a wealth of possibilities to define mechanisms of protein-protein assembly under natural intracellular conditions. For example, by manipulating the structures of various proteins it will be possible to develop particular antigen delivery systems for the presentation of immunogenic epitopes against other infectious diseases.

Towards this end, we have manipulated the VP7 protein of CLP to receive foreign immunogenic epitopes. Initially a chimeric VP7 protein containing 48

Table 5. Vaccination of horses with recombinant AHSV proteins

Horse no.	Day 0 inoculum	Day 21 booster	Pre-challenge titre									Challenge	Post-challenge titre		
			Day 0	Day 21	Day 28	Day 35	Day 74	Day 105	Day 133	Day 164	Day 194	(1 ml subcutaneously) Day 204	Day 225	Day 234	Day 257
1	VP2 (5 µg)	VP2 (5 µg)	Neg	–	160	160	80	80	30	40	40	+	>2560	>20480	40960
2	VP2 (10 µg)	VP2 (10 µg)	Neg	20	320	320	160	160	60	40	40	–	NT	NT	40
3	VP2 (20 µg)	VP2 (20 µg)	Neg	120	640	640	640	80	240	160	160	+	>2560	2560	1280
4	VP2 (40 µg)	VP2 (40 µg)	Neg	60	320	480	640	320	20	20	20	+	>2560	2560	640
5	VP2 (80 µg)	V2 (80 µg)	Neg	80	>640	>640	640	160	30	40	colic[a]	–	–	–	–
6	VP2 (5 µg) + VP5 (0.5 µg)	VP2 (5 µg) + VP5 (0.5 µg)	Neg	–	–	20	80	160	30	20	20	–	NT	NT	20
7	VP2 (10 µg) + VP5 (1 µg)	VP2 (10 µg) + VP5 (1 µg)	Neg	60	640	106	640	80	160	120	80	–	NT	NT	40
8	VP2 (20 µg) + VP5 (2 µg)	VP2 (20 µg) + VP5 (2 µg)	Neg	320	640	480	640	320	160	80	80	+	>2560	5120	2560
9	VP2 (40 µg) + VP5 (4 µg)	VP2 (40 µg) + VP5 (4 µg)	Neg	20	320	640	640	2560	160	80	80	–	NT	NT	320
10	VP2 (80 µg) + VP5 (8 µg)	VP2 (80 µg) + VP5 (8 µg)	Neg	480	>640	>640	1280	320	240	160	160	+	>2560	5120	1280
11 (control)	Adjuvant and saline	Adjuvant and saline	Neg	Neg	Neg	Neg	Neg	Neg	Neg	Neg	Neg	+	Died		

[a] One horse died from colic

Table 6. Temperature reactions of challenged horses

Horse no.	Inoculum protein (g)	Day post challenge												
		1	3	4	5	6	7	8	9	10	11	12	13	14
1	VP2 (5 µg)	38.5 38.0	39.1 39.4	38.7	38.1	40.2 38.7	40.5 39.2	39.9 39.1	39.8 39.1	38.6	38.0	38.2	37.1 38.5	38.5 38.7
3	VP2 (20 µg)	37.8 37.5	38.5 38.3	38.3	37.5	38.2 38.0	38.1 38.0	38.3 38.0	38.1 38.0	38.0	38.0	37.7	37.7 38.2	38.1 38.0
4	VP2 (40 µg)	38.1 38.0	38.4 38.5	38.4	37.5	38.4 37.9	38.3 37.9	38.3 37.9	38.8 38.4	38.0	38.0	37.5	37.0 38.0	37.8 38.2
8	VP2 (20 µg) + VP5 (2 µg)	38.6 38.2	38.3 38.3	38.1	37.9	38.9 37.6	39.1 38.0	38.7 38.0	38.3 38.0	38.1	38.1	37.8	37.4 38.1	38.5 38.2
10	VP2 (80 µg) + VP5 (8 µg)	38.2 38.1	38.9 38.2	38.5	38.5	38.8 38.1	38.7 38.5	38.8 38.3	38.5 38.1	38.2	38.2	38.2	37.1 38.6	38.4 38.2
11 (control)	–	38.6 38.2	38.4 38.3	38.8	39.0	40.6 39.8	40.9 39.9	41.1 40.5	41.0	Died	–	–	–	–

amino acid residues of the hepatitis B virus (HBV) pre-S_2 region placed at the amino terminus of VP7 was expressed by recombinant baculovirus and chimeric CLP were synthesized in the presence of recombinant VP3 protein [3]. The location of pre S_2-VP7 protein on the particle was analysed by immunoelectron microscopy using gold labelled anti HBV pre-S_2 serum.

Subsequently, to analyse the three dimensional structure of VP7 molecule, baculovirus expressed VP7 protein has been purified and crystallized. The recent X ray crystallographic analysis revealed that the VP7 molecules exists as trimers and each molecule has two domains. The three VP7 molecules interact side to side to form the trimer [2, 12, 13]. The upper domain of the molecule mainly contains β-sheet and β-strands with a number of loops. The lower domain, on the other hand is mainly composed of tight α helices which cannot be disturbed. However, there are several sites within the various loops of the upper domain that can be manipulated for insertion of foreign epitopes as indicated in the ribbon diagram of the trimer (Fig. 10).

As indicated in the VP7 diagram, a number of regions appeared to be accessible to the surface of the molecule including an Arg-Gly-Asp (RGD: aa. 168–170) motif. Therefore, this site as well as two other sites (Ala aa. 145 and Gly aa. 238) were chosen for insertion of foreign epitopes. All three sites were manipulated and restriction sites were created for receiving foreign epitopes. The RGD site was used for insertion of the Hepatitis B virus surface antigen (HBsAg). Chimeric HBsAg-VP7 protein was expressed and tested for its ability to form CLP. Indeed the chimeric protein could form CLP which were similar to

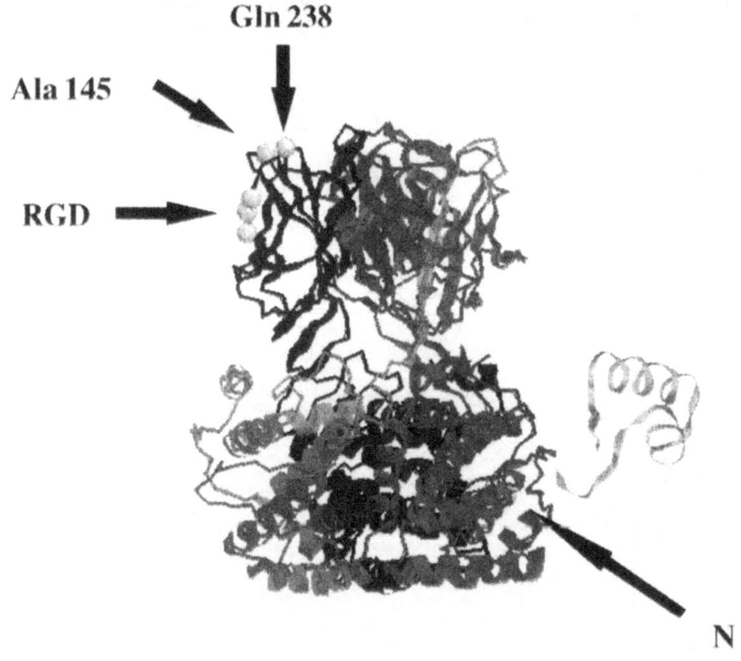

Fig. 10. BTV VP7 trimer showing the sites used for epitope insertion

CLP produced with the normal VP7. They had complete or nearly complete set of VP7 spikes and appeared to be as stable as non-chimeric CLP. The other sites were used to insert immunogenic regions of surface glycoprotein gp51 of bovine leukemia virus (BLV). Both sites appeared to be suitable for insertion of extra amino acid residues without disturbing CLP formation (unpubl. obs.). The immunogenic and protective properties of these chimeric particles are now under investigation.

Conclusion

The baculovirus expression system has provided a method to understand how viral proteins interact and which of them assemble to form virus particles. This approach has also provided information on the overall sequence of particle assembly. It seems that the viral genome and the three minor proteins interact with each other to form the innermost core particle. Subsequently, each of the components interact with the VP3 to form the subcore. The 120 molecules of VP3 then serve as scaffold for the attachment of the VP7 trimers and thus complete a stable and transcriptionally active icosahedral core particle. At a later stage of virus replication the two proteins of the outer capsid are simultaneously layered onto the surface of the VP7 trimers.

Virus structures, which contain multiple polypeptide species encoded by separate genes (mRNA species) and which involve non equimolar ratios of proteins, such as those of BTV, present one of the most difficult and challenging subjects to study. Synthesis of subviral particles in eukaryotic cells by an expression system opens up a wealth of possibilities for defining the mechanism of protein-protein assembly under natural intracellular conditions. For example, by manipulating the structures of various BTV proteins it will be possible to define the essential morphogenic interactions of the seven structural and four non structural proteins.

In addition, it is anticipated that the availability of large quantities of viral components, and of CLP and VLP, will not only facilitate an understanding of their three dimensional structure at the atomic level, but due to the authenticity in their biological and immunological properties, the subunit VP2 (for AHSV) and the assembled VLP (for BTV) are ideal as viral vaccines, as they lack any genetic material and can not replicate. Curiously, for AHSV only 5 μg of VP2 appears to be sufficient for protection against virulent virus challenge.

Since CLP and VLP are highly immunogenic and elicit protective immunity in vertebrate hosts, BTV CLP and VLP are now being developed to deliver multiple peptide components representing viral epitopes in order to elicit protective immunity. Since BTV CLP and VLP are large multi protein structures, they can incorporate alternative forms of the structural proteins (e.g., chimeric VP7). In addition, there is a possibility to use more than one of the proteins to deliver antigens (e.g., VP2, VP5 and VP7). This feature has important implications since to decrease the chance of non responsiveness, vaccines designed to elicit immunity should not be based on a single viral sequence or antigen, but should include as many immunogenic sequences as possible. Moreover, VLP based on alternative BTV serotypes can be used for successive immunizations in order to evade the anti BTV response elicited in a primary vaccination. This is another novel feature of the BTV system, not currently available in other antigen delivery systems (e.g., vaccinia, polio, Ty particles, etc).

BTV CLP and VLP offer several particular advantages over other systems. First, as noted above, large quantities can be produced due to the expression capabilities of baculovirus vectors (ca. 20–30 mg per litre culture, produced in serum free medium, stable to freeze drying, etc). Second, CLP and VLP can be purified using a one step generic protocol based on the physical properties of the particle (gradient centrifugation of cell lysates). Third, they are devoid of any detectable amount of insect, or baculovirus proteins, RNAs, or DNAs. Fourth, the purification procedure is gentle enough to maintain the morphological structure of the particles in their native conformations. Lastly, and most importantly, VLP have inherent properties of inducing both B cell and T cell responses in vertebrate hosts.

References

1. Basak AK, Stuart DI, Roy P (1992) Preliminary crystallographic study of bluetongue virus capsid protein VP7. J Mol Biol 228: 687–689
2. Basak AK, Grimes, JM, Gouet P, Roy P, Stuart DI (1997) Structures of orbivirus VP7: implications for the role of this protein in the viral life cycle. Structure 5: 871–883
3. Belyaev AS, Roy P (1992) Presentation of hepatitis B virus preS2 epitope on bluetongue virus core-like particles. Virology 190: 840–844
4. Belyaev AS, Roy P (1993) Development of baculovirus triple and quadruple expression vectors: co-expression of three or four bluetongue virus proteins and the synthesis of bluetongue virus-like particles in insect cells. Nucleic Acids Res 21: 1219–1223
5. Belyaev AS, Hails RS, Roy P (1992) High-level expression of five foreign genes by a single recombinant baculovirus. Gene 156: 229–233
6. Bishop DHL (1992) Baculovirus expression vectors. Semin Virol 3: 253–264
7. Burroughs JN, O'Hara RS, Smale CJ, Hamblin C, Walton A, Armstrong R, Mertens PPC (1994) Purification and properties of virus particles, infectious sub viral particles, cores and VP7 crystals of African horse sickness virus serotype 9. J Gen Virol 75: 1849–1857
8. Chuma T, LeBlois H, Sanchez-Vizcaino JM, Diaz-Laviada M, Roy P (1992) Expression of the major core antigen VP7 of African horsesickness virus by a recombinant baculovirus and its use as a group-specific diagnostic reagent. J Gen Virol 73: 925–931
9. French TJ, Roy P (1990) Synthesis of bluetongue virus (BTV) core-like particles by a recombinant baculovirus expressing the two major structural core proteins of BTV. J Virol 64: 1530–1536
10. French TJ, Marshall JA, Roy P (1990) Assembly of double-shelled, virus-like particles of bluetongue virus by the simultaneous expression of four structural proteins. J Virol 64: 5695–5700
11. Fukusho A, Yu Y, Yamaguchi S, Roy P (1989) Completion of the sequence of bluetongue virus serotype 10 by the characterization of the structural protein, VP6, and a non-structural protein, NS2. J Gen Virol 70: 1677–1689
12. Grimes JM, Jakana J, Ghosh M, Basak AK, Roy P, Chiu W, Stuart DI, Prasad BVV (1994) An atomic model of the outer layer of the bluetongue virus core derived from X-ray crystallography and electron microscopy. Structure 5: 885– 893

13. Grimes J, Basak, AK, Roy P, Stuart D (1995) The crystal structure of bluetongue virus VP7. Nature 373: 167–170
14. Hewat EA, Booth TF, Roy P (1992) Structure of bluetongue virus particle by cryo-electron microscopy. J Struct Biol 109: 61–69
15. Hewat EA, Booth TF, Loudon PT, Roy P (1992) 3-D reconstruction of bluetongue virus core-like particles by cryo-electron microscopy. Virology 189: 10–20
16. Hirasawa T, Roy P (1990) The complete nucleotide sequence of VP5 of a strain of bluetongue virus of serotype 2 isolated in the USA reveals its close relationship with a virus of serotype 1 isolated in Australia. Virus Res 15: 107–112
17. Huismans H, Van Der Walt NT, Cloete M, Erasmus BJ (1987) Isolation of a capsid protein of bluetongue virus that induces a protective immune response in sheep. Virology 157: 172–179
18. Inumaru S, Roy P (1987) Production and characterisation of the neutralisation antigen VP2 of bluetongue virus serotype 10 using a baculovirus expression vector. Virology 157: 472–479
19. Iwata H, Chuma T, Roy P (1992) Characterisation of genes encoding two of the major capsid proteins of epizootic haemorrhagic disease virus indicates a close genetic relationship to bluetongue virus. J Gen Virol 73: 915–924
20. Iwata H, Hirasawa T, Roy P (1991) Complete nucleotide sequence of segment 6 of epizootic haemorrhagic disease virus; the outer capsid protein VP5 is homologous to that of bluetongue virus. Virus Res 20: 273–281
21. Iwata H, Yamagawa M, Roy P (1992) Evolutionary relationships among the gnat-transmitted orbiviruses that cause African horsesickness, bluetongue and epizootic hemorrhagic disease as evidenced by their capsid proteins. Virology 191: 251–261
22. LeBlois H, Roy P (1993) A single point mutation in the VP7 major core protein of bluetongue virus prevents the formation of core-like particles. J Virol 67: 353–359
23. LeBlois H, Fayard B, Urakawa T, Roy P (1991) Synthesis and characterisation of chimaeric particles between epizootic haemorrhagic disease virus and bluetongue virus: functional domains are conserved on the VP3 protein. J Virol 65: 4821–4831
24. Loudon PT, Roy P (1991) Assembly of five bluetongue virus proteins expressed by recombinant baculoviruses: inclusion of the largest protein VP1 in the core and virus-like particles. Virology 180: 798–802
25. Loudon PT, Roy P (1992) Interaction of nucleic acids with core-like and subcore-like particles of bluetongue virus. Virology 191: 231–236
26. Loudon PT, Hirasawa T, Oldfield S, Murphy M, Roy P (1991) Expression of outer capsid protein VP5 of two bluetongue viruses, and synthesis of chimeric double-shelled virus-like particles using combinations of recombinant baculoviruses. Virology 182: 793–801
27. Martinez-Torrecuadrada J, Iwata H, Venteo A, Casal I, Roy P (1994) Expression and characterisation of the two outercapsid proteins of African horsesickness virus. The role of VP2 in virus neutralisation. Virology 202: 348–359
28. Monastyrskaya K, Staeuber N, Sutton G, Roy P (1997) Effects of domain-switching and site-directed mutagenesis on the properties and functions of the VP7 proteins of two orbiviruses. Virology 237: 217–227
29. O'Reilly DR, Miller LK, Luckow VA (1992) (eds) Baculovirus expression vectors: a laboratory manual. WH Freeman, London
30. Oldfield S, Hirasawa T, Roy P (1990) Sequence conservation of the outer capsid protein, VP5, of bluetongue virus, a contrasting feature to the outer capsid protein VP2. J Gen Virol 72: 449–451
31. Prasad BVV, Yamaguchi S, Roy P (1992) Three-dimensional structure of single-shelled BTV. J Virol 66: 2135–2142

32. Roy P, Gorman BM (1990) Bluetongue viruses. Curr Top Microbiol Immunol 162
33. Roy P, Bishop DHL, LeBlois H, Erasmus BJ (1994) Long-lasting protection of sheep against bluetongue challenge after vaccination with virus-like particles: evidence for homologous and partial heterologous protection. Vaccine 12: 805–811
34. Roy P, Callis J, Erasmus BJ (1994) Protection of sheep against bluetongue disease after vaccination with core-like and virus-like particles: evidence for homologous and partial heterologous protection. In: Proceedings for 97th Annual Meeting, USA Health Association. La Vegas, pp 88–97
35. Roy P, French TJ, Erasmus BJ (1992) Protective efficacy of virus-like particles for bluetongue disease. Vaccine 10: 28–32
36. Roy P, Hirasawa T, Fernandez M, Blinov VM, Sanchez-Vizcaino JM (1991) The complete sequence of the group-specific antigen, VP7, of African horsesickness disease virus serotype 4 reveals a close relationship to bluetongue virus. J Gen Virol 72: 1237–1241
37. Roy P, Marshall JA, French TJ (1990) Structure of bluetongue virus genome and its encoded proteins. Curr Top Microbiol Immunol 162: 43–87
38. Stuart DI, Gouet P, Grimes J, Malby R, Diprose J, Zientara S, Burroughs JN, Mertens PPC (1998) Structural studies of orbivirus particles. In: Mellor PS, Baylis M, Hamblin C, Calisher CH, Mertens PPC (eds) African horse sickness. Springer, Wien New York, pp 235–250 (Arch Virol [Suppl] 14)
39. Urakawa T, French TJ, Adachi Y, Fukusho A, LeBlois H, Flamand M, Mertens PPC, Roy P (1994) Synthesis of recombinant baculoviruses expressing the outer capsid protein VP2 of five BTV serotypes and the induction of neutralising antibodies to homologous and heterologous BTV serotypes. Virus Res 31: 149–161
40. Verwoerd DW, Els HJ, De Villiers E-M, Huismans H (1972) Structure of the bluetongue virus capsid. J Virol 110: 783–794
41. Yang YY, Li JK (1992) Complete genomic sequences of the VP5 protein gene of bluetongue virus serotype 11 and 17. Virus Res 23: 163–171

Authors' address: Dr. P. Roy, Department of Biochemistry, University of Oxford, Mansfield Road, Oxford, OXI 35R, U.K.

Expression of the major core structural proteins VP3 and VP7 of African horse sickness virus, and production of core-like particles

S. Maree[1], **S. Durbach**[2], **F. F. Maree**[3], **F. Vreede**[1], and **H. Huismans**[3]

[1] Biochemistry Division, Onderstepoort Veterinary Institute, Onderstepoort, South Africa
[2] Molecular Biology Unit, South African Institute for Medical Research, Johannesburg, South Africa
[3] Department of Genetics, University of Pretoria, Pretoria, South Africa

Summary. The genome segments encoding the seven structural proteins of African horse sickness virus (AHSV), including the largest coding for VP1, were cloned and sequenced. Analysis of the VP1 sequence supports the putative identity of this protein as an RNA polymerase. The genes encoding the two major core proteins, VP3 and VP7, were also cloned and expressed by both in vitro translation and by means of recombinant baculoviruses. Co-infection of insect cells with VP3 and VP7 recombinant baculoviruses resulted in the intracellular formation of multimeric particles with a diameter of 72 nm, which structurally resembled authentic AHSV cores (core like particles: CLP). The complete genome of AHSV has now been cloned and sequenced.

*

The etiological agent of African horse sickness is a double stranded RNA virus in the genus *Orbivirus*, family *Reoviridae* [9]. The ten double stranded (ds) RNA genome segments of these viruses encode 7 structural proteins (VP1–7) and at least three distinct nonstructural proteins (NS1–3). The genome is enclosed in an icosahedral core particle composed primarily of two major proteins, VP3 and VP7. The knob-like protrusions on the surface of this core are composed of VP7 trimers, while VP3 forms an internal scaffold, on which the VP7 layer and the three internal minor proteins, VP1, VP4 and VP6 are assembled [7, 8, 17]. The 3 minor proteins are all presumed to be associated with enzymatic activities of the core particle, which include the dsRNA dependent ssRNA polymerase and RNA capping activities. VP1, encoded by genome segment 1, is the prime candidate for the RNA polymerase. Motifs characteristic of RNA polymerases have been identified in the predicted VP1 amino acid sequence of BTV [11]. The core particle is surrounded by an outer capsid, composed of two major proteins, VP2 and VP5, of which VP2 is located on the surface of the virion. VP2 is also the most variable of the capsid proteins and it is the major serotype-specific antigen [3, 14, 22].

BTV is the first orbivirus from which all 10 different genome segments were fully characterized. Most of what is known about the structure and function of orbivirus proteins therefore relates to BTV [18, 19]. Research on the structural proteins of BTV can be subdivided in a number of major focus areas, which include the structure and assembly mechanism of viral particles, the enzymatic functions of minor core proteins and the role of the structural proteins in eliciting a protective immune response. In much of this work the baculovirus expression system has been exploited extensively for the synthesis of individual proteins as well as for production of core-like particles (CLPs) and virus-like particles (VLPs). These expression systems have been particularly valuable for structural analyses and research on potential subunit vaccines [1, 5, 6].

Much less is known about the structural proteins of AHSV and most of the focus has been on protective antigens and the development of subunit vaccines [15]. Until very recently a number of the genes that encode the capsid proteins had either not yet been cloned or had not been expressed. No work on the assembly of either AHSV CLPs or VLPs has been published. One of the aims in our laboratory has been to clone, characterize and express all the genes that encode AHSV structural proteins. We are interested in identifying potentially important differences between BTV and AHSV capsid proteins and to reveal common structural elements in cognate orbiviruses, that can provide us with a better understanding of how the structure of individual proteins relates to their function. This paper reports that all ten the different genes of AHSV, including the VP1, VP3 and VP6 genes have now been cloned and sequenced. We have also expressed most of these genes and were able to demonstrate that co-expression of VP7 and VP3 in insect cells results in the self-primed assembly of CLPs.

The different structural genes were all cloned by a modified version of the strategy described by Bremer et al. [2]. Using these modifications, dsRNA was not denatured before polyadenylation and full length cDNA fragments of each genome segment were isolated by gel electrophoresis. These two refinements resulted in a very large increase in the percentage of full-sized genomic clones that were obtained. All cDNA fragments were cloned into the *PstI* site of pBR322 via the G/C-tailing method and putative clones were characterized by *PstI* digestion, dot-blot hybridization and terminal sequencing. All seven structural genes, of a number of different AHSV serotypes, were cloned with this approach. The genome segments which encode VP1, VP2, VP5, VP6 and VP7 were fully sequenced, as cDNA copies. The others (coding for VP3 and VP4) were only partially sequenced, because the sequence was already known. Details about the size and proposed function of the different genes that have been cloned and sequenced either by ourselves or others is summarized in Table 1.

Sequencing of the VP1 and VP6 genes of AHSV confirmed some of the earlier observations that were made with respect to the cognate BTV genes. Genome segment 1 of AHSV-9 is 3965 nucleotides in length. Translation of the largest open reading frame yields a protein comprising 1305 amino acids with a number of motifs that show some similarity to other RNA dependent RNA polymerases. The VP6 protein on the other hand, has motifs associated with a helicase [20]. We

Table 1. AHSV genome segments and the encoded proteins

Genome segment[a]	Serotype	Segment length	Terminal sequences of coding strand (fully conserved bases in bold underlined)		5' UTR[b]	3' UTR[b]	Viral protein	Protein length	Predicted mol wt	Location[c]
		(bp)	5'	3'	(bp)	(bp)		(aa)		
1[1]	9	3965	**GUU**UAU...	...ACU**UAC**	13	34	VP1	1305	150292	IC
2[2]	3	3221	**GUU**UAA...	...ACU**UAC**	12	35	VP2	1057	123063	C
3[3]	4	2792	**GUU**AAU...	...ACU**UAC**	26	48	VP3	905	103269	OC
4[4]	4	1978	**GUU**UAU...	...CCU**UAC**	11	38	VP4	642	75826	IC
5[5]	9	1566	**GUU**UAU...	...ACA**UAC**	19	29	VP5	505	56771	C
6[6]	9	1748	**GUU**AAA...	...ACU**UAC**	35	66	NS1	548	63377	NS
7[7]	9	1167	**GUU**UAA...	...ACU**UAC**	17	103	VP7	349	37916	OC
8[8]	9	1166	**GUU**UAA...	...ACA**UAC**	22	46	NS2	365	41193	NS
9[9]	3	1169	**GUU**AAA...	...ACU**UAC**	17	42	VP6	369	38464	IC
10[10]	9	756	**GUU**UAA...	...ACU**UAC**	18	84	NS3	217	23659	NS
							NS3A	206	22481	NS

[a] The relevant Genbank accession numbers and source references are footnoted: [1] U94887 (present study); [2] U01832 [25]; [3] M94681 [7]; [4] D14402 [14]; [5] U74489 [4]; [6] U01069 [15]; [7] U90337 [13]; [8] M69090 [22]; [9] U19881 [19]; [10] D12480 [21]

[b] Untranslated region

[c] *IC* Inner core; *OC* outer core; *C* outer capsid; *NS* non-structural

have also cloned and expressed the VP4 gene of AHSV-9. Partial sequencing has indicated very high similarity to the VP4 gene of AHSV-4 [16]. The VP2 and VP5 genes of both AHSV-9 and AHSV-3 have been cloned and expressed and are currently under investigation with respect to their ability to induce a protective immune response.

Although the VP3 gene of AHSV-4 has been sequenced [10], as yet there have been no other reports on expression of the gene or any attempts to co-express VP3 and VP7 genes with the aim of producing AHSV CLPs. Genes that were expressed in this investigation were first modified by a PCR method to remove the flanking homopolymeric tails and to introduce restriction enzyme recognition sites at the flanking ends of the coding regions [13]. These modified genes were then cloned into the pBS transcription vector to produce VP3 and VP7 mRNAs which were subsequently translated in a cell-free rabbit reticulocyte system (Amersham). The [^{35}S]-methionine labelled translation products were analyzed by SDS-PAGE and autoradiography (Fig. 1a and b). Bands corresponding to the predicted size of VP3 (103 k Da) and VP7 (38 k Da) were observed. These bands were absent from mock-infected BHK cells and co-migrated with the VP3 or VP7 proteins synthesized in AHSV-9 infected BHK cells. The yield of VP7 was much higher than that of VP3.

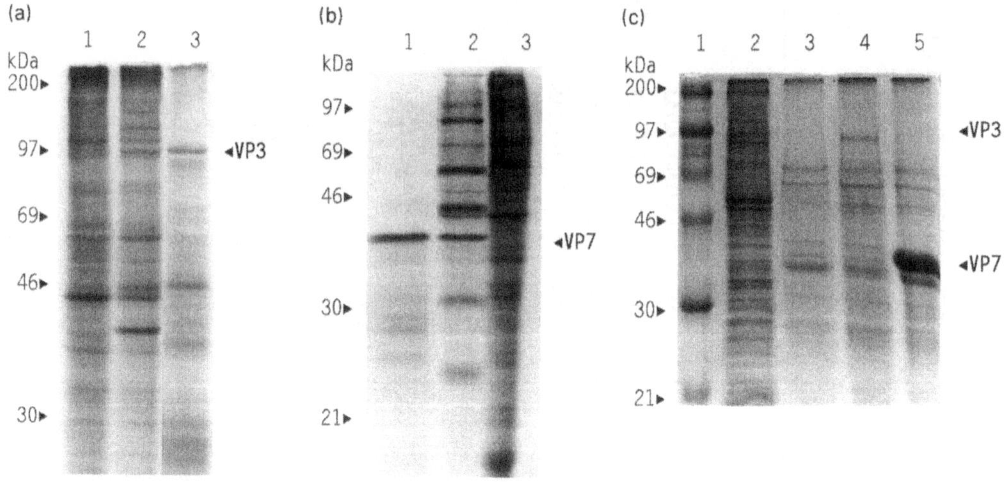

Fig. 1. a Autoradiograph of radio labelled translation products of VP3 messenger RNA, produced in a cell-free expression system. *1* Mock-infected CER cells; *2* AHSV-9 infected CER cells; *3* in vitro translation products of VP3 mRNA, transcribed from the genome segment 3 (S3) PCR/cDNA hybrid gene. **b** Autoradiograph of the radio labelled VP7 protein expressed in vitro. *1* In vitro translation products of VP7 mRNA derived form the PCR-tailored genome segment 7 (S7) cDNA; *2* AHSV-9 infected CER cells; *3* mock-infected CER cells. **c** Expression of the AHSV VP3 or VP7, in Sf9 cells by recombinant baculoviruses. Protein samples were resolved by SDS-PAGE and visualised by staining with Coommassie blue. Bands in *1* represent standard protein size markers. *S. frugiperda* cells were mock-infected (*2*), infected with wildtype baculovirus (*3*), BAC S3 Hyb (*4*), or BAC S7 PCR (*5*). The sizes of the molecular weight markers in kDa are indicated

The AHSV-9 VP3 and VP7 genes were then individually expressed in insect cells by means of recombinant baculoviruses. The BAC-to-BAC baculovirus expression system was used to construct the recombinant baculovirus DNA (bacmid) as described by Luckow et al. [12]. Sf9 cells were transfected with the composite VP3 and VP7 bacmid DNA and recombinant viral stocks were prepared by plaque selection. The recombinant viruses were used to infect insect cells and the proteins were then analyzed by SDS-PAGE. The results (Fig. 1c) indicate that proteins corresponding to the estimated sizes of VP3 or VP7 were synthesized in the respective infected cells and that they were absent in the wild-type or mock-infected Sf9 cells. The yield of VP7 was again much higher compared to that of VP3 and the protein aggregated into large crystals identical to the crystalline structures found in AHSV-4 infected cells, when observed under a light microscope [4].

In order to establish whether AHSV VP3 and VP7 were able to interact and assemble into CLPs, the two genes were co-expressed by infecting Sf9 cells with the VP3 and VP7 recombinant baculoviruses at multiplicities of 5 p.f.u./cell for each virus. VP3–VP7 complexes, with 225 (± 25) S values, were isolated at 48 h p.i. by sucrose gradient centrifugation and analyzed by conventional electron microscopy. The negative stained particles (Fig. 2a) largely resembled authentic but

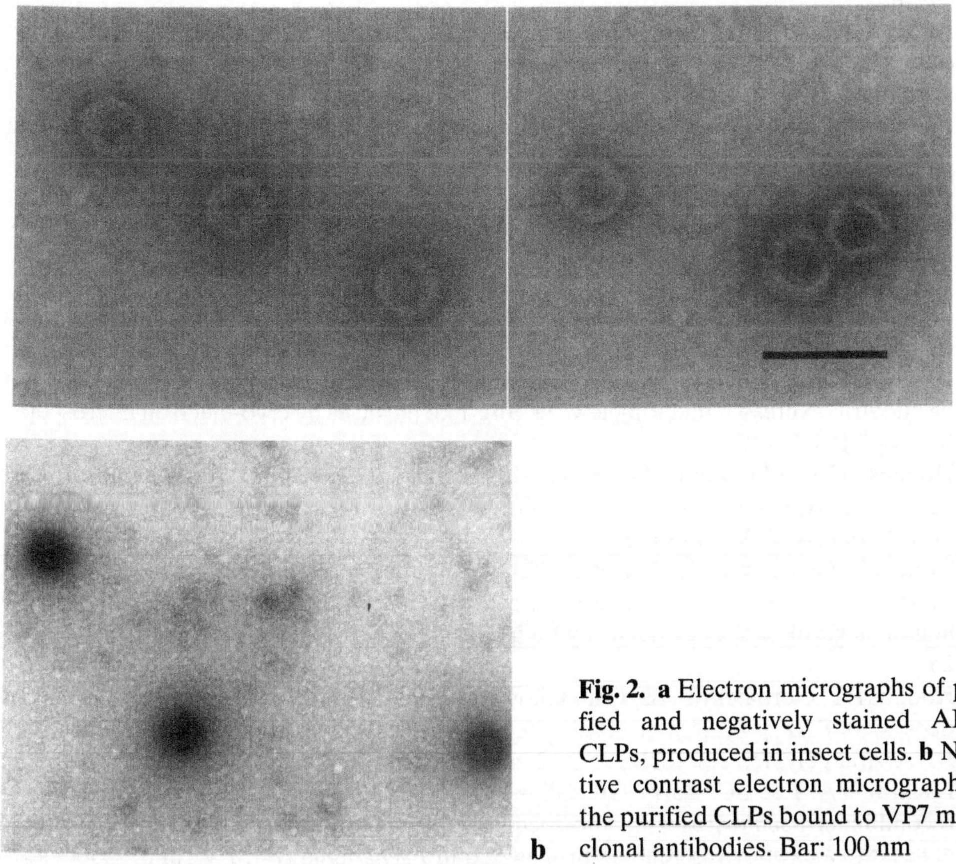

a

b

Fig. 2. a Electron micrographs of purified and negatively stained AHSV CLPs, produced in insect cells. **b** Negative contrast electron micrographs of the purified CLPs bound to VP7 monoclonal antibodies. Bar: 100 nm empty AHSV core particles. The 72 nm particles were characterized by a dark central region absent in the normal RNA containing core particles. The region of higher contrast displays icosahedral symmetry surrounded by a less electron dense layer with knobby morphological protrusions extending outwards. The identity of the CLPs was verified by immuno electron microscopy with VP7 monoclonal antibodies prepared by Van Wyngaardt et al. [21]. Electron micrographs revealed that each particle of interest was masked by a dark shadow which could be attributed to the specific interaction of monoclonal VP7 antibodies with the VP7 outer shell (Fig. 2b). A few contaminating baculovirus tubules were present which did not bind the AHSV, VP7 specific monoclonal antibodies and therefore served as negative controls.

References

1. Belyaev AS, Roy P (1993) Development of baculovirus triple and quadruple expression vectors: co-expression of three or four bluetongue virus proteins and synthesis of bluetongue virus-like particles in insect cells. Nucleic Acids Res 21: 1219–1223
2. Bremer CW, Huismans H, Van Dijk AA (1990) Characterization and cloning of the African horse sickness virus genome. J Gen Virol 71: 793–799
3. Burrage TG, Trevejo R, Stone-Marschat M, Laegreid WW (1993) Neutralizing epitopes of African horse sickness virus serotype 4 are located on VP2. Virology 196: 799–803
4. Chuma T, Le Blois H, Sanchez-Vizcaino JM, Diaz-Laviada M, Roy P (1992) Expression of the major core antigen VP7 of African horse sickness virus by a recombinant baculovirus and its use as a group-specific diagnostic agent. J Gen Virol 73: 925–931
5. French TJ, Roy P (1990) Synthesis of bluetongue virus (BTV) core like particles by a recombinant baculovirus expressing the two major structural core proteins of BTV. J Virol 64: 1530–1536
6. French TJ, Marshall JJA, Roy P (1990) Assembly of double-shelled virus like particles of bluetongue virus by simultaneous expression of four structural proteins. J Virol 64: 5695–5700
7. Grimes J, Basak AK, Roy P, Stuart DI (1995) The crystal structure of bluetongue virus VP7. Nature 373: 167–170
8. Hewat EA, Booth TF, Loudon PT, Roy P (1992) Three dimensional reconstruction of baculovirus expressed bluetongue virus core-like particles by cryo-electron microscopy. Virology 189: 10–20
9. Holmes IH (1991) Family *Reoviridae*. In: Francki RIB, Fauquet CM, Knudson DL, Brown F (eds) Classification and Nomenclature of Viruses. Springer, Wien New York, pp 186–199 (Arch Virol [Suppl] 2)
10. Iwata H, Yamagawa M, Roy P (1992) Evolutionary relationships among the gnat transmitted orbiviruses that cause African horse sickness, Bluetongue and epizootic hemorrhagic disease as evidence by their capsid protein sequences. Virology 191: 251–261
11. Koonin EV, Gorbalenya AE, Chumakov KM (1989) Tentative identification of RNA-dependent RNA polymerases of dsRNA viruses and their relationship to positive strand RNA viral polymerases. FEBS 252: 42–46
12. Luckow VA, Lee SC, Barry GF, Olins PO (1993) Efficient generation of infectious recombinant baculoviruses by site-specific transposon mediated insertion of foreign genes into a baculovirus genome propagated in *Escherichia coli*. J Virol 67: 4566–4579

13. Maree FF, Huismans H (1997) Characterization of tubular structures composed of nonstructural protein NS1 of African horse sickness virus expressed in insect cells. J Gen Virol 78: 1077–1082
14. Martinez-Torrecuadrada JL, Casal JI (1995) Identification of a linear neutralization domain in the protein VP2 of African horse sickness virus. Virology 210: 391–399
15. Martinez-Torrecuadrada JL, Díaz-Laviada M, Roy P, Sánchez C, Vela C, Sánchez-Vizcaino JM, Casal JI (1996) Full protection against African horse sickness virus serotype 4 VP2, VP5 and VP7. J Gen Virol 77: 1211–1221
16. Mizukoshi N, Sakamoto K, Iwata A, Tsuchiya Y, Ueda S, Apiwatnakorn B, Kamada M, Fukusho A (1993) The complete nucleotide sequence of African horse sickness virus serotype 4 (vaccine strain) segment 4, which encodes the minor core protein VP4. Virus Res 28: 299–306
17. Prasad BVV, Yamaguchi S, Roy P (1992) Three-dimensional structure of single-shelled bluetongue virus. J Virol 66: 2135–2142
18. Roy P (1989) Bluetongue virus genetics and genome structure. Virus Res 13: 179–206
19. Roy P (1992) Bluetongue virus proteins. J Gen Virol 73: 3051–3064
20. Turnbull PJ, Cormack SB, Huismans H (1996) Characterization of the gene encoding core protein VP6 of two African horse sickness virus serotypes. J Gen Virol 77: 1421–1423
21. Van Wyngaardt W, Du Plessis DH, Van Wyngaardt S, Verschoor JA (1992) Production and properties of monoclonal antibodies against African horse sickness virus, serotype 3. Onderstepoort J Vet Res 59: 129–133
22. Vreede FT, Huismans H (1994) Cloning, characterization and expression of the gene that encodes the major neutralization-specific antigen of African horse sickness virus serotype 3. J Gen Virol 75: 3629–3633

Authors' address: Prof. H. Huismans, Department of Genetics, University of Pretoria, Pretoria 0002, South Africa.

VP7 from African horse sickness virus serotype 9 protects mice against a lethal, heterologous serotype challenge

A. M. Wade-Evans*, **L. Pullen**, **C. Hamblin**, **R. S. O'Hara**, **J. N. Burroughs**, and **P. P. C. Mertens**

Institute for Animal Health, Pirbright Laboratory, Woking, Surrey, U.K.

Summary. An established mouse model system was used to evaluate the effectiveness of the major outer core protein VP7 of African horse sickness virus (AHSV) serotype 9 as a subunit vaccine. Balb C mice were immunised with VP7 crystals purified from AHSV infected BHK cells. In groups of mice, each of which was immunised with ≥ 1.5 μg of the protein in Freund's adjuvant, ≥ 80% of mice survived challenge with a virulent strain of a heterologous AHSV serotype (AHSV 7), that killed ≥ 80% of the mice in the uninoculated control groups. This level of protection was significantly greater than that observed in mice inoculated with equivalent amounts of either denatured VP7 (50% survival), or GST/VP7 fusion protein (50–70% survival), or which were vaccinated with AHSV 9 (40–50% survival). The VP7 protein folding, or its assembly into crystals, are thought to play some role in the effectiveness of the protective response observed. Titres of circulating antibodies against AHSV VP7 were determined by competitive ELISA but did not appear to correlate with the levels of protection observed. Passive transfer of these antibodies to syngeneic recipients also failed to protect Balb C mice from the AHSV 7 challenge. The observed protection is therefore unlikely to be due to an antibody mediated immune response.

Introduction

African horse sickness virus (AHSV) is a ten segmented dsRNA virus belonging to the genus *Orbivirus* within the family *Reoviridae*. AHSV, which is enzootic in a large area of sub-Saharan Africa, is transmitted by species of *Culicoides* and can infect all species of equidae. The virus is highly infectious but causes a non contagious disease (African horse sickness: AHS), which is particularly severe in serologically naive horses, causing mortality that can often exceed 95%. In contrast the death rates observed in mules donkeys and zebra are usually considerably lower (50%).

---

* Present address: N. I. B. S. C. Potters Bar, Hertfordshire, U.K.

9 distinct serotypes of AHSV and 24 serotypes of bluetongue virus (BTV: the prototype *Orbivirus*) are currently recognised, which cross neutralise poorly if at all in serum neutralisation assays [3, 8, 10]. None of the vaccines made from a single virus serotype (that are currently available) would offer effective protection against a heterologous serotype challenge [15] and there is frequently a requirement in the field to protect animals against multiple virus types. Although AHSV contains fewer serotypes and therefore poses less of a problem than BTV, the serological complexity of both virus species increases the complexity of vaccination and vaccine design. Historically vaccination against BTV has depended on the use of multiple inoculations containing several BTV serotypes (24 serotypes in total). The AHSV vaccine normally used in South Africa incorporates attenuated viruses of eight serotypes [21], which are given as two doses of 4 serotypes each. However, the use of live, attenuated, multivalent vaccines introduces several potentially important problems [9, 19]. Even limited replication of these vaccine viruses is likely to result in production of antibodies to all of the virus structural and non-structural proteins, making it difficult to distinguish between vaccinated and infected animals by serological methods [1]. Reassortment between serotypes, or between vaccine and wild type strains of virus can readily occur, resulting in the generation of progeny virus strains which may have novel serological properties and could have modified or even increased virulence characteristics [5, 13, 14, 16–18]. There is therefore a need to develop a safe, economical and effective vaccine to protect susceptible equidae against multiple serotypes of AHSV.

It has previously been reported that in the absence of neutralising antibodies, cell mediated immunity can play an important part in development of a protective immune response to orbiviruses [11]. Based on these findings and reports that VP7 of BTV (the outer core protein and major serogroup (virus species) specific antigen) contains immunodominant, serotype cross-reactive T-cell epitopes [2], it was decided to investigate the protective potential of AHSV VP7 as a subunit vaccine.

Material and methods

Synthesis of bacterial GST-VP7 fusion protein

The glutathione-S-transferase / VP7 fusion protein (GST/VP7) was expressed in *E. coli* and the bacterial inclusion bodies purified as described previously [22].

Purification of VP7 crystals

Crystals of VP7, synthesised in BHK cells infected with AHSV serotype 9 (AHSV 9), were purified as previously described [4] and were heat treated in 0.1 M tris/HCl pH 8.0 at 56°C for 1 h prior to use. VP7 from crystals was denatured by boiling for 30 min prior to inoculation.

Determination of the virulence of AHSV 4, 7 and 9 and their suitability for use as challenge virus (in adult mice)

Eight week old adult female Balb C mice (five mice / group) were inoculated intravenously or intraperitoneally with: 50 µl, 100 µl, 175 µl, or 200 µl of AHSV 4, 7 or 9 at concentrations of 1.4, 1.2 or 2.0 x 10^7 $TCID_{50}$/ml respectively [17]. The mice were observed for a period of 21 days post-infection. Any individuals showing clinical signs, including shivering, fever, paralysis or wasting, were immediately euthanized.

Stock viruses

AHSV 7 (isolate Karen from 29/6/1960) was supplied as a high titre mouse brain (MB) passage, which was plaque purified three times in BHK cells and grown for a further three passages in BHK cells (BHK 3). Stock tissue culture preparations (unpurified) of AHSV 7, AHSV 9 (isolate 90/61 from 2/5/1973, MB3, BHK6) and AHSV 4 (isolate Ricarda from 21/4/1989, MB1, BHK 6) were all titrated in BHK cells. Virus titres were calculated using the method of Kärber [12].

Immunisation of mice

Twelve groups of ten female Balb C mice (eight weeks old) were inoculated subcutaneously with the following amounts of protein.

Group 1: control group (no protein inoculation).
Group 2: 3 doses of 1.5 µg (100 µl) each of purified GST: on day 1 in Freunds complete adjuvant (FCA); on day 21 in Freunds incomplete adjuvant (FIA); on day 42 in FIA.
Group 3: 3 x 1.5 µg (100 µl) each of purified GST / VP7: on day 1 in FCA; on days 21 and 42 in FIA.
Group 4: 3 x 1.5 µg (100 µl) each of purified and denatured VP7 crystals: on day 1 in FCA; on days 21 and 42 in FIA.
Group 5: 3 x 1.5 µg (100 µl) each of purified VP7 crystals: on day 1 in FCA; on days 21 and 42 in FIA.
Group 6: 2 x 1.5 µg (100 µl) each of purified VP7 crystals: on day 1 in FCA; on day 21 in FIA.
Group 7: 1 x 1.5 µg (100 µl) of purified VP7 crystals: on day 1 in FCA.
Group 8: 1 x 1.0 µg (100 µl) of purified VP7 crystals: on day 1 in FCA.
Group 9: 1 x 0.5 µg (100 µl) of purified VP7 crystals: on day 1 in FCA.
Group 10: 1 x 0.25 µg (100 µl) of purified VP7 crystals: on day 1 in FCA.
Group 11: 1 x 1.5 µg (100 µl) of purified VP7 crystals: on day 1, no adjuvant.
Group 12: 2.0 x 10^7 $TCID_{50}$ of AHSV 9 (200 µl) on day 1, no adjuvant.

Test bleeds were taken from the tail vein of all the mice prior to each inoculation. Mice were challenged with AHSV 7 on day 49 (as described below). All studies were repeated in triplicate.

AHSV 7 challenge

On day 49 all of the mice were challenged by intravenous inoculation with 200 µl of AHSV 7 (1.8 x 10^7 $TCID_{50}$/ml). The mice were subsequently examined twice each day for any clinical signs, over a period of 21 days post challenge (pc).

Anti VP7 antibody titres

Titres of antibodies against AHSV VP7 were assayed by competitive ELISA, as previously described [6].

Viraemia and viral antigen titres in murine tissues post challenge

Thirty female Balb C mice were given a single inoculation of 1.5 μg (100 μl) of purified VP7 crystals in 50% FCA on day 1. Another thirty female mice were kept as a control group. All sixty mice were challenged with AHSV 7 as described above on day 49 post inoculation. Three mice from each group were sacrificed on days 1, 2, 3, 4, 5, 6, 8, 10, 12 and 14. Heparinised blood samples were collected and washed three times in phosphate buffered saline, then titrated on BHK cell monolayers. Heart, liver, spleen, brain, inguinal lymph node and lungs were collected from each mouse. Homogenates of each organ were assayed by sandwich ELISA for the presence of antigen [7].

Passive antibody transfer

A group of ten female Balb C mice were inoculated on day 1, 21, and 42 with 1.5 μg of purified VP7 crystals, as described previously for group 5. A control group of ten mice received PBS in 50% FCA. The mice were sacrificed on day 49 and the serum collected. The anti-VP7 antibody titre of each serum sample was assayed by competitive ELISA [6]. The antibody titres from the immunised mice were all greater than 2.9 ($\log_{10}$) and these sera were subsequently pooled. The pooled serum was retitrated and 150 μl was injected intravenously into each of ten syngeneic Balb C recipient mice. An additional ten mice received 150 μl of VP7-antibody negative serum. A third group of ten mice were left untreated. Each of the thirty mice was bled after a further 5 h, in order to measure the levels of circulating antibodies to VP7. The mice were then challenged with AHSV 7 and observed for a further 21 days.

Results

Intravenous injection of AHSV 7 into the tail vein of adult Balb C mice, was found to be a more effective route of inoculation, in terms of inducing clinical signs than the intraperitoneal route (Table 1). In the experiments described here, the virus isolate that induced the highest mortality rate in the Balb C mice, was AHSV 7, which reproducibly resulted in ≥ 80% mortality. A stock AHSV 7 virus preparation, with a final titre of 1.8×10^7 $TCID_{50}$/ml, was used in all of the challenge experiments described.

The percentage mortality of mice in the vaccinated and control groups, caused by intravenous challenge with a normally lethal dose of AHSV 7, are summarised

Table 1. Titration of three AHSV isolates to determine their pathogenicity using either intravenous or intraperitoneal routes of inoculation. Percentage survival rates

Virus serotype	50 μl i.v.	100 μl i.v.	175 μl i.v.	200 μl i.v.
AHSV 4 1.4×10^7 $TCID_{50}$/ml	100%	100%	100%	100%
AHSV 7 1.2×10^7 $TCID_{50}$/ml	80%	60%	60%	20%
AHSV 9 2.0×10^7 $TCID_{50}$/ml	100%	100%	100%	100%
Virus serotype	**50 μl i.p.**	**100 μl i.p.**	**175 μl i.p.**	**200 μl i.p.**
AHSV 4 1.4×10^7 $TCID_{50}$/ml	100%	100%	100%	100%
AHSV 7 1.2×10^7 $TCID_{50}$/ml	100%	100%	100%	100%
AHSV 9 2.0×10^7 $TCID_{50}$/ml	100%	100%	100%	100%

in Table 2. Between 80 and 100% of the mice which had received ≥ 1.5 μg of purified VP7 crystals in the presence of adjuvant, survived the challenge. Even in the absence of adjuvants, 70% of the mice which had received 1.5 μg also survived. However, in the control groups (Groups 1 and 2) only 0 to 20% survived the AHSV 7 challenge. In the groups which had received a single inoculation, 90–100% of those which had received 1.5 μg of purified VP7 crystals survived (Group 7); 60% of the mice that received 1.0 μg survived (group 8); but only 0–10% survived in the groups of mice which had received 0.5 μg or 0.25 μg of VP7 (Groups 9 and 10). There was no evidence of any protection in these last two groups when they were compared to the controls (Groups 1 and 2: 0 to 20% and 20 to 30% survival) (0.5 μg, P = 0.492 and 0.25 μg, P = 1.0).

In groups of mice that had been inoculated three times with 1.5 μg of the bacterial GST/ VP7 protein, 50 to 70% survived the subsequent AHSV 7 challenge. Relatively similar survival levels were also observed in those groups that had received either denatured VP7 (Group 4: 50% survival) or which were vaccinated with AHSV 9 prior to challenge with AHSV 7 (Group 12: 40–50% survival).

Analyses of the serum samples collected from test bleeds of VP7 innoculated mice are shown in Fig. 1. Not all of the protected mice had high titres of antibodies to VP7 prior to challenge. In addition, some of the individual mice that became sick or died, had reasonably high levels of anti-VP7 antibodies. The viraemia observed in unvaccinated mice (1.66–2.66 $TCID_{50}$/ml) persisted from 3 to 8 days pc. In contrast, in the mice immunised with VP7 the duration of viraemia was shorter (4 to 6 days pc) (Fig. 2). The distribution of virus, in tissues from VP7 vaccinated and the unvaccinated control mice, was similar after the challenge, with antigen detected only in the heart and brain tissues (data not shown).

Table 2. Percentage survival of innoculated mice after challenge with AHSV 7

Group	Inoculum	% Survival (range)	Days post challenge, over which death occurred
1	zero protein controls	0–20%	6 to 14
2	GST (3 x 1.5 μg)	20–30%	7 to 12
3	GST/VP7 (3 x 1.5 μg)	70–50%	6 to 14
4	denatured VP7 (3 x 1.5 μg)	50%	6 to 15
5	VP7 crystals (3 x 1.5 μg)	80–90%	7 to 14
6	VP7 crystals (2 x 1.5 μg)	90%	11
7	VP7 crystals (1 x 1.5 μg)	90–100%	10
8	VP7 crystals (1 x 1.0 μg)	60%	8 to 13
9	VP7 crystals (1 x 0.5 μg)	0%	7 to 13
10	VP7 crystals (1 x 0.25 μg)	0–10%	7 to 14
11	VP7 crystals (1 x 1.5 μg, NA[a])	70%	4 to 14
12	AHSV 9 (2.0 x 10^7 $TCID_{50}$)	40–50%	8 to 13

[a] *NA* No adjuvant

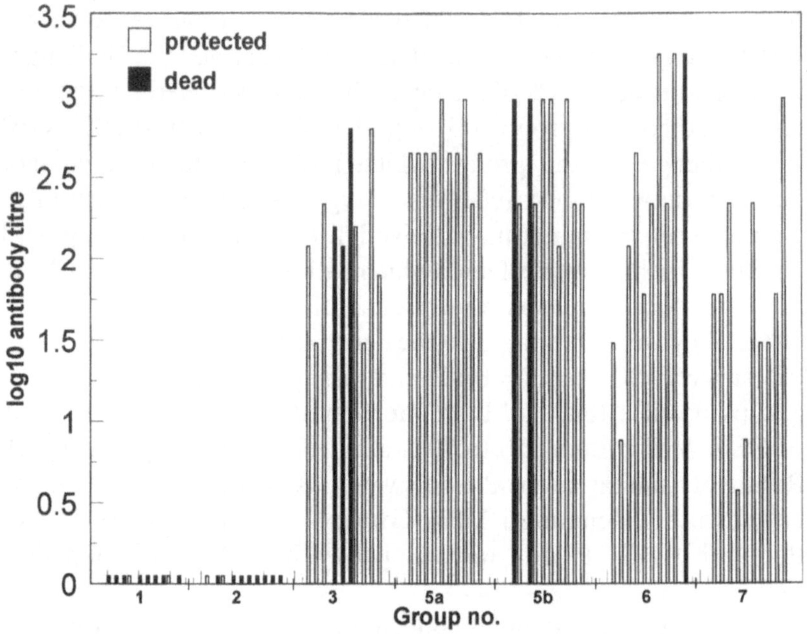

Fig. 1. Anti-AHSV VP7 antibody titres in serum samples taken from mice, prior to heterologous serotype challenge with AHSV 7. Groups of mice were: uninoculated (*1*); or inoculated with GST (*2*); with GST/VP7 (*3*); with denatured VP7 (*4*); or with varying amounts (4.5 μg, 3.0 μg, or 1.5 μg) of purified AHSV 9 VP7 crystals (*5–7*), as described in detail in Material and methods and Table 2. Serum samples were taken from individual mice on days 1, 21 and 42 and analysed by competitive ELISA [6], to determine the titre of circulating antibodies to AHSV VP7 prior to challenge with AHSV 7. The results given are for the test bleeds taken on day 42. Groups 5a and 5b show the results recorded from two separate experiments with identical vaccination schedules

Discussion

The experimental data presented demonstrates that in a mouse model system AHSV 9 VP7 is effective, as a subunit vaccine, against challenge with a heterologous AHSV 7. Up to 90% protection was achieved, even from a single inoculation of 1.5 μg of purified VP7 crystals. The results also indicate that the duration of viraemia but not necessarily its level could be reduced by immunisation with VP7. However, VP7 crystals which had been denatured at 100°C, or a bacterial expressed GST/VP7 fusion protein, gave significantly less protection when administered in directly comparable amounts. In both cases the reduced level of protection was comparable to that produced by vaccination with live AHSV 9 virus. The conformation and possibly the assembly of the VP7 into crystals therefore appears to be important in the mechanism of protection against the heterologous serotype challenge.

Cells infected with AHSV do not provide an entirely satisfactory source of VP7 crystals for vaccination purposes, since they inevitably present the risk of contamination with infectious virus. It has been reported that AHSV 4 VP7 also

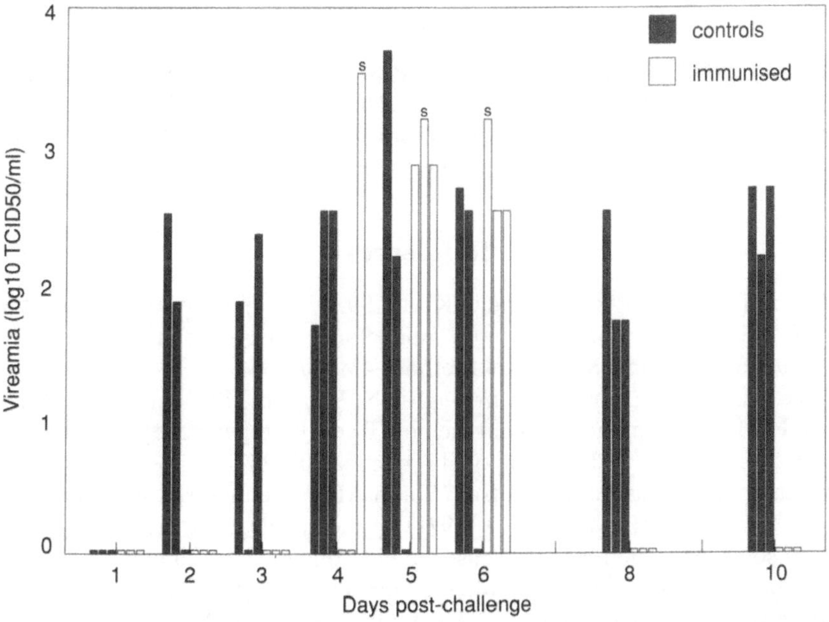

Fig. 2. The levels of viraemia in immunised and control mice following challenge with AHSV 7. Two groups of thirty mice, which were either inoculated with 1.5 μg of AHSV 9 VP7 crystals, or kept as uninoculated controls, were challenged with AHSV 7 at 49 days post inoculation. Three mice were selected at random from each group on days 1, 2, 3, 4, 5, 6, 8 and 10 post challenge. The amount of infectious virus in the blood samples was determined by titration on BHK cells. The shaded columns represent the control mice, the open columns represent the vaccinated animals. "S" above a column indicates that the mouse was showing clinical signs of infection at the time of sampling

forms crystalline structures when expressed in insect cell cultures by a recombinant baculovirus containing a cDNA copy of AHSV 4 genome segment 7 [20].

Passive transfer of antibodies from immunised Balb C mice failed to protect syngeneic recipients from AHSV 7 challenge (data not shown). This indicates that an antibody response is unlikely to represent the primary mechanism involved in the protection induced by vaccination with VP7 crystals. It is possible that immunisation with VP7 induces a protective T-cell response but further experiments are required to confirm this and to determine which T-cell subsets are involved.

This is the first study with any of the orbiviruses to demonstrate that a high level of protection can be generated against a heterologous virus serotype, using purified VP7 as a subunit vaçcine. It may therefore make a significant contribution to the design of subunit vaccines for these viruses in the future. There are clearly limits on the extrapolation and application of these data to horses, since all of the results collected were generated using a mouse model system. However, the results indicate the potential value of even a single inoculation of VP7 crystals as a prophylactic vaccine, which may be effective against all nine serotypes of AHSV. Future studies will include a trial of VP7 as a vaccine in horses (the most severely affected and economically important host for AHSV).

Acknowledgements

The authors would like to thank S. Rathbone, P. Warder, and S. Wilsden for the day to day care of the mice and Andrew Walton for preparation of the AHSV 9 VP7 crystals. This work was jointly funded by the Ministry of Agriculture Fisheries and Food UK and the Commission of the European Community (grant number 8001-CT91-0211).

References

1. Anderson J, Mertens PPC, Herniman KAJ (1993) A competitive ELISA for the detection of anti-tubule antibodies using a monoclonal antibody against bluetongue virus non-structural protein NS1. J Virol Methods 43: 167–176
2. Angove HL, Wade-Evans AM, Burroughs JN, Mertens PPC, Parkhouse RME, Takamatsu H (1998) Immunodominant helper T-cell epitopes recognised by sheep are arranged as a cluster in the amino-acid sequence 202–253 of outer capsid protein VP2 from bluetongue virus serotype 1 from South Africa (manuscript in preparation)
3. Borden EC, Shope RE, Murphy FA (1971) Physiochemical and morphological relationships of some arthropod bourne viruses to bluetongue virus – a new taxonomic group. Physiochemical and serological studies. J Gen Virol 13: 261–271
4. Burroughs JN, O'Hara RS, Smale CJ, Hamblin C, Walton A, Armstrong R, Mertens PPC (1994) Purification and properties of virus particles, infectious sub viral particles, cores and VP7 crystals of African Horse sickness virus serotype 9. J Gen Virol 75: 1849–1857
5. Cowley JA, Gorman BM (1989) Cross neutralisation of genetic reassortants of bluetongue virus serotypes 20 and 21. Vet Microbiol 19: 37–51
6. Hamblin C, Graham SC, Anderson EC, Crowther JR (1990) A competitive ELISA for the detection of group specific antibodies to African horse sickness virus. Epidemiol Infect 104: 303–312
7. Hamblin C, Mertens PPC, Mellor PS, Burroughs JN, Crowther JR (1991) A serotype specific enzyme-linked immunosorbent assay (ELISA) for the detection and identification of African horse sickness viruses. J Virol Methods 31: 285–292
8. Holmes IH, Boccardo G, Estes MK, Furuichi MK, Hoshino Y, Joklik WK, McCrae M, Mertens PPC, Milne RG, Samal KSK, Shikata E, Winton JR, Uyeda I (1995) Family *Reoviridae.* In: Murphy FA, Fauquet CM, Bishop DHL, Ghabrial SA, Jarvis AW, Martelli GP, Mayo MA, Summers MD (eds) Virus Taxonomy. Classification and Nomenclature of Viruses. Sixth Report of the International Committee on Taxonomy of Viruses. Springer, Wien New York, pp 208–237 (Arch Virol [Suppl] 10)
9. House JA (1998) Future international management of African horse sickness vaccines. In: Mellor PS, Baylis M, Hamblin C, Calisher CH, Mertens PPC (eds) African horse sickness. Springer, Wien New York (Arch Virol [Suppl] 14)
10. Howell PG (1962) The isolation and identification of further antigenic types of African horse sickness virus. Onderstepoort J Vet Res 29: 139–149
11. Jeggo MH, Wardley RC, Brownlie J (1984) A study of the role of cell-mediated immunity in bluetongue virus infection of sheep, using cellular adoptive transfer techniques. Immunology 52: 403–410
12. Kärber G (1931) Beitrag zur kollektiven Behandlung pharmakologischer Reihenversuche. Arch Exp Pathol Pharmakol 162: 480–483
13. Martin LA, Meyer AJ, O'Hara RS, Fu H, Knowles NJ, Mertens PPC (1998) Phylogenetic analysis of African horse sickness virus genome segment 10: sequence variation involved in virus release and virulence characteristics. In: Mellor PS, Baylis M, Hamblin C, Calisher CH, Mertens PPC (eds) African horse sickness. Springer, Wien New York (Arch Virol [Suppl] 14)

14. Mertens PPC, Pedley S, Cowley J, Burroughs JN, Corteyn AH, Jeggo MH, Jennings AM, Gorman BM (1989) Analysis of the roles of bluetongue virus outer capsid proteins VP2 and VP5 in determination of virus serotype. Virology 170: 561–565
15. McIntosh BM (1958) Immunological types of horse sickness virus and their significance in immunisation. Onderstepoort J Vet Res 27: 465–538
16. Nuttal PA, Jacobs SC, Jones LD, Carey D, Moss SR (1992) Enhanced neurovirulence of tick bourne orbiviruses resulting from genetic modulation. Virology 187: 407–412
17. O'Hara RS (1994) Identification of the genome segments and proteins controlling the virulence of African horse sickness virus. Thesis submitted in part for the degree of doctor of Philosophy, University of Reading
18. O'Hara RS, Meyer AJ, Pullen L, Martin LA, Mertens PPC (1998) Development of a mouse model system and identification of the individual genome segments of African horse sickness virus serotypes 3 and 8 involved in determination of virulence. In: Mellor PS, Baylis M, Hamblin C, Calisher CH, Mertens PPC (eds) African horse sickness. Springer, Wien New York, pp 259–279 (Arch Virol [Suppl] 14)
19. Roy P (1995) Towards the control of African horse sickness by recombinant technology. In: Nakajima H, Plowright W (eds) Equine infectious diseases VII (Proceedings of the Seventh International Conference). R&W Publications, Newmarket, pp 65–70
20. Roy P, Hirasawa T, Fernandez M, Blinov VM, Sanchez-Vizcaino JM (1991) The complete sequence of the group-specific antigen, VP7, of African horse sickness disease virus serotype 4 reveals a close relationship to bluetongue virus. J Gen Virol 72: 1237–1241
21. Taylor MB, van der Meyden CH, Erasmus BJ, Reid RR, Labuschagne JH, Dreyer L, Prozesky OW (1992) Encephalitis and chorioretinitis associated with neurotropic African horse sickness virus infection in laboratory workers. Part IV. Experimental infection of the veruet monkey *Cercopithecus pygerythrus*. S Afr Med J 81: 462–467
22. Wade-Evans AM, Woolhouse T, O'Hara R, Hamblin C(1993) The use of African horse sickness virus VP7 antigen, synthesised in bacteria, and anti-VP7 monoclonal antibodies in a competitive ELISA. J Virol Methods 45: 179–188

Authors' address: Dr. A. M. Wade-Evans, N.I.B.S.C., Blanche Lane, South Mimms, Potters Bar, Hertfordshire, EN63QG, U.K.

Molecular epidemiology of African horse sickness virus based on analyses and comparisons of genome segments 7 and 10

S. Zientara[1], C. Sailleau[1], E. Plateau[1], S. Moulay[1], P. P. C. Mertens[2], and C. Crucière[1]

[1] Centre National d'Etudes Vétérinaires et Alimentaires, Laboratoire Central de Recherches Vétérinaires, Maisons Alfort, France
[2] The Institute for Animal Health, Pirbright Laboratory, Pirbright, Woking, U.K.

Summary. This paper describes a method to rapidly identify African horse sickness virus (AHSV), using a single tube reverse transcription polymerase chain reaction (PCR). This method was used to amplify cDNA copies of genome segments 7 and 10 from several different AHSV strains, of different serotypes, which were then analysed by sequencing and/or endonuclease digestion. AHSV VP7 (encoded by genome segment 7) is one of the two major capsid proteins of the inner capsid layer, forming the outer surface of the core particle. VP7 is highly conserved and is the major serogroup specific antigen common to all nine AHSV serotypes. Digestion of the 1179 bp cDNA with restriction enzymes, allowed differentiation of several strains of different serotypes and identified six distinct groups containing AHSV-1, 3, 6 and 8; AHSV-2; AHSV-4; AHSV-5; AHSV-7; and AHSV-9. Differences were detected between wild type viruses and vaccine strains that had been attenuated by multiple passage in suckling mouse brain or in tissue cultures.

RFLP analysis was also used to study variation the 758 bp cDNA copies of AHSV genome segment 10, which encodes the two small non-structural membrane proteins NS3 and NS3a. In this way it was possible to distinguish each of the strains tested, except AHSV 4 (USDA) and AHSV 9 (USDA). However, these isolates could be distinguished by RFLP analysis of genome segment 7 cDNA.

Using sequence analysis of genome segment 10 we were able to classify the virus isolates into three groups: AHSV-1, 2 and 8; AHSV-3 and 7; AHSV 4, 5, 6 and 9. These studies confirmed that the virus which first appeared in central Spain in July 1987, subsequently spread into northern Morocco in October 1989.

Introduction

African horse sickness (AHS) is an infectious disease of *Equidae* associated with high morbidity and mortality. AHS is caused by African horse sickness virus

(AHSV), which is a member of the genus *Orbivirus*, within the family *Reoviridae* [3, 12, 32]. The virus is transmitted by biting midges of *Culicoides* vector species, viz. *C. imicola* [7], although it may also be mechanically transmitted by other biting insects, e.g. mosquitoes. AHS occurs mainly in southern, eastern and central Africa, although it has also occurred in the middle east, parts of Asia and recently in Spain, Portugal and Morocco.

AHSV was introduced into Spain in July 1987 (to 1990), by importation from Namibia of infected zebras (in which clinical signs of AHSV infection are less apparent than in horses). AHS which was caused by AHSV 4, spread to Portugal in 1989 and to Morocco (1989–1991).

There are seven AHSV structural proteins, which surround the ten dsRNA molecules of the AHSV genome [1, 22] forming a double shelled virus particle. The outer capsid, which is involved in virus neutralisation by antibodies, is composed of two major proteins (VP2 and VP5) that are responsible for antigenic variability within the virus serogroup (the AHSV species) [19]. The inner capsid, or core particle, is composed of two additional major proteins (VP3 and VP7) and three minor proteins (VP1, VP4 and VP6). VP3 and VP7 are highly conserved among the nine serotypes of AHSV [2, 21, 23]. At least three non structural proteins (NS1, NS2 and NS3/NS3a) have also been identified in AHSV infected cells [5, 11, 13, 30]. NS3 and NS3a are encoded by genome segment 10 [30].

The classical methods of direct diagnosis of African horse sickness (AHS), are intra cerebral inoculation of suckling mice, or infection of cell cultures (BHK 21 or Vero cells). Serological diagnosis was based on the complement fixation test [16], although recently ELISA for detection of both AHSV group specific antigens and antibodies have been developed and are now also in routine use [9, 10, 34]. Virus serotype is currently determined by serum neutralisation assays [23]. However these assays are time consuming and the rapid course of the disease often results in death of horses before all of the diagnostic tests and identification of the serotype can be carried out. In order to improve AHS diagnosis, molecular tools have been developed for the identification of AHSV (molecular probes, reverse transcriptase-PCR) [2, 19, 20, 26, 29, 31, 37].

This paper describes the analysis of restriction fragment length polymorphism (RFLP) in cDNA of genome segments 7 and 10 (encoding VP7 and NS3/3a respectively). These cDNA were derived from different strains of AHSV of different serotypes. Vaccine and wild type strains were also compared. Nucleotide sequence data was also generated for segment 10 of two strains, a virulent strain (AHSV 4 (M)) and a vaccine strain (AHSV 4 (USDA)).

Materials and methods

Viruses and cells

Vaccine strains of the nine AHSV serotypes (designated as USDA strains: Table 1) were kindly provided by Dr. J. House (US Department of Agriculture [USDA], Foreign Animal Disease Diagnostic Laboratory, Greenport, New York. USA). Strains of the different serotypes were also obtained from South Africa (Dr B. Erasmus, Onderstepoort Veterinary

Table 1. Year of isolation, origin and passage history of AHSV isolates used in RT-PCR

Virus isolate	Date of isolation	Origin	Viral strain	Passage history (passage number)
AHSV 1 (USDA)	1968	South Africa	vaccine	sm p100[a]
AHSV 2 (USDA)	1968	South Africa	vaccine	sm p100
AHSV 3 (USDA)	1968	South Africa	vaccine	sm p100
AHSV 4 (USDA)	1968	South Africa	vaccine	sm p100
AHSV 4 (SA)	1968	South Africa	vaccine	Vero 10
AHSV 4 (Sp)	1990	Spain	wild type	Vero p1
AHSV 4 (M)	1990	Morocco	wild type	Vero p1
AHSV 5 (USDA)	1968	South Africa	vaccine	sm p100
AHSV 6 (USDA)	1968	South Africa	vaccine	sm p100
AHSV 7 (USDA)	1968	South Africa	vaccine	sm p100
AHSV 8 (USDA)	1968	Iran	vaccine	sm p100
AHSV 9 (USDA)	1968	Iran	vaccine	sm p100
AHSV 9 (SA)	1968	South Africa	vaccine	Vero p10

[a] *sm* Suckling mice

Research Institute, Onderstepoort, South Africa). The isolates of AHSV 8 and 9, designated as USDA strains, were isolated in Iran (Table 1). The USDA AHSV strains were prepared in 1968 by passaging in suckling mouse using intracranial inoculation. Commercially available vaccine strains of AHSV 4 and AHSV 9 (Onderstepoort Veterinary Research Institute, Onderstepoort, South Africa) were also used in this study (Table 1). The Spanish strain of AHSV 4 (Sp) was kindly provided by Dr H. Hooghuis from the European AHSV Reference Laboratory, Algete (Madrid). The Moroccan strain of AHSV 4 (M) was isolated from a spleen sample from an infected horse, by direct inoculation of Vero cell cultures (this sample was provided by Dr. El-Harrack, Biopharma Laboratory, Rabat).

All of the AHSV strains were propagated in Vero cells (American Type Culture Collection (ATCC) reference number, CCL 81), grown in 175 cm^2 flasks and RPMI 1640 growth medium, supplemented with 8% foetal calf serum, 100 μg/ml streptomycin and 100 UI/ml penicillin. The infected monolayers showed cytopathic effects 48–96 h later, depending on the virus strain. The 50% end point titre was determined using the methods of Reed and Muench [23] and varied from 10^5–10^7 $TCID_{50}$/ml.

Nucleic acid sample preparation

Total RNA was extracted from infected cell cultures or clinical samples, using the acid guanidinium thiocyanate-phenol-chloroform, single extraction method of Chomczynski and Sacchi [4]. Cell debris was remove by low speed centrifugation (10 min at 500 g). Material in the supernatants was then pelleted by ultracentrifugation at 100000 **g**, for 3 h at 4°C (Beckman SW28 rotor). The pellets were resuspended in 400 μl of solution D (4 M guanidinium thiocyanate, 25 mM sodium citrate (pH 7.0), 0.5% sarcosyl, 0.1% β mercaptoethanol) in an Eppendorf tube. 50 μl of 2 M sodium acetate (pH 4.0), 500 μl of water saturated phenol and 100 μl of chloroform-isoamyl alcohol (24:1), were added. The homogenate was mixed by inversion after the addition of each reagent and finally shaken vigourously for 10 sec, cooled on ice for 15 min and then centrifuged at 13 000 rpm for 10 min (Biofuge A,

Heraeus). 20 UI of RNasin (40 U/μl, Boehringer Mannheim) was added to the aqueous phase, which contained the RNA. The RNA was precipitated with isopropanol at –20°C overnight. The RNA was pelleted at 13000 rpm for 10 min at 4°C, and resuspended in 1.0 ml of 70% ethanol, pelleted, vacuum dried and dissolved in 10 μl of diethylpyrocarbonate (DEPC) treated water. The same methods were used to prepare the RNA from uninfected tissue culture cells, in order to provide negative controls for RT-PCR assays.

Primers: genome segment 7

Two primers were selected from the published sequence of AHSV 4 segment 7 [24] (Table 2). Using these primers in the RT PCR, allowed the amplification of the whole of genome segment 7 of AHSV (1179 bp), which includes the open reading frame (ORF) for VP7.

Primers: genome segment 10

Primers were selected from the published sequence of AHSV 3 segment 10 [30] (Table 2). AHSV 3 genome segment 10 is 758 bp in length, with a dsRNA Mr of 45 x 10^3 Dalton [30].

Reverse transcriptase and polymerase chain reaction

2.5 μl of the resuspended RNA pellet (containing the viral dsRNA) was denatured by adding an equal volume of 20 mM methyl mercuric hydroxide (ten minutes at room temperature), prior to being used as a template for cDNA synthesis. The methyl mercuric hydroxide was reduced by addition of 1.0 μl of 0.7 M β mercaptoethanol (5 min at room temperature), followed by addition of 2.0 μl of RNasin (40 UI/μl) [33]. All of the components required for both the reverse transcriptase and PCR were combined in a single reaction in a 0.5 ml Eppendorf tube. Both reactions were then run one after the other, in a single uninterrupted thermal cycling programme, as previously described by Sellner et al. [28]. 4.0 μl of the denatured RNA solution was added to 100 μl of 10 mM tris pH 8.8, 50 mM KCl, 200 mM of each dNTP, 1.5 mM $MgCl_2$, 0.1 mg/ml of gelatine, containing 10 pmol of each primer, 10 U of avian myeloblastosis virus reverse transcriptase (AMV-RT-Appligene, Illkirch, France) and 7 U of *Taq* DNA polymerase (Boerhinger Mannheim). The RT-*Taq* polymerase ratio was less than 3:2 as recommended by Sellner et al. [27]. The tubes were over laid with 100 μl of paraffin oil and incubated in a thermal cycler PTC 100/60 (Prolabo), according to the following protocol: 1 h at 37°C, 5 min at 95°C; 40 cycles of 55°C for 1 min, 70°C for 2 min, and 95°C for 1 min; then a terminal extension step at 70°C for 8 min.

Table 2. Primers used for the amplification of AHSV genome segments 7 and 10 in RT PCR

Primer	Sequence (5'–3')	Location (bp)
Segment 7 (1179 bp in length)		
P1 (forward)	GTTAAAATTCGGTTAGGATG	1–20
P2 (reverse)	GTAAGTGTATTCGGTATTGA	1179–1159
Segment 10 (758 bp in length (AHSV 3))		
P1 (forward)	GTTTAAATTATCCCTTGTCA	1–21
P2 (reverse)	GTAAGTCGTTATCCCGGCTC	758–738

Table 3. Restriction enzymes used to analyse the PCR amplified cDNA of AHSV 4 (Sp) segment 7, or AHSV 3 segment 10, with a summary of restriction sites and cleavage products

AHSV 4 segment 7			
Restriction enzyme (cleavage site)		Number of predicted cleavage sites in AHSV 4 segment 7 cDNA	Size of cleavage products (bp)
Asn I	(ATTAAT)	3	432, 348, 273, 126
BamH I	(GGATCC)	1	1087, 92
Hinf I	(GANTC)	1	1030, 149
Hph I	(GGTGA)	3	444, 397, 258, 80
Pvu II	(CAGCTG)	0	uncleaved
Sac I	(GAGCTC)	2	759, 292, 128
Sph I	(GCATGC)	1	1122, 57
Cfo I	(GCGC)	5	511, 243, 158, 150, 97, 20
AHSV 3 segment 10			
Restriction enzyme (cleavage site)		Number of predicted cleavage sites in AHSV 3 segment 7 cDNA	Size of cleavage products (bp)
Hinf I	(GANTC)	4	332, 168, 127, 109, 22
Pvu II	(CAGCTG)	0	uncleaved
Sph I	(GCATGC)	0	uncleaved
Hph I	(GGTGA)	1	618, 140
Cfo I	(GCGC)	0	uncleaved
Rsa I	(GTAC)	3	315, 254, 109, 80

Analysis of restriction fragment length polymorphism (RFLP)

The sequences of AHSV 4 segment 7 cDNA [24] and of AHSV 3 segment 10 cDNA [30] were analysed for restriction sites using the "IntelliGenetics PC/Gene Restri" software. Eight enzymes were used; *Asn* I ; *BamH* I; *Hinf* I; *Hph* I; *Pvu* II; *Sac* I; *Sph* I; *Cfo* I. Six of these enzymes were selected to confirm the identity of the amplified segment 10 cDNA products from the different virus strains *Hinf* I; *Hph* I; *Rsa* I; *Pvu* II; *Sph* I; *Cfo* I (Tables 3 to 5). 10 µl aliquots of the amplified PCR mixtures from each AHSV strain were hydrolysed with each of the restriction enzymes at 37°C for 1 h (10 U per reaction) in a total volume of 20 µl (using the buffers recommended by the manufacturers). The resulting cleavage fragments, along with pBR328 / *Bgl* I and pBR328 / *Hinf* I DNA standards, were separated by electrophoresis using a 2% horizontal agarose slab gel (100 V for 1 h), stained with ethidium bromide and visualised under UV illumination. In order to more easily detect mutations which affect the restriction sites, five restriction enzymes which did not cleave the AHSV 3 segment 10 cDNA were not selected for use in these analyses.

Determination of nucleotide sequences

The amplified cDNA products from segment 10 of AHSV 4 virulent strain and AHSV 4 vaccine strain were ligated into *Srf* I digested (blunt end site) PCR script SK(+) vector

(Stratagene) and the mixture was used to transform competent *E. coli* XL-Blue cells. Both the individual clones and the PCR products and from AHSV 2, 5, 6 and 7 were sequenced using ABI PRISM "Dye terminator, Cycle Sequencing Ready Reaction" kits with AmpliTaq DNA polymerase, FS and a 373 DNA sequencer (Applied Biosystems).

Computer analysis of nucleotide and amino acid sequences

Sequence data for segment 10 of AHSV 4 (USDA) (vaccine) and AHSV 4 (M) (virulent strain) have been deposited with GenBank (accession numbers Z48734 and Z48735, respectively). Amino acid sequences for NS3/3a of each serotype were deduced from the nucleotide sequence. The GenBank accession numbers for the nucleotide sequence of genome segment 10 of the strains AHSV 2, 5, 6, and 7 are: U59279; U60188; U60189; U60190, respectively.

Phylogenetic trees for the amino acid sequences were also calculated from the pairwise alignment scores provided by CLUSTAL, within the PC/GENE suite of programmes.

Results

Table 4 summarises the restriction patterns of segment 7 cDNAs from AHSV strains belonging to the nine serotypes. The restriction patterns of the AHSV 4 (SA) strain (obtained with eight different enzymes) are identical to those predicted from the published nucleotide sequence of genome segment 7 from the Spanish isolate (AHSV 4 [Sp]) [24]. The segment 7 cDNAs from different AHSV strains (of the nine serotypes), all had the same pattern of *BamH* I and *Sac* I cleavage sites, producing 2 and 3 cleavage fragments respectively with these enzymes. *Asn* I generates two bands of fragments from segment 7 cDNA from AHSV 1, 3, 6 and 8, however, it generates two different bands with the cDNA of isolates AHSV 4 (USDA) and AHSV 9 (USDA) and three bands with AHSV 2 and 5. *Asn* I also gave unique patterns with segment 7 cDNA of AHSV 7 and AHSV4 (SA). *Hph* I does not cut the segment 7 cDNA from the strains of AHSV 2, 4 (USDA), or 5 but it did cleave the cDNA of AHSV 1, 3, 4 (SA), 6, 8, 9 (SA) and 9 (USDA) giving similar patterns of fragments in each case. *Pvu* II gave two similar fragments after digestion of AHSV 2, 5, 9 (SA) and 9 (USDA) segment 7 cDNA but this enzyme did not cleave the cDNA of the other serotypes. *Cfo* I generates one pattern of restriction fragments for segment 7 cDNA from strains AHSV 1, 3, 6, 7, 8, 9 (USDA) and a different pattern for AHSV 4 (USDA) and 5 but unique patterns for AHSV 2, 4 (SA) and 9 (SA). *Cfo* I gives a cleavage fragment of 550 bp with AHSV 9 (USDA) segment 7 cDNA, which is not generated by cleavage of AHSV 9 (SA).

Restriction endonuclease analysis of segment 10 cDNAs from different strains of AHSV

Genome segment 10 of AHSV was chosen for amplification and further study, because unlike the homologous segment of bluetongue virus, cross hybridisation studies have demonstrated that it is not highly conserved [2]. However the 3' and 5' terminal regions do have common sequences [18]. As predicted from the

Table 4. Summary of RFLP analysis data for segment 7 cDNAs of different AHSV isolates

Restriction enzymes	Isolates with identical cleavage patterns of segment 7 cDNA	Isolates with unique cleavage patterns
Asn I	AHSV 1, 3, 6 and 8. AHSV 2 and 5. AHSV 4 (USDA), 9 (USDA) and 9 (SA).	AHSV 4 (SA) AHSV 7
BamH I	No differences were detected in any isolates from the nine serotypes.	
Hinf I	AHSV 1, 3, 6, 7, 8, 9 (USDA) and 9 (SA) {no cleavage sites}.	AHSV 4 (SA)
Hph I	AHSV 1, 3, 4 (SA), 6, 8, 9 (USDA) and 9 (SA). AHSV 2, 4 (USDA) and 5 {no cleavage sites}.	AHSV 7
Pvu II	AHSV 1, 3, 4 (USDA), 4 (SA), 6, 7 and 8 {no cleavage sites}. AHSV 2, 5, 9 (USDA), and 9 (SA).	
Sac I	No differences were detected in any isolates from the nine serotypes.	
Sph I	AHSV 1, 2, 3, 4 (USDA), 4 (SA), 5, 6, 7 and 8. AHSV 9 (USDA), and 9 (SA) {no cleavage sites}.	
Cfo I	AHSV 1, 3, 6, 7, 8 and 9 (USDA). AHSV 4 (USDA) and 5.	AHSV 2 AHSV 4 (SA) AHSV 9 (SA)

sequence of segment 10 cDNA from AHSV 3 (USDA), its restriction patterns match those of AHSV3 (SA) [30]. Table 5 summarises the RFLP analysis of segment 10 from the different AHSV strains. *Hinf* I cuts the cDNA of all of the different strains analysed, giving similar patterns for the four different strains of AHSV 4 and AHSV 9 (USDA). *Pvu* II cuts the segment 10 cDNAs of AHSV 1, 2 and 8 (giving similar patterns) but does not cleave those of the other strains. *Sph* I only cuts the segment 10 cDNA of AHSV 4 (SA), 4 (Sp) and 4 (M). *Hph* I did not cut the segment 10 cDNAs of AHSV 5, 8, 9 or the strains of type 4, but did generate fragments which were of differing sizes for the strains of the other serotypes. *Cfo* I cuts segment 10 cDNA from the strains of AHSV 1, 2, 4, 5, 6, 8 and 9, while *Rsa* I cuts the cDNA of all of the strains tested.

The four AHSV 4 strains (USDA, SA, Sp and M) produce similar patterns of segment 10 restriction fragments by electrophoresis, after digestion with each of the selected restriction enzymes, except *Sph* I. This enzymes cuts the cDNA of the

Table 5. RFLP analysis. Restriction enzymes which give different patterns of cleavage with AHSV segment 10 cDNA

	AHSV 1	AHSV 2	AHSV 2	AHSV 4 (SA, M, SP)	AHSV 4 (USDA)	AHSV 5	AHSV 6	AHSV 7	AHSV 8	AHSV 9 (USDA)	AHSV 1 (SA)
AHSV 1	–										
AHSV 2	C, R	–									
AHSV 2	H, P, S, C, R	H, P, C	–								
AHSV 4 (SA,M,SP)	H, P, S, Hp, C, R	H, P, S, Hp, C	H, S, Hp, C	–							
AHSV 4 (USDA)	H, P, Hp, C, R	H, P, Hp, C	H, Hp, C	S	–						
AHSV 5	H, P, Hp, C, R	H, P, Hp, C, R	H, Hp, C, R	H, S, R	H, R	–					
AHSV 6	H, P, Hp, C, R	H, P, Hp, R	H, Hp, C, R	H, S, Hp, C, R	H, Hp, C, R	H, Hp, C, R	–				
AHSV 7	H, P, C, R	H, P, C	H	H, Hp, C,	H, Hp, C	H, Hp, C, R	H, Hp, C, R	–			
AHSV 8	Hp, C	Hp, R	H, P, Hp, C, R	H, P, C, R	H, P, C, R	H, P, S, C, R	H, P, Hp, R	H, P, Hp, C, R	–		
AHSV 9 (USDA)	H, P, Hp, C, R	H, P, Hp, C	H, Hp, C	S	–	H, R	H, Hp, C, R	H, Hp, C	H, P, C, R	–	
AHSV 1 (SA)	H, P, Hp, C, R	H, P, Hp, C	H, Hp, C	H, S	H	H, R	H, Hp, C, R	Hp, C	H, P, C, R	H	–

Abbreviations for restriction endonucleases: *Hinf* I = H; *Pvu* II = P; *Sph* I = S; *Hph* I = Hp; *Rsa* I = R; *Cfo* I = C); identical restriction fragment patterns (–)

Table 6. Location of the putative initiation codons in genome segment 10 of isolates AHSV 4 (M) and AHSV 4 (USDA)

AHSV isolate	Initiation codons (bp number)	Termination codon (bp number)	Total number of codons
AHSV 4 (M)	19–21 52–54	669–671	217
AHSV 4 (USDA)	20–22 53–55	670–672	217

South African, Moroccan, and Spanish strains of AHSV 4 but not that of AHSV 4 (USDA). However, AHSV 4 (USDA) and AHSV 9 (USDA) gave identical RFLP patterns from segment 10 cDNA when digested with the six selected restriction enzymes. The two AHSV 9 strains differ in their *Hinf* I restriction sites in segment 10. However, the cleavage pattern for segment 10 of AHSV 9 (USDA) is identical to that of an AHSV field isolate from Senegal (data not shown). The cleavage patterns of viruses isolated from spleen samples are identical to those of the Spanish and Moroccan strains of AHSV 4. The segment 10 cDNA restriction fragment pattern forAHSV 3 and 7 were different only when produced using *Hinf* I.

Nucleotide sequence of segment 10 of AHSV 4 (M)

Segment 10 of AHSV 4 (M) and AHSV 4 (USDA) was found to be 747 and 753 bp in length respectively. As with segment 10 of other orbiviruses [6, 30], there is a relatively long 3' untranslated region of 76 bp for AHSV 4 (M) and 81 bp for AHSV 4 (USDA). The location of the putative initiation and termination codons, as well as the number of codons for each of the two sequences are shown in Table 6.

Discussion

RFLP analyses of the AHSV segment 7 cDNAs has allowed us to place the different virus strains analysed into a number of different groups. Using *Sph* I, similar segment 7 patterns can be obtained for each of the AHSV strains except those of type 9, which were not cleaved by this enzyme. When the patterns of the segment 7 cDNA fragments were analysed collectively, it is possible to distinguish the strains of AHSV 2, 4, 5, 7, and 9. The four isolates of AHSV 1, 3, 6 and 8 produced similar patterns of segment 7 cDNA fragments after analysis with each of the eight enzymes used (Table 4) but collectively could be distinguished from any of the other strains. The strains that we have analysed can be grouped by RFLP analysis of genome segment 7, but further data is required, from analyses of other isolates, before definitive conclusions can be drawn concerning the reproducibility of these groupings, in relation to virus serotype or other viral properties. AHSV 2 and 5 strains gave identical segment 7 cDNA cleavage patterns with each of the enzymes except *Cfo* I. With this enzyme AHSV 2 gave a unique pattern, while the AHSV 5 strain gave a pattern similar to that obtained with AHSV 4 (USDA). Setting aside the data obtained with *Cfo* I, the strains AHSV 2 and 5 could be distinguished from the other viruses tested. With the data obtained using *Cfo* I they could also be distinguished from each other. AHSV 7 segment 7 cDNA gave a unique pattern of RFLP with *Asn* I and *Hph* I. This virus strain could therefore be distinguished from the other strains analysed in this paper. The two strains of AHSV 9 (USDA and SA) can be differentiated from the other AHSV strains examined by RFLP analysis of segment 7 cDNA using *Sph* I and were identical by RFLP analysis of segment 7 with each enzyme except *Cfo* I. The single difference detected by RFLP analysis between the two strains of AHSV 9 might reflect different passage histories of the two strains in non equine cell lines or in suckling mouse brain.

RFLP analysis of the segment 10 cDNA also demonstrated that it is possible to differentiate between the majority of the different strains of AHSV that were analysed using this method (Table 5) and to place them within groups. By comparison with BTV and in view of the non structural nature of the NS3 and NS3a proteins it encodes, the level of variation that exists in genome segment 10 of AHSV is surprising [6, 8, 37]. Overall, RFLP analysis of AHSV 3 and 7 gave similar patterns for segment 10, except when digested with *Hinf* I, suggesting a relatively close relationship between these viruses. This correlates with results of a cross-hybridisation study by Bremer et al. [2], who reported that an AHSV 3 segment 10 cDNA probe would only cross hybridise to that of AHSV 7, even under relatively low stringency conditions (< 75% homology). Despite the similarities observed in segment 10 of these strains they can also be distinguished by RFLP analysis of genome segment 7 (see above) [35].

Three of the four AHSV 4 strains (SA, Sp and M), gave identical restriction fragments from segment 10 cDNA, which differed from those of AHSV 4 (USDA) only with *Sph* I. In addition, no differences were detected in the patterns of cleavage for segment 7 cDNA between the South African vaccine strain and the field isolate from Spain [35], but four of the enzymes gave different cleavage patterns with AHSV 4 (USDA). Differences in the sequences of genome segments 7 and 10 of AHSV 4 (USDA) could reflect the high passage number of this strain in mice, which may have induced mutations during adaptation to this non equine system.

The cDNAs of segment 7 of AHSV 9 (USDA) and AHSV 9 (SA) were similar by RFLP analysis and segment 10 differed only in their *Hinf* I cleavage sites.

Overall higher variation were observed by RFLP analysis, in AHSV genome segment 10 than in segment 7 [35]. This reflects a higher than expected level of sequence variation in segment 10, as reported by De Sa et al. [6] for isolates of serotypes 1, 4 and 8. Our RFLP analysis data indicatc close relationships between the strains analysed from serotypes 4 and 9, as well as between AHSV 1 and 8. There is also evidence that AHSV 3 is relative close to the AHSV 4+9 group.

Using sequence analysis of genome segment 10, we were able to classify the virus isolates into three groups: AHSV-1, 2 and 8; AHSV-3 and 7; AHSV 4, 5, 6 and 9 (see also reference 15). Segment 10 of the Spanish and Moroccan strains of

Table 7. Matrix showing the percentage divergence (top right) and the percentage similarity (bottom left) between pairs of nucleotide sequences from different AHSV isolates

	Percentage divergence												
	AHSV 1	AHSV 2	AHSV 3	AHSV 4 (USDA)	AHSV 4 (M)	AHSV 4 (SP)	AHSV 4 SA)	AHSV 5	AHSV 6 (UP)	AHSV 6	AHSV 7	AHSV 8	AHSV 9
AHSV 1	–	2.1		30.8	29.5			30.3	30.6	31	29.3		
AHSV 2	97.9	–	28.5	30.1	30	32.1	30.7	30.3		30.9	28.6	2.7	30.7
AHSV 3		71.8	–	30.8	23.7			29.9	6.5	25.7	2.1		
AHSV 4 (USDA)	69.2	69.1	76.2	–	5.4	6.3	7.6	6.3	2.5	13.8	23.9	29.9	5.4
AHSV 4 (M)	70.5	70	76.3	94.6	–	0.7	7.8	2.7		13.5	2.4	29.2	4.5
AHSV 4 (SP)		67.9		93.7	99.3	–		2.1		15.1	25.2		
AHSV 4 (SA)		69.3		92.4	92.2		–	7.2		15.1	23.9		
AHSV 5	69.7	69.7	70.1	93.7	97.3	97.9	92.8	–	2.9	13.1	24.5	29.9	5.6
AHSV 6 (UP)		69.4		93.5	97.5			97.1	–	13.8	23.8		
AHSV 6	69	69.1	74.3	86.2	86.5	84.9	84.9	86.9	86.2	–	26.8	30.9	12.5
AHSV 7	70.7	71.4	97.9	76.1	76	74.8	76.1	75.5	76.2	73.2	–	28.8	24.6
AHSV 8		97.3		70.1	70.8			70.1		69.1	71.2	–	
AHSV 9		69.3		94.6	95.5			94.4		87.5	75.4		–
	Percentage homology												

The nucleotide sequence of the isolates from AHSV serotypes 3 and 9 were determined by Van Staden and Huismans [30]. The sequences of isolates from AHSV serotypes 1, 4 (SA) and 8, were determined by De Sa et al. [6]. The sequence of the Spanish strain of AHSV 4 (SP), was by Laviada et al. [14] and the sequence of the University of Pretoria isolate of AHSV 6 (UP) was by Stoltz et al. (direct submission to GenBank, accession number U26171)

AHSV showed only 0.7% sequence divergence (Table 7). This very high level of similarity confirms that these isolates are very closely related and further supports the epidemiological data that the virus which first appeared in Spain in July 1987, spread into Morocco in October 1989. These data disprove previous suggestions that the AHSV present in Southern Spain and in Morocco were generated by reversion from the vaccine [17].

The nucleotide sequence of the vaccine strain AHSV 4 (USDA) segment 10 shows 5.4%, 6.3% and 7.6% divergence when compared with the virulent strains of AHSV 4 (M), AHSV 4 (Sp) and AHSV 4 (SA: unknown virulence) respectively [14, 25]. These levels of variation may also correlate with the high passage level of AHSV 4 (USDA) in suckling mouse brains. An in depth comparison of variation in genome segment 10 of AHSV between strains from the nine different serotypes, is presented by Sailleau et al. [25] and by Martin et al. [15].

The PCR technique is able to detect AHSV genome segment 7, 8 and 10 [29, 36] in tissues from horses infected with AHSV. The PCR method associated with RFLP analysis is a sensitive and valuable technique to compare different isolates for epidemiological studies. However, the principal serotype specific antigen of AHSV is the outer capsid protein VP2, which is encoded by genome segment 2. These viruses have the ability to efficiently exchange genetic material, by a process of genome segment reassortment [22]. These observations demonstrate that although the analyses of RFLP and sequences of genome segments 7 and 10 are valuable for comparing different vaccine and wild type AHSV strains, the differences observed cannot have any absolute correlation with virus serotype [15, 22]. The development of a PCR RFLP based method of analysing virus serotype, possibly involving genome segment 2, would depend on obtaining RFLP data from a larger number of isolates of each serotype. This would allow an assessment of the stability of the restriction sites involved and of the impact of intra serotype genetic variability in segment 2, on this method.

Acknowledgement

This work was funded by an EC grant, contract number 8001-CT91-0211.

References

1. Bremer CW (1976) A gel electrophoretic study of the proteins and nucleic acid components of African horse sickness virus. Onderstepoort J Vet Res 43: 193–200
2. Bremer CW, Huismans H, Van Dijk AA (1990) Characterisation and cloning of the African horse sickness virus genome. J Gen Virol 72: 793–799
3. Borden EC, Shope RF, Murphy FA (1971) Physicochemical and morphological relationships of some arthropod bourne viruses to bluetongue virus – a new taxonomic group, physicochemical and serological studies. J Gen Virol 13: 261–271
4. Chomczynski P, Saachi N (1978) A single step method for RNA isolation by acid guanidiniun thiocyanate-phenol-chloroform extraction. Anal Biochem 162: 156–159
5. Devaney MA, Kendall J, Grubman KC (19988) Characterization of a non structural phosphoprotein of two orbiviruses. Virus Res 11: 151–164

6. De Sa R, Zellner M, Grubman MJ (1994) Phylogenetic analysis of segment 10 from African horse sickness virus and cognate genes from other orbiviruses. Virus Res 33: 157–165
7. Du Toit RM (1944) the transmission of bluetongue and horse sickness by *Culicoides.* Onderstepoort J Vet Res 560: 536–541
8. Grubman MJ, Lewis SA (1992) Identification and characterization of the structural and non-structural proteins of African horse sickness virus and determination of the genome coding assignments. Virology 92: 397–406
9. Hamblin C, Mertens PPC, Mellor PS, Burroughs JN, Crowther JR (1991) A serotype specific enzyme-linked immunosorbent assay (ELISA) for the detection and identification of African horse sickness viruses. J Virol Methods 31: 285–292
10. Hamblin C, Anderson EC, Mellor PS, Graham SD, Mertens PPC, Burroughs JN (1992) The detection of African horse sickness virus antigens and antibodies in young equidae. Epidemiol Infect 108: 193–201
11. Huismans H, Els HJ (1979) Characterization of tubules associated with the replication of the different orbiviruses. Virology 92: 397–406
12. Holmes IH, Boccardo G, Estes MK, Furuichi MK, Hoshino Y, Joklik WK, McCrae M, Mertens PPC, Milne RG, Samal KSK, Shikata E, Winton JR, Uyeda I (1995) Family *Reoviridae.* In: Murphy FA, Fauquet CM, Bishop DHL, Ghabrial SA, Jarvis AW, Martelli GP, Mayo MA, Summers MD (eds) Virus Taxonomy. Classification and Nomenclature of Viruses. Sixth Report of the International Committee on Taxonomy of Viruses. Springer, Wien New York, pp 208–237 (Arch Virol [Suppl] 10)
13. Laviada MD, Arias M, Sanchez-Viscaino JM (1993) Characterisation of African horse sickness virus serotype 4-induced polypeptides in Vero cells and their reactivity in Western immuno blotting. J Gen Virol 74: 81–87
14. Laviada MD, Roy P, Sanchez Viscaino JM, Casal JI (1995) The use of African horse sickness NS3 protein, as a marker to differentiate infected from vaccinated horses. Virus Res 38: 205–218
15. Martin L-A, Meyer AJ, O'Hara RS, Fu H, Knowles NJ, Mertens PPC (1998) Phylogenetic analysis of African horse sickness virus genome segment 10: sequence variation and virus release. In: Mellor PS, Baylis M, Hamblin C, Calisher CH, Mertens PPC (eds) African horse sickness. Springer, Wien New York, pp 281–293 (Arch Virol [Suppl] 14)
16. McIntosh BM (1958) Immunological types of horse sickness virus and their significance in immunisation. Onderstepoort J Vet Res 27: 465–538
17. Mellor PS (1994) Epizootiology and vectors of African horse sickness. Virus Comp Imm Microbiol 173: 287–296
18. Mertens PPC, Sangar DV (1985) Analysis of the terminal sequences of the genome segments of four orbiviruses. Virology 140: 55–67
19. Mizukoski N, Sakamoto K, Iwata A, Tsuchiya T, Ueda S, Watanabe T, Kamada M, Fukusho A (1992) The complete sequence of African horse sickness virus serotype 4 (vaccine strain) RNA segment 5 and its predicted polypeptide compared to NS1 of bluetongue virus. J Gen Virol 73: 2425–2428
20. Moulay S, Zientara S, Sailleau C, Crucière C (1995) Detection of African horse sickness viruses by dot blot hybridisation using a digoxigenin labelled probe. Mol Cell Probes 9: 233–237
21. Oellerman RA, Els HJ, Erasmus BJ (1970) Characterisation of African horse sickness virus. Arch Ges Virusforsch 29: 163–174
22. O'Hara RS, Meyer AJ, Pullen L, Martin L-A, Mertens PPC (1998) Development of a mouse model system and identification of individual genome segments of African horse sickness virus, serotypes 3 and 8 involved in determination of virulence. In: Mellor PS,

Baylis M, Hamblin C, Calisher CH, Mertens PPC (eds) African horse sickness. Springer, Wien New York, pp 259–279 (Arch Virol [Suppl] 14)

23. Reed LJ, Muench H (1938) Simple method of estimating 50 percent end points. Am J Hyg 27: 493
24. Roy P, Hirasawa T, Fernandez M, Blinov VM, Sanchez-Vizcaino JM (1991) The complete sequence of the group-specific antigen, VP7, of African horse sickness disease virus serotype 4 reveals a close relationship to bluetongue virus. J Gen Virol 72: 1237–1241
25. Sailleau C, Moulay S, Zientara S (1997) Nucleotide sequence comparison of the segments S10 of the nine African horse sickness virus serotypes. Arch Virol 142: 965–978
26. Sakamoto K, Punyahotra R, Mizukoshi N, Ueda S, Imagawa H, Sugiura T, Kamada M, Fukusho A (1994) Rapid detection of African horse sickness by the reverse transcriptase polymerase chain reaction (RT-PCR) using the amplimer for segment 3 (VP3 gene). Arch Virol 136: 87–97
27. Sellner LN, Coellen RJ, Mackenzie JS (1992) Reverse transcriptase inhibits Taq polymerase activity. Nucleic Acids Res 20: 1487–1490
28. Sellner LN, Coellen RJ, Mackenzie JS (1992) A one tube one manipulation RT-PCR reaction for detection of Ross River virus. J Virol Methods 40: 255–264
29. Stone-Marschat M, Carville A, Skowronek A, Laegreid AA (1994) Detection of African horse sickness virus by reverse transcription PCR. J Clin Microbiol 32: 697–700
30. Van Staden V, Huismans H (1991) A comparison of the genes which encode non structural protein NS3 of different orbiviruses. J Gen Virol 72: 1073–1079
31. Venter EH, Van Dijk AA, Huismans H, Van der Lugt JJ (1992) The detection of bluetongue and African horse sickness using dot-spot and conventional in situ hybridisation. In: Walton TE, Osburn BI (eds) Bluetongue African horse sickness and related orbiviruses. CRC Press, Boca Raton, pp 671–678
32. Verwoerd DW, Huismans H, Erasmus BJ (1979) Orbiviruses. Compr Virol 14: 285–345
33. Wade-Evans AM, Mertens PPC, Bostock CJ (1990) Development of the polymerase chain reaction for the detection of bluetongue virus in tissue samples. J Virol Methods 30: 15–24
34. Wade-Evans AM, Woolhouse T, O'Hara R, Hamblin C (1993) The use of African horse sickness virus VP7 antigen, synthesised in bacteria, and anti-VP7 monoclonal antibodies in a competitive ELISA. J Virol Methods 45: 179–188
35. Zientara S, Sailleau C, Moulay S, Crucière C (1993) Diagnosis and molecular epidemiology of the African horse sickness virus by the polymerase chain reaction and restriction patterns. Vet Res 24: 385–395
36. Zientara S, Sailleau C, Moulay S, Crucière C (1994) Diagnosis of the African horse sickness virus serotype 4 by a one tube, one manipulation RT PCR reaction from infected organs. J Virol Methods 46: 179–188
37. Zientara S, Sailleau C, Moulay S, Crucière C (1995) Differentiation of African horse sickness viruses by polymerase reaction and segment 10 restriction patterns. Vet Microbiol 47: 365–375

Authors' address: Dr. S. Zientara, Centre National d'Etudes Vétérinaires et Alimentaires, Laboratoire Central de Recherches Vétérinaires, 22 rue Pierre Curie, B.P. 67, F-94703 Maisons Alfort Cedex, France.

Structural studies of orbivirus particles

D. I. Stuart[1,2], **P. Gouet**[1], **J. Grimes**[1], **R. Malby**[1], **J. Diprose**[1], **S. Zientara**[3], **J. N. Burroughs**[4], and **P. P. C. Mertens**[4]

[1] Laboratory of Molecular Biophysics, Oxford, U.K.
[2] The Oxford Centre for Molecular Sciences, New Chemistry Laboratory, Oxford, U.K.
[3] Centre National d'Études Vétérinaires et Alimentaires, Laboratoire Central de Recherches Vétérinaires, Maisons-Alfort, France
[4] Pirbright Laboratory, Institute for Animal Health, Pirbright, Surrey, U.K.

Summary. We are using crystallographic methods to investigate the structure of AHSV and BTV. Our initial approach was to investigate the structure of the major protein component of the viral core, VP7(T13). This trimeric protein has been studied in several crystal forms from both orbiviruses and reveals a structure made up of conserved domains, capable of conformational changes and possessing a cleavage site. Further crystallographic analyses of native particles have provided a picture of the VP7(T13) and VP3(T2) layers of the BTV core. The VP7(T13) layer consists of 260 trimers arranged rather symmetrically and possessing very similar structures, thereby following the rules of quasi equivalence. The VP3(T2) layer is thin and contains 120 copies of 100 kDa protein arranged as 60 approximate dimers. This type of icosahedral construction has not been observed before and appears to contain a genome which is highly ordered. We anticipate that all of these features will be common to AHSV.

Introduction

Structural studies have provided insights into the biological functions of many viruses [10, 13, 22, 27, 29, 30, 33, 37–39, 48, 62]. In comparison to smaller viruses that have already been studied in atomic detail, orbiviruses and the other members of the family *Reoviridae* [11, 32, 46], have additional constraints on their life cycle, which are the consequence of their double stranded RNA genome [35]. Since dsRNA would trigger the host's defence mechanisms if it were released into the cytoplasm of an infected cell, the orbivirus genome must be retained within the viral capsid structure. In the absence of host cell transcriptases which can use dsRNA as template for mRNA synthesis, the naked virus genome would also be transcriptionally and translationally inert within the cell cytoplasm and would therefore be unable to initiate replication. The orbivirus core, which contains the ten genome segments, also provides a compartment within which they can be repeatedly transcribed by core associated enzymes. mRNA copies of each of the segments are synthesised simultaneously and released from the core particle into the host cell cytoplasm, to function as templates for translation, as well as for negative strand viral RNA synthesis within nascent progeny virus particles [35].

The intact virion possesses seven structural proteins [9, 33, 43, 60]. Three of these are present in relatively smaller numbers in each particle (minor proteins: VP1, VP4 and VP6) and have enzymatic functions associated with the synthesis and capping of the viral RNA [16, 43, 44, 55, 58, 59]. Two relatively more abundant proteins (major proteins: VP2 and VP5) form the outer shell of the intact virus particle [9, 33, 43, 60]. Under suitable conditions (eg. 0.1 M tris / HCl pH 8.0) the virion is relatively stable, but treatment with $MgCl_2$ (BTV and AHSV), or CsCl (BTV), or reduction in pH, releases the components of the outer capsid from the surface of the core [9, 33, 43, 60]. The outer capsid layer can also be modified by treatment with proteases (such as trypsin, chymotrypsin, or plasmin) which results in cleavage of VP2 in the outer capsid and some modification of the biological properties and solubility of the particle [9, 43, 45]. The remaining two major proteins (VP3(T2) and VP7(T13)) form the inner and outer layers of the core particle (respectively) which is not only central to the process of transcription and genome replication but, we believe, is also the cornerstone of the virus particle assembly process [6, 17, 19]. Our studies have focussed primarily on BTV, which provides a paradigm for AHSV and other orbiviruses.

Materials and methods

Analysis and enumeration of the virion structural proteins

The proteins of BTV serotype 1 from South Africa (BTV1 SA) were labelled with [^{3}H] leucine or [^{35}S] methionine and purified as described previously [9, 43]. The individual proteins were separated by SDS PAGE, in 11% gels , stained with Coomassie Blue, then excised. The strips of polyacrylamide were dried at 37°C in 'low background' glass scintillation vials. Solubilization solution (made up of 95 ml 30% hydrogen peroxide and 5 ml of concentrated (880) ammonia solution) was added (0.5 ml) to each dried gel slice, followed by a thin layer of paraffin oil. The vials were then held at 37°C for 6 h prior to addition of 50 µl of glacial acetic acid. The vials were capped and shaken vigorously, then "counted" in a Packard 460 liquid scintillation counter using a dpm program. Using values for the percentages of leucine and methionine and for the molecular weight of each protein [20, 53, 54] molar ratios were calculated for each of the proteins. Assuming 780 copies of VP7 per particle [24–26, 50] the numbers of the other virion structural proteins could be estimated.

Crystallographic analysis of core particles of BTV1 SA

The core particles of BTV1 SA were purified and crystallised, using standard techniques, as described by Mertens et al. [43] and Burroughs et al. [8]. The crystals grew to between 0.3–0.5 mm in length (Fig. 1a), and were premounted in quartz capillaries within the secure laboratories at IAH Pirbright, prior to transportation to synchrotron radiation sources at the SRS and the ESRF. This was done in accordance with disease security regulations laid down by MAFF. Data were collected, initially to low resolution at the SRS Daresbury, and later to higher resolution at the ESRF at Grenoble, France. Only 1 exposure of 0.25° was possible per position at ~8°C. due to radiation damage, although usually more than one position was possible per crystal. The crystals belong to space group $P2_12_12$, with unit cell dimensions of a = 796 Å, b = 822 Å and c = 753 Å, and the asymmetric unit contained ½ particle. Data have been integrated using DENZO [49] and the reflections scaled and merged together using SCALEPACK [49]. In total over 8,000,000 data have been measured, giving a unique data set of some 2,800,000 reflections. The structure of the core has been solved by molecular replacement, using a combination of cryo-EM data and crystallographic data [26]. The search model was a Cα model of the outer layer of the core, derived by fitting VP7 into a cryoelectron microscopy reconstruction of the core [25]. A self-rotation, using X-PLOR, gave the orientation of the particle around the z-axis, and a translation function then gave the position of the particle along z. Rigid body refinement of the model yielded phases to 6 Å resolution. At this point real space density modification was used to refine the crude model phases, by averaging and solvent flattening of the electron density maps. The result revealed the structure not only of the VP7(T13) layer, but also the VP3(T2) layer. A Cα model for the units of secondary structure was built and refined, ultimately against the 60-3.8 Å data. The NCS operators were then optimised and the map modification resumed. The final electron density map was of sufficient quality for the unambiguous chain tracing and sequence alignment of the 2 molecules of VP3(T2). The final correlation coefficient, a measure of the extent to which the data and electron density are consistent with each other, was 0.89 (a value of 1 indicating perfect agreement).

Crystallographic analysis of core particles of BTV10 US

BTV10 USA (the CA-8 strain from North America) was grown and viral cores prepared using established procedures [8, 43]. Crystals were grown at 20°C in micro bridges [28]. The largest crystals reached a diameter of 0.4 mm (Fig. 1b) and were more fragile than BTV 1 core crystals [8]. The BTV 10 core crystals belong to space group $P4_12_12$, with unit cell dimensions of a = b = 1,115 Å, c = 1,584 Å. This unit cell contains one core particle in each of its 8 asymmetric units. Data collection was at the high brilliance undulator beam line ID2 (BL4) at the European Synchrotron Radiation Facility (ESRF), Grenoble, France. The highly parallel nature of the beam, in conjunction with aperture limiting the beam to as little as 50 µm, allowed the diffraction orders to be resolved at a wavelength of some 1 Å, using a commercially available (MAResearch) imaging plate detector. Careful experimental design allowed relatively low angle diffraction (60 Å resolution) to be collected. Crystals were cooled to about 8°C and randomly oriented. Oscillation ranges per image were 0.15–0.3°. It was possible to obtain 2 images from some crystals. A total of 100 images were used, from 75 crystals (although many more crystals were examined), Images were indexed and integrated with the program DENZO [49], keeping overlapping reflections. Overlapped reflections were deconvoluted using PrOW [5]. Data were scaled and merged with SCALEPACK [49]. A unique data set of just greater than 1 million reflections was obtained covering the resolution range 60 to 6.5 Å, with a completeness of 54% The phase problem was solved and space group determined using the program X-PLOR [7]. The particle orientation was determined by inspection of a self-rotation function and then refined using published methods [7]. The space group and particle position were determined more quickly by factoring the space group symmetry operators to reduce the problem to 1-D and 2-D searches (since the search model, was a complete BTV core, these calculations were computationally demanding). Rigid body refinement was combined with a grid search refinement of the unit cell dimensions to give the final estimates quoted above. A correlation coefficient (described above) of 0.72 was obtained for the model, providing phases which were refined by cyclic averaging and solvent flattening to yield a final correlation coefficient of 0.90. A cyclic averaging procedure was performed using both

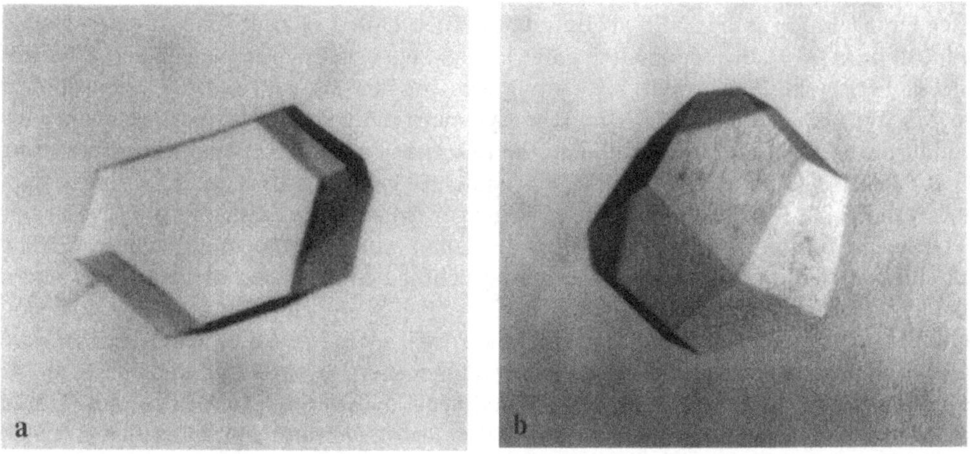

Fig. 1. Crystals of BTV cores. (**a**) shows a crystal of BTV1 SA cores (orthorhombic symmetry, these crystals can grow up to 0.8 mm in the longest direction). (**b**) shows a crystal of BTV10 USA (tetragonal symmetry, maximum dimension 0.4 mm)

BTV1 (100-6.5 Å subset, 98.9% complete) and BTV10 data, with the two averaged maps being cross-averaged during the cycle in order to dissect out structural features conserved between the somewhat disordered interiors of the two serotypes (Figs. 7, 8). Segments of dsRNA were modelled into the density of the protomeric unit using the graphics program TURBO-FRODO [52]. Figures have been drawn using TURBO-FRODO [52] (Figs. 4 and 8), EXPLORER [21] (Fig. 7) and BOBSCRIPT [18] plus RASTER3D [1, 42] (Figs. 2, 3, 5, and 6).

Results and discussion

The VP7(T13) protein

As part of the on-going work on the structure of the orbivirus core, we have undertaken structural studies (in collaboration with Prof. P. Roy) on individual protein components of the core, such as VP7(T13) from BTV and AHSV. Crystallographic analyses of both VP7(T13) from AHSV and BTV at high resolution reveal that the molecular unit that makes up the building block of the virus is a tightly assembled trimer. Work on different crystal forms also reveals that the protein has some conformational flexibility and is susceptible to proteolytic cleavage. In our analyses we have looked at 4 different crystal forms, to provide a total of 16 independent views of the molecule of VP7 [3, 4, 24–26]. In both viruses (AHSV and BTV) the monomer of VP7(T13) has a 2 domain structure composed an outer and an inner domain (where the "outer" domain lies at a greater radius from the centre of the core particle: Fig. 2) . The outer domain consists of about 130 residues, representing the central portion of the molecule's amino acid chain and has a 'jelly-roll' topology common to many viral structural proteins [57]. The lower, inner domain is entirely α-helical, and has a topology not seen in other viral structural proteins. The relative disposition of the these domains is such that they wrap around the molecular 3-fold axis in a right handed sense (Fig. 2).

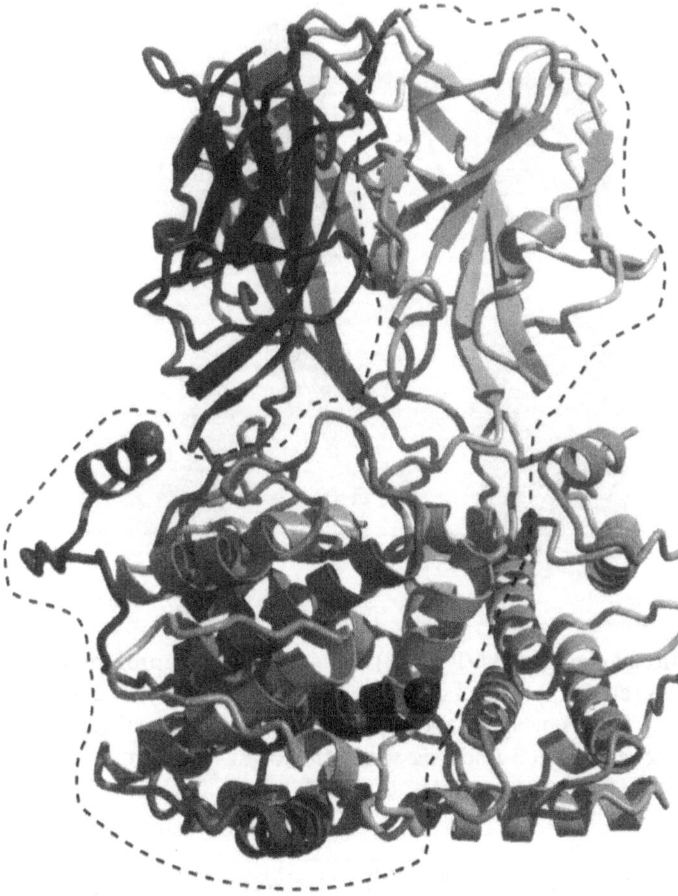

Fig. 2. Trimer of orbivirus VP7(T13). The protein from BTV is shown, although AHSV VP7(T13) would be virtually indistinguishable. The 3-fold symmetry axis of the trimer is disposed vertically so that the flat base at the bottom of the image is the surface which adheres to the VP3(T2) subcore. Note the clear division into two domains, the upper domain contains mainly β-strands (shown as arrows), whilst the lower domain is constructed of a number of closely packed helices. One subunit is outlined with a dashed line to clarify the twist of the subunit with respect to the molecular 3-fold axis. The N and C termini of the outlined molecule are represented by small balls (the lower ball being the N-terminus)

VP7(T13) from both BTV and AHSV has a tendency to undergo proteolytic cleavage, at a point between the 2 domains [3, 4]. This is clearly seen in some crystals of VP7(T13) from both BTV and AHSV, where only a trimeric fragment, comprising the top domain, is present in the crystal. Comparison of these two fragments shows that they are structurally very similar. They can be superimposed with an rms deviation in equivalent Cα positions of 1.2 Å, with the major differences occurring at the molecular surface. It is unclear whether this cleavage has some biological significance, or is as a result of the inter domain region being more available to proteases, due to inherent conformational flexibility. Certainly

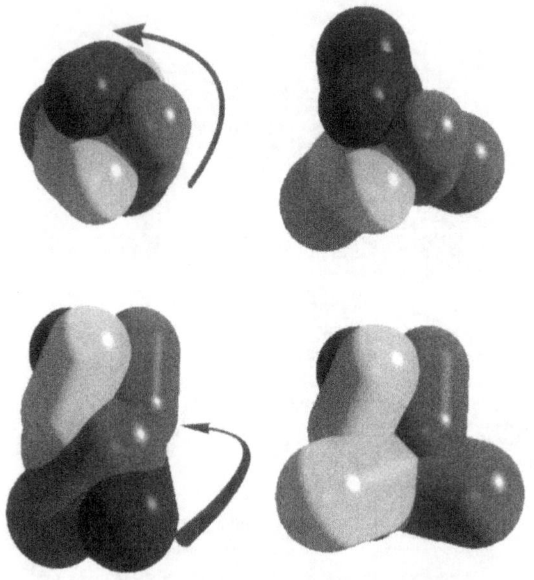

Fig. 3. Conformational changes in VP7(T13). The 4 cartoons represent views of both the normal structures, as seen on the intact core (the leftmost 2 images). The unwound structure (images on the right). Each sausage shaped object represents a domain and the upper views are looking down the molecular 3-fold axis, whilst the lower views show a view equivalent to that of Fig. 2

there is evidence for considerable flexibility between the 2 domains. Thus in one crystal form of BTV VP7(T13), the structure of which could only be determined at low resolution (6 Å), the α-helical domains had moved substantially, compared to the upper jelly-roll domains, which were fixed. This movement results in a massive repositioning of the lower domains, which we have interpreted as an unwinding of the structure, by a rotation of 160° (Fig. 3). The alternative model is where the trimer is "wound up", resulting in a reduced but still large movement. However, this model would result in steric hindrance and on a general structural basis seems less feasible.

Protein composition of the orbivirus particle

Virus particles of BTV1 SA, were labelled with either [^{3}H] leucine or [^{35}S] methionine and purified as described by Mertens et al. [43]. The different proteins were separated by SDS PAGE and the amount of label incorporated into each band was measured. Using the amino acid composition of the different proteins (and consequently the numbers of leucines or methionines in each [20, 53, 54]) it has been possible to calculate their relative abundance in preparations of the purified virus particles. Using the number of VP7(T13) molecules (780 molecules per particle: as previously determined [24–26, 50]) to provide a "standard", we have calculated the number of each of the other structural proteins in the individual virus particle (Table 1). The averaged number of 120 VP3(T2) molecules per particle fits exactly with the number determined in our structural studies [26]. The number of molecules of VP5 (approximately 360 copies per particle) is significantly larger than previously suggested by cryoelectron microscopy (120 copies per particle [29, 30]), indicating that VP5 in the outer capsid layer could be organised as trimers. This conclusion is supported by the cryo-EM reconstruction of whole particles of Broadhaven virus, where VP5 is clearly also seen as a trimer [56]. The number of VP2 molecules may be an under estimate since it is very sensitive to proteolytic digestion [9, 43, 45] and there is a tendency for VP2 in preparations of purified virus to become cleaved during storage. However, the results of cryo-EM reconstruction of BTV and Broadhaven virus again indicate that this molecule may be trimeric [30, 56] forming 60 "triskelion" like structures on the outer particle surface.

The numbers of the minor core proteins in virus particles could be related either to the 10 dsRNA genome segments, or to the icosahedral structure of the protein capsid (we have evidence that these are clustered close to the 12 icosahedral 5-fold axes). It is therefore considered possible that there could be 10 or 12 copies (respectively) of VP1 (the RNA polymerase [54, 55, 57, 59]). VP4,

Table 1. Analysis of the structural proteins of purified bluetongue virus particles (serotype 1 from South Africa[a]) by electrophoresis and estimation of their numbers per particle

Protein[b]	Number of copies of each protein per virus particle[c]				Average[e] copy number	Adjusted value[f]
	Expt 1[d]	Expt 2	Expt 3	Expt 4		
VP1 (Pol)				7.9	7.9	12
VP2[g]	162	159	189	152	165.5	180
VP3 (T2)	119	112	137	112	120	120
VP4 (Cap)				19.6	19.6	24
VP5	324	343	437	376	370	360
VP6 (Hel)				72	72	72
VP7 (T13)	780	780	780	780	780	780

[a] Virus particles of BTV1 SA were purified as described by Mertens et al. [43]

[b] *Cap* capping enzyme (guanylyltransferase and methyltransferase); *Pol* RNA Polymerase; *Hel* Helicase

[c] Means from different numbers of gel slices. For the major proteins (VP2, VP3(T2), VP5, VP7(T13)) 14 gel slices were counted in four separate experimental series. For the minor proteins 5 gel slices were counted in a single experimental series

[d] In experiment 1, virus labelled with [^{35}S] methionine was used. In experiments 2 to 4 virus labelled with [^{3}H] leucine was used

[e] Average values were calculated from the different experimental series

[f] Values adjusted to the nearest number consistent with icosahedral virus symmetry

[g] Because of the sensitivity of VP2 to proteases and its partial cleavage in some virus preparations, values for its relative abundance in the virus particle may be underestimates

which is the transmethylase 1 and 2 [16, 44] as well as the guanylyltransferase [44], forms dimers [16] and may therefore be present as either 20 or 24 copies per particle. VP6 is reported to be a helicase [58] and may form hexamers. It could therefore be present as 60 or 72 copies per particle.

Studies of the native orbivirus core structure

A structural analysis of the core of BTV is crystallographically very challenging, due to the relatively large size of the particle (6.7 x 10^7 Da [45, 46]). A full report of the structure has been presented [23, 26]. The core is made up of an outer surface composed of 260 trimers of VP7(T13) (38 kDa), sitting on a scaffold of 120 copies of VP3(T2) (100 kDa), which form a "sub core" layer. The trimers are arranged on a pseudo-hexagonal T = 13 lattice and are all very similar to each other, suggesting that they conform to the ideas of quasi-equivalence [12, 34]. These quasi-equivalent trimers build up the core surface layer, such that they form pairs related by almost exact 2-fold symmetry, so that the local contacts are indeed quasi-equivalent in nature (Fig. 4). This is extremely unusual and is achieved by the unique architecture of this layer. Unlike smaller viral capsids, where the surface of interaction between the capsid proteins extends over a considerable radial depth, the contact regions between the different VP7(T13) trimers are located in a thin band within the lower α-helical domain and are essentially 1-dimensional in nature. This region is formed in large part by a short hydophobic α-helix, that seems to act as a 'greasy hinge', about which the trimers are able to roll, as they pack down onto the surface of the inner core of VP3(T2) (Fig. 5). This rolling enables the same modes of interaction to be used between the different trimers (P-T), with minimal distortion.

The inner core protein VP3(T2) is arranged as 120 copies per particle, with icosahedral symmetry, on a T = 2 lattice, as shown in Fig. 6. VP3(T2) is highly unusual with the 901 amino acids being organised into a structure that resembles a triangular wedge (130 Å by 75 Å) with a thickness varying between 13 Å and 35 Å. There are 3 domains that make up the VP3(T2) molecule: an apical domain that sits nearest to the icosahedral 5-fold axis; a carapace domain, which forms a rigid plate; and a dimerisation domain that is involved in forming the quasi-2 fold interaction between the different molecules of VP3(T2). As far as we are aware none of these domains resembles any other protein structure yet seen.

The innermost protein shell of the core (the "sub core"), is believed to be of particular significance in the process of orbivirus replication and morphogenesis. VP3(T2) is the only viral protein which can assemble when expressed by itself, to form icosahedral particles similar in appearance to the authentic sub core ([47], P. Roy, pers. comm.). The sub core layer is also sufficiently stable that it can retain its the particle size and structure, after removal of the outer capsid layer and the VP7(T13) layer of the core surface [33, 40, 41]. During assembly of progeny virus particles within the viral inclusion bodies (VIB) in infected cells, the smallest particles, which are thought to contain viral RNA are equivalent in size to sub cores [6, 17, 19]. The sub core is therefore thought to be involved both in genome packaging and in defining the size and overall morphology of the outer layers of the capsid which are added later.

The icosahedral building block of the sub core shell contains two molecules of VP3(T2) (120 copies per particle) which although chemically identical perform different structural roles and are therefore denoted as A and B. These two molecules fit together in a manner previously unseen. Conceptually the 60 'A' subunits define the size of the subcore by forming a continuous scaffold of pairs, which span icosahedral 2-fold axes and link adjacent 5-fold axes. This scaffold is sealed off by the 60 'B' subunits, which cluster at the icosahedral 3-fold axes to form triangular plugs (Fig. 6). A tightly packed, almost hermetic, shell is achieved by distortion of these VP3(T2) molecular building blocks. The conformational changes occur primarily at 2 places in the molecules; the dimerisation domain, which is intimately involved in forming the quasi 2-fold interface between molecules A and B, and between the carapace and apical domains. There is a modest pivoting of the dimerisation domain, which combines with a slight rotation of the apical domain against the carapace domain, to swing the tip of the apical domain some 35 Å. There are additional small changes which combine to ensure that the molecules fit tightly around each 5-fold axis, in the form of a decamer. The subtle rotation of the dimerisation domains is such that there is an almost exact local 2-fold relationship between dimerisation domains of the A and B molecules in adjacent decamers, so that an A molecule clips onto a B molecule in the neighbouring decamer and vice versa. This quasi 2-fold relationship breaks down beyond the dimerisation domains. This almost exact quasi 2-fold interaction is a characteristic of the 120 molecule architecture, which may be thought of as a sub-triangulation of the icosahedron into a T = 2 structure.

The VP7(T13) and VP3(T2) layers interact through flattish, predominantly hydrophobic surfaces. As a direct consequence of the symmetry mismatch between the two layers of the core there are 13 different sets of contacts between the 13 copies of VP7(T13) packed onto the 2 copies of VP3(T2). It is remarkable that both VP3(T2) A and VP3(T2) B molecules make completely different yet extensive contacts with the various VP7(T13) molecules, whose structures are virtually identical to one another and to the structure which was solved using the isolated protein.

The VP7(T13) subunits form themselves into trimers, even when expressed as a single protein, which (AHSV) may then assemble into flat sheets of hexameric rings [9]. One may think of the completion of the outer core layer as the crystallisation of 260 preformed trimers of VP7(T13) onto a fragile subcore made up of 120 molecules of VP3(T2). The driving force for this assembly of the outer core layer would be the adhesion of the relatively flat surfaces, with the relative orientation of adjacent trimers of VP7(T13) being determined by their side-to-side interactions with each other. This two-dimensional crystallisation model avoids the need for conformational switching (a mechanism used by less complex viruses for controlled capsid assembly).

The atomic structure of the principle proteins of the core was derived from the analysis of BTV1 SA, however there are other components of the core whose

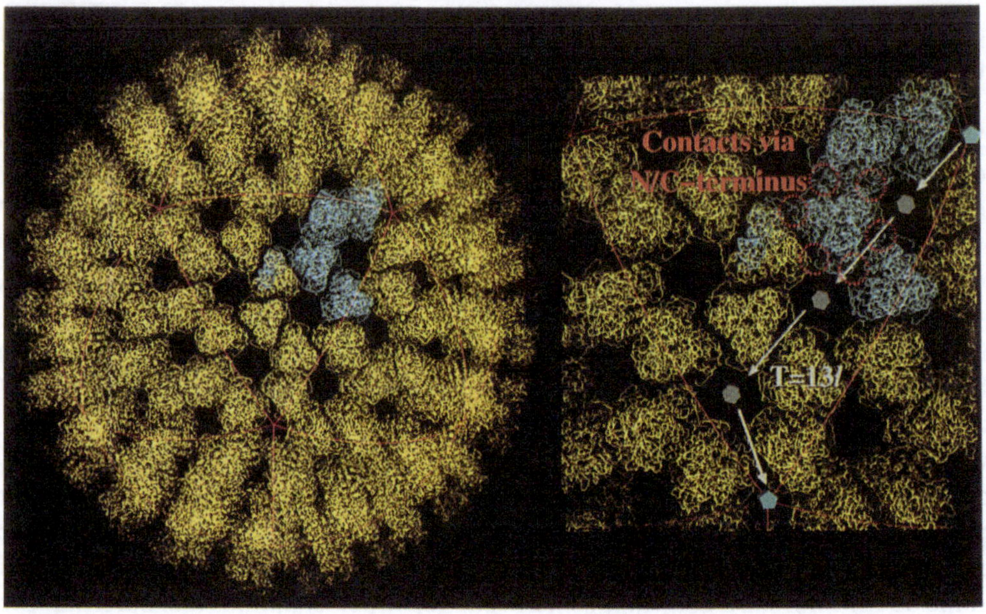

Fig. 4. The architecture of the VP7(T13) layer of BTV. Each fuzzy yellow object is a trimer of VP7(T13) with the organisation of the trimers over the surface of the core being clarified by picking out the icosahedral building block in blue (red lines join icosahedral 5-fold axes to help guide the eye). The outer diameter is 730 Å. The right hand panel shows a close-up, with some of the local 6-fold axes (a consequence of the T = 13 arrangement of the subunits) shown in grey and two adjacent icosahedral 5-fold axes shown in blue. The dotted red circles indicate points of contact between neighbouring trimers

Fig. 5. VP7(T13) trimer-trimer contacts in the BTV core. The semi-transparent green surface encloses one trimer whilst the other (related by a 2-fold axis running vertically out of the page) is shown as wires and tubes to represent the course of the polypeptide chain. It should be apparent that a single helix is central to the contact and the contact is largely through short hydrophobic side-chains

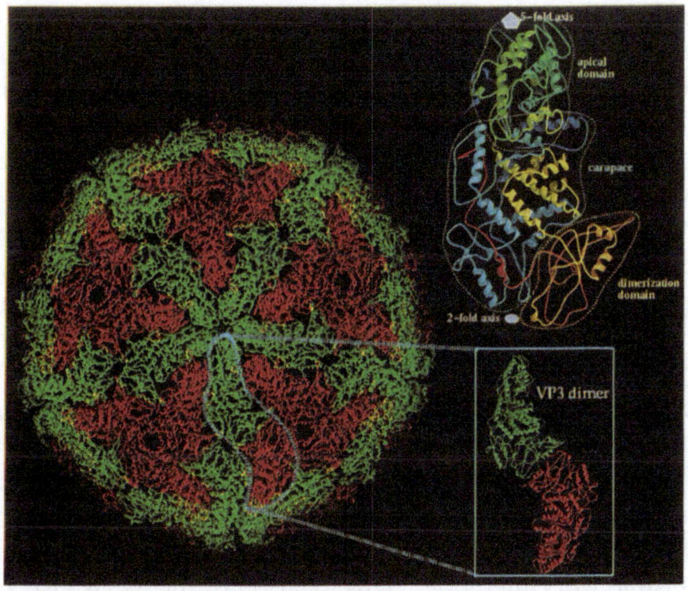

Fig. 6. The architecture of the VP3(T2) layer of BTV. The A and B molecules (the distinction is discussed in the text) are shown in green and red respectively. Each of these 120 molecules has a structure similar to that illustrated on right of the image

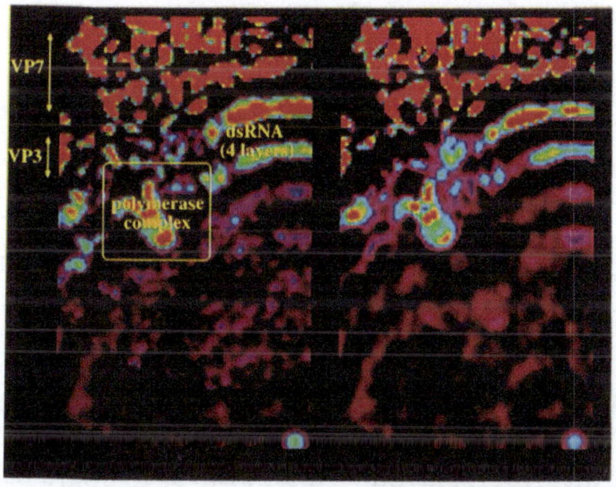

Fig. 7. Sections through the electron density for BTV1-SA and BTV10-USA. The centre of the particle is marked as a blue dot at the bottom of the picture and the various radial layers are labelled on the left hand panel

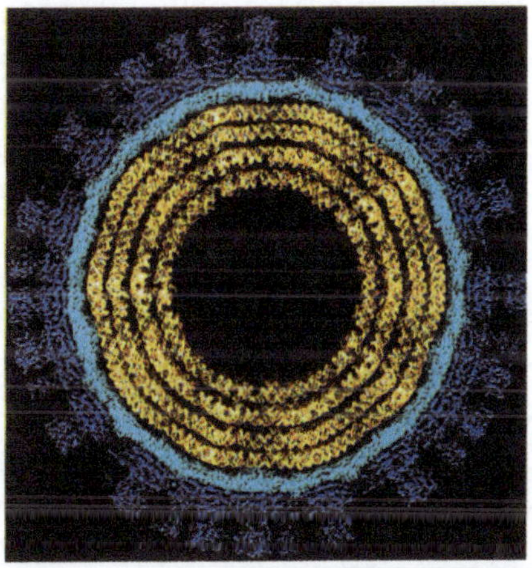

Fig. 8. A thin section through the core of BTV. The VP7(T13) layer is shown in dark blue, the thin VP3(T2) layer in pale blue and the four layers of genomic dsRNA are largely yellow

structure cannot fully obey icosahedral symmetry since the number of molecules in each particle is not a multiple of 60. To our initial surprise we detected some electron density within the VP3(T2) protein layer and we have commenced a careful analysis of this by determining the structure of BTV10 USA, in addition to BTV1 SA. This analysis is at lower resolution, but this is appropriate since the features we wish to see are more blurred. Figure 7 shows the electron density that we find within the two viruses. There is density positioned directly under the icosahedral 5-fold axes, which we interpret as the 'minor proteins' associated with transcription of the viral genome, which we term the transcriptase complex. (TC). These proteins are reported to possess helicase [58], polymerase [55, 59] and RNA capping activity [16, 44]. We estimate (Table 1) that the TC is formed from 1 molecule of VP1 (polymerase), a dimer of VP4 (capping enzyme) and the hexameric VP6 (helicase). We assume that 12 such complexes are arranged at the 5-fold apices of the icosahedrally symmetric sub-core during assembly of the virus. Although there is no direct evidence for this, from the close association of the TC with the VP3(T2) layer and from the absence of large holes in that layer, it seems a plausible working model. This location of the TC is also supported by observation of transcriptionally active particles of other members of the family *Reoviridae* [2, 14, 22, 37, 51, 61]. In the orbivirus core, derived from intact virus, 10 of these complexes will be intimately associated with one of the 10 segments of genomic dsRNA. This is in agreement with the data for both BTV1 and BTV10 shown in Fig. 7 where four layers of electron density line the VP3(T2) shell and are closely associated with the density we assign to the TC (the overall layout of the RNA within these four layers is shown in Fig. 8). This demonstrates that genomic RNA is packaged in a rather structured way which follows, to a considerable extent, the icosahedral symmetry of the core. This order is imposed by the VP3(T2) layer which possesses, on its inner surface, shallow grooves that seem to form tracks along which tubes of dsRNA lie. These grooves are chemically 'bland', and there is a paucity of specific interactions with the dsRNA. However, the peculiar architecture of the VP3(T2) layer means that the A and B molecules of VP3(T2) are situated at different distances from each of the 5-fold apices. This appears to generate a spiral of grooves for the RNA around each 5-fold apex.

The distance between the layers of RNA is 26–30 Å and although we cannot produce a full 'wiring diagram' for the arrangement of the genome within the core we can suggest a plausible model for the general layout. This allows us to begin to understand how RNA can be tightly packed within the core and yet remain sufficiently fluid for efficient, repeated, and independent transcription of the 10 segments of dsRNA (totalling about 20,000 base pairs of RNA). We suggest that each RNA segment is associated with a TC complex at the 5-fold apices and is wound around this TC, until it hits a neighbouring segment. At this point due to steric hindrance it may flip down to form the next layer, and wind back in towards the TC, like rope coiled up on a ships deck, orderly stored and yet ready for action. This model is also in agreement with some recent observations of other members of the family *Reoviridae* made by cryo electron microscopy [31, 47, 51].

Conclusions

As the most complex macromolecular assembly yet visualised in atomic detail the BTV core may provide a paradigm for other members of the family *Reoviridae* [35, 36]. BTV has found a solution to the problem of assembling a structure without exact symmetry and without conformational switching in the VP7(T13) layer. The core is segregated into two layers. The thin inner layer is formed from 120 large triangular plates of VP3(T2), constructed using an architecture previously unseen, to form a complete shell whilst minimising the inevitable conformational distortion. This fragile internal scaffold defines the size of the virus and is rigidified by the cloak of 260 essentially identical trimers of VP7(T13), which assembles following the principles of quasi-equivalence. Extensive non-specific interactions, ameliorate the symmetry mismatch between the two layers to produce a robust shell. Such a mixture of assembly modes provides a mechanism appropriate for the construction of even larger macromolecular complexes. It is clear from our analyses of VP7(T13) and from mapping amino acid sequence data onto the structural information for the BTV core that BTV provides a good working model for ASHV and the orbiviruses.

Acknowledgements

We thank the SRS, ESRF & EMBL for their assistance in data collection, in particular the staff of ID2 at the ESRF; the disease security officers D. Goodridge, S. Williams, A. Meyer & P. Wilkinson; M. Pickford for technical assistance; R. Bryan, K. Measures & R. Esnouf for computing; S. Lee for assistance with figures; P. Roy and E. Y. Jones for discussion. This work has been given long-term support by the BBSRC and MRC; RM is supported by a CJ Martin Fellowship from NHMRC Australia; PPCM and JNB are supported by MAFF.

References

1. Bacon D, Anderson WF (1988) A fast algorithm for rendering space-filling molecule pictures. J Mol Graphics 6: 219–220
2. Bartlett NM, Gillies SC, Bullivant S, Bellamy AR (1974) Electron-microscopy of reovirus reaction cores. J Virol 14: 315–326
3. Basak AK, Grimes JM, Gouet P, Roy P, Stuart DI (1997) Structures of orbivirus VP7: implications for the role of this protein in the viral life cycle. Structure 5: 871–883
4. Basak AK, Gouet P, Grimes J, Roy P, Stuart D (1996) Crystal-structure of the top domain of African horse sickness virus VP7 – comparisons with bluetongue virus VP7. J Virol 70: 3797–3806
5. Bourgeois D, Nurizzo D, Kahn R, Cambillau C (1998) An integration routine based on profile fitting with optimized fitting area for the evaluation of weak and/or overlapped two-dimensional Laue or monochromatic patterns. J Appl Cryst 31 (in press)
6. Brookes SM, Hyatt AD, Eaton BT (1993) Characterization of virus inclusion bodies in bluetongue virus-infected cells. J Gen Virol 74: 525–530
7. Brunger AT (1992) X-PLOR Version 3.1 (Yale University Press)
8. Burroughs JN, Grimes JM, Mertens PPC, Stuart DI (1995) Crystallization and preliminary X-ray analysis of the core particle of bluetongue virus. Virology 210: 217–220
9. Burroughs JN, O'Hara RS, Smale CJ, Hamblin C, Walton A, Armstrong R, Mertens PPC (1994) Purification and properties of virus particles, infectious sub viral particles,

cores and VP7 crystals of African horse sickness virus serotype 9. J Gen Virol 75: 1849–1857
10. Butcher SJ, Dokland T, Ojala PM, Bamford DH, Fuller SD (1997) Intermediates in the assembly pathway of the double-stranded RNA virus phi 6. EMBO J 16: 4477–4487
11. Calisher CH, Mertens PPC (1998) Taxonomy of African horse sickness virus. In: Mellor PS, Baylis M, Hamblin C, Calisher CH, Mertens PPC (eds) African horse sickness. Springer, Wien New York, pp 3–11 (Arch Virol [Suppl] 14)
12. Caspar DLD, Klug A (1962) Physical principles in the construction of regular viruses. Cold Spring Harbor Sym Quant Biol 27: 1–24
13. Chiu W, Burnett RM, Garcea RL (eds) (1997) Structural biology of viruses. Oxford University Press, New York
14. Dai R, Wu A, Shen X, Qian L, Sun Y (1982) Isolation of the genome-enzyme complex from cytoplasmic polyhedrosis virus of silkworm Bombyx mori. Sci Sin B 25: 29–35
15. Delano WL, Brunger AT (1995) the direct rotation function – rotational Patterson correlationb search applied to molecular replacement. Acta Crystallogr D 51: 740–748
16. Devi R, Burroughs JN, Mertens PPC, Roy P (1998) Transmethylase 1 and transmethylase 2 activities associated with the Cap protein (VP4) of BTV, in native cores, core like particles and as a purified, baculovirus expressed protein. Proc Natl Acad Sci USA (in press)
17. Eaton BT, Hyatt AD, Brookes SM (1990) The replication of bluetongue virus. Curr Top Microbiol Immunol 162: 89–118
18. Esnouf RM (1997) An extensively modified version of MolScript that includes greatly enhanced colouring capabilities. J Mol Graphics 15: 133–138
19. Fu H, Bland AP, Wade-Evans A M, Mertens PPC, Mellor PS (1998) Ultra structural comparison of bluetongue virus replication in mammalian and Culicoides cells. (Manuscript submitted)
20. Fukusho A, Yu Y, Yamaguchi S, Roy P (1989) Completion of the sequence of bluetongue virus serotype 10 by the characterization of the structural protein, VP6, and a non-structural protein, NS2. J Gen Virol 70: 1677–1689
21. Foulser D (1995) IRIS EXPLORER; A framework for investigation. Comp Graphics 29: 13–16
22. Gillies S, Bullivant S, Bellamy AR (1971) Viral RNA polymerases: electron microscopy of reovirus reaction cores. Science 174: 694–696
23. Gouet P, Grimes JM, Diprose JM, Malby R, Burroughs JN, Zeintara S, Stuart, DI, Mertens PPC (1998) Crystallographic studies show ordered double-stranded RNA in bluetongue virus. (submitted)
24. Grimes JM, Basak AK, Roy P, Stuart DI (1995) The crystal structure of bluetongue virus VP7. Nature 373: 167–170
25. Grimes JM, Jakana J, Ghosh M, Basak AK, Roy P, Chiu W, Stuart DI, Prasad BVV (1997) An atomic model of the outer layer of the bluetongue virus core derived from X-ray crystallography and electron cryomicroscopy. Structure 5: 885–893
26. Grimes JM, Burroughs JN, Gouet P, Diprose JM, Malby R, Zeintara S, Mertens PPC, Stuart DI (1998) The atomic structure of the bluetongue virus core. Nature (in press)
27. Hadfield AT, Lee WM, Zhao R, Oliveira MA, Minor I, Rueckert RR, Rossmann MG (1997) The refined structure of human rhinovirus 16 at 2.15 Å resolution: Implications for the viral life cycle. Structure 5: 427–441
28. Harlos K (1992) Micro-bridges for sitting-drop crystallizations J Appl Crystallogr 25: 536–538
29. Hewat EA, Booth TF, Loudon PT, Roy P (1992) Three-dimensional reconstruction of baculovirus expressed bluetongue virus core-like particles by cryo-electron microscopy. Virology 189: 10–20

30. Hewat EA, Booth TF, Roy P (1992) Structure of bluetongue virus particles by cryoelectron microscopy. J Struct Biol 109: 61–69
31. Hill, CL (1997) Analysis of the structure and function of a cytoplasmic polyhedrosis virus. Ph.D. Thesis, Oxford University
32. Holmes IH, Boccardo G, Estes MK, Furuichi MK, Hoshino Y, Joklik WK, McCrae M, Mertens PPC, Milne RG, Samal KSK, Shiata E, Winton JR, Uyeda I (1995) Family Reoviridae. In: Murphy FA, Fauquet CM, Bishop DHL, Ghabrial SA, Jarvis AW, Martelli GP, Mayo MA, Summers MD (eds) Virus Taxonomy. Classification and Nomenclature of Viruses. Sixth Report of the International Committee on Taxonomy of Viruses. Springer, Wien New York, pp 208–237 (Arch Virol [Suppl] 10)
33. Huismans H, Van Dijk AA, Els HJ (1987) Uncoating of parental bluetongue virus to core and subcore particles in infected L cells. Virology 157: 180–188
34. Johnson JE, Speir JA (1997) Quasi-equivalent viruses: a paradigm for protein assemblies. J Mol Biol 269: 665–675
35. Joklik WK (1983) In: Fraenkel-Conrat H, Wagner RR (eds) The Reoviridae. Plenum Press, New York
36. Juuti JT, Ravantti JJ, Bamford DH (1997) The common and uncommon among dsRNA viruses. In: Arias CF, Lopez S (eds) 6th International Symposium on dsRNA Viruses. Cocoyoc, Mexico, pp A1–A10
37. Lawton JA, Estes MK, Prasad BV (1997) Three-dimensional visualisation of mRNA release from actively transcribing rotavirus particles. Nature Struct Biol 4: 118–121
38. Lawton JA, Zeng CQY, Mukherjee SK, Cohen J, Estes MK, Prasad BVV (1997) Three-dimensional structural analysis of recombinant rotavirus-like particles with intact and amino-terminal deleted VP2: implications for the architecture of the VP2 capsid layer. J Virol 71: 7363–7360
39. Liddington RC, Yan Y, Moulai J, Sahli R, Benjamin TL, Harrison SC (1991) Structure of simian virus 40 at 3.8-Å resolution. Nature 354: 278–284
40. Loudon PT, Roy P (1991) Assembly of five bluetongue virus proteins expressed by recombinant baculoviruses: inclusion of the largest protein VP1 in the core and virus-like particles. Virology 180: 798–802
41. Loudon PT, Roy P (1992) Interaction of nucleic acids with core-like and subcore-like particles of bluetongue virus. Virology 191: 231–236
42. Merritt EA, Murphy MEP (1994) Raster 3D version 2.0. A program for photorealistic molecular graphics. Acta Crystallog D 50: 869–873
43. Mertens PPC, Burroughs JN, Anderson J (1987) Purification and properties of virus particles, infectious subviral particles, and cores of bluetongue virus serotypes 1 and 4. Virology 157: 375–386
44. Mertens PPC, Burroughs JN, Wade-Evans AM, Le Blois H, Oldfield S, Basak A, Loudon P, Roy P (1992) Analysis of guanylyltransferase and transmethylase activities associated with bluetongue virus cores and recombinant baculovirus-expressed core-like particles. Bluetongue, African horse sickness and related orbiviruses. In: Walton TE, Osburn BI (eds) Second International Symposium on bluetongue, African horse sickness and related orbiviruse. CRC Press, Boca Raton, pp 404–415
45. Mertens PPC, Burroughs JN, Walton A, Wellby MP, Fu H, O'Hara RS, Brookes SM, Mellor PS (1996) Enhanced infectivity of modified bluetongue virus particles for two insect cell lines and for two Culicoides vector species. Virology 217: 582–593
46. Mertens PPC et al (in preparation) Family Reoviridae. In: Seventh Report of the International Committee on Taxonomy of Viruses
47. Moss SR, Nuttall PA (1994) Subcore- and core-like particles of Broadhaven virus (BRDV), a tick-bourne orbivirus, synthesized from baculovirus expressed VP2 and VP7, the major core proteins of BRDV. Virus Res 32: 401–407

48. Oliveira MA, Zhao R, Lee WM, Kremer MJ, Minor I, Rueckert RR, Diana GD, Pevear DC, Dutko FJ, Mckinlay MA, Rossmann MG (1993) The structure of human rhinovirus 16. Structure 1: 51–68
49. Otwinoski ZO (1993) DENZO. In: Saywer L, Isaacs N, Bailey S (eds) Proceedings of CCP4 study weekend: Data collection and reduction
50. Prasad BVV, Yamaguchi S, Roy P (1992) Three-dimensional structure of single-shelled bluetongue virus. J Virol 66: 2135–2142
51. Prasad BVV, Rothnagel R, Zeng CQY, Jakana J, Lawton JA, Chiu W, Estes MK (1996) Visualisation of ordered genomic RNA and localisation of transcriptional complexes in rotavirus. Nature 382: 471–473
52. Roussel A, Inisan AG, Cambillau C (1991) AFMB and biographics. Marseille, France
53. Roy P (1989) Bluetongue virus genetics and genome structure. Virus Res 13: 179–206
54. Roy P (1989) Bluetongue virus proteins. J Gen Virol 73: 3051–3064
55. Roy P, Fukusho A, Ritter D G, Lyons D (1988) Evidence for genetic relationship between RNA and DNA viruses from the sequence homology of a putative polymerase gene of bluetongue virus with that of vaccinia virus: conservation of RNA polymerase genes from diverse species. Nucleic Acids Res 16: 11759–11767
56. Schoehn G, Moss SR, Nuttall PA, Hewat EA (1997) Structure of Broadhaven virus by cryoelectron microscopy: correlation of structural and antigenic properties of Broadhaven virus and bluetongue virus outer capsid proteins. Virology 235: 191–200
57. Stuart D (1993) Viruses. Curr Opin Struct Biol 3: 167–174
58. Stauber N, Martinez-Costas J, Sutton G, Monastyrskaya K, Roy P (1997) Bluetongue virus VP6 protein binds ATP and exhibits an RNA-dependent ATPase function and a helicase activity that catalyze the unwinding of double-stranded RNA substrates. J Virol 71: 7220–7226
59. Urakawa T, Ritter DG, Roy P (1989) Expression of largest RNA segment and synthesis of VP1 protein of bluetongue virus in insect cells by recombinant baculovirus; association of VP1 protein with RNA polymerase activity. Nucleic Acids Res 17: 7395–7401
60. Van Dijk AA, Huismans H (1980) The in vitro activation and further characterization of the bluetongue virus-associated transcriptase. Virology 104: 347–356
61. Yazaki K, Miura K (1980) Relation of the structure of cytoplasmic polyhedrosis virus and the synthesis of its messenger RNA. Virology 105: 467–479
62. Zhu Y, Hemmings AM, Iwasaki K, Fujiyoshi Y, Zhong B, Yan J, Isogai M, Omura T (1997) Details of the arrangement of the outer capsid of rice dwarf phytoreovirus, as visualized by two-dimensional crystallography. J Virol 71: 8899–8901

Authors' address: Dr. P. P. C. Mertens, Institute for Animal Health, Pirbright Laboratory, Ash Road, Pirbright, Woking Surrey GU24 ONF, U.K.

Characterization of two African horse sickness virus nonstructural proteins, NS1 and NS3

V. van Staden, C. C. Smit, M. A. Stoltz, F. F. Maree, and **H. Huismans**

Department of Genetics, University of Pretoria, Pretoria, South Africa

Summary. Each of the ten segments of the African horse sickness virus (AHSV) genome encodes at least one viral polypeptide. This report focuses on the non-structural proteins NS1 and NS3, which are encoded by genome segments 5 and 10 respectively. The NS1 protein assembles into tubular structures, which are characteristically produced during orbivirus replication in infected cells. NS1 expressed by a recombinant baculovirus in Sf9 cells also forms tubules, which were analysed electron microscopically. These tubules had an average diameter of 23 ± 2 nm, which is less than half the width of the corresponding bluetongue virus (BTV) tubules. They were also more fragile at high salt concentrations or pH. The cytotoxic effects produced by NS3 were examined by constructing of mutated versions and expressing them in insect cells. Substitution of amino acids 76–81 in a conserved region (highly conserved amongst all AHSV NS3 proteins, as well as other orbiviruses) with similar amino acids, did not influence the cytotoxicity of the mutant protein. However, mutation of four amino acids, from hydrophobic to charged amino residues, (aa 165–168) in a predicted transmembrane region of NS3, largely abolished its cytotoxic effect. It is considered likely that the mutant protein is unable to interact with cellular membrane components, thereby reducing its toxicity.

Introduction

African horse sickness virus (AHSV) is a member of the genus *Orbivirus* (family *Reoviridae*), of which bluetongue virus (BTV) is the prototype member. The ten double-stranded (ds)RNA genome segments of these viruses encode at least 10 virus-specified proteins, three of which are the nonstructural proteins, NS1, NS2 and NS3 [10, 13]. The exact role of these nonstructural proteins in virus replication is not clear, and this applies in particular to AHSV. As part of a long-term study of the function of AHSV nonstructural proteins we have recently focused our attention on two of these proteins, namely NS1 and NS3. The first part of this paper concerns the structure of AHSV NS1 tubules, as synthesized in insect cells infected with an "AHSV NS1" baculovirus recombinant. The second part of the paper focuses on the different conserved structural elements of NS3 and their contribution to its observed cytotoxicity.

Structure of NS1 tubules

It has previously been shown that the unique virus-specified tubular structures observed in the cytoplasm of AHSV-infected cells are composed of NS1 [3]. These tubules are characteristically produced in infected cells, during the replication of orbiviruses. When the NS1 genes of either BTV, or epizootic hemorrhagic disease virus (EHDV) are expressed in insect cells, the NS1 protein self-assembles into tubular structures similar to those observed in virus-infected cells [15, 12]. The BTV NS1 tubules have a diameter of 52.3 nm and are composed of a coiled ribbon of NS1 dimers with 22 dimers per helix turn [2]. Very little is known about the structure of the AHSV tubules except that they appear to be morphologically different from those of EHDV and BTV [3]. To obtain more information on the characteristics of AHSV NS1 we have recently cloned, sequenced and expressed the gene encoding NS1 from AHSV serotype 6 [8]. This has enabled us to produce large amounts of NS1 tubules in *Spodoptera frugiperda* (Sf9) insect cells. AHSV NS1 tubules were subsequently purified from cellular lysates of cells infected with the "NS1 recombinant" baculovirus, by sucrose gradient centrifugation [8]. Fractions containing material sedimenting in the 200–400 S region were pooled. The particulate material was then pelleted by centrifugation (at 90000 **g** for 2 h), resuspended in 0.15 M STE buffer (0.15 M NaCl; 0.01 M Tris pH 7.4; 0.001 M EDTA) and analysed by electron microscopy, using negative staining with 2% (w/v) uranyl acetate. The results (Fig. 1) indicated the presence of tubules with an average diameter of 23 ± 2 nm, which varied in length up to 4 μm. These AHSV tubules had an internal structure with a fine reticular "cross-weave" appearance, which is very different in appearance to that of BTV and EHDV tubules. There was no evidence of the ladder-like or segmented appearance found in the much wider BTV (68 nm) and EHDV (52 nm) tubules. Hollow, circular structures were also observed which could represent cross-sections of tubules or very short subsections of larger tubules. The effect of different biophysical conditions on the morphology of the tubules was investigated. They were found to be unstable, or significantly reduced in length, at high salt concentrations (> 1 M NaCl, or > 0.2 M $CaCl_2$). AHSV tubules were also unstable at a higher pH range (pH 8.5 or above) but like those of BTV [9] were relatively resistant to lower pH, although they appeared to be denatured at pH 5.0 or less. Most of the results suggested that the AHSV tubules are more fragile than BTV tubules.

Mutation analysis of NS3

Proteins NS3 and NS3A of AHSV are encoded from two in-phase overlapping reading frames of segment 10, so that they differ only with respect to the ten addi-

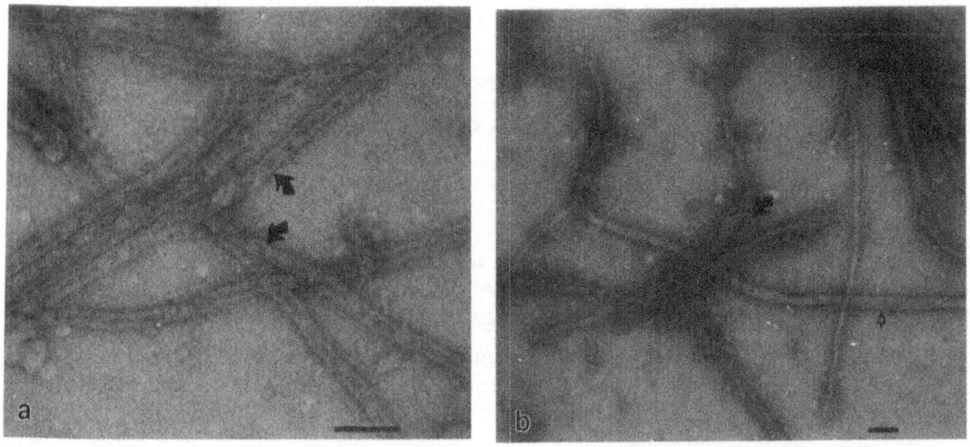

Fig. 1. Negative contrast electron micrographs. The recombinant baculovirus-expressed NS1 tubules were purified by sucrose gradient centrifugation and stained with 2% uranyl acetate (**a**). The purified tubules were decorated with anti-AHSV-6 antiserum before staining with uranyl acetate (**b**). The NS1 tubules, indicated by solid arrows, could be distinguished from the baculovirus tubule contaminants (open arrows), which are unlabelled. Bar markers represent 100 nm

tional amino acids present at the N terminus of NS3 [16]. The exact function of these proteins in the viral replication cycle is unclear. However, the NS3 protein has been shown to be present in membrane components of infected cells. Immunogold labelling has been used to locate NS3/NS3A in areas of the plasma membrane associated with the release of viral particles from AHSV-infected Vero cells, as well as in membranes of cells expressing NS3 from a recombinant baculovirus [14]. This strongly suggests that AHSV NS3 is involved in the final stages of viral morphogenesis, in a manner similar to that previously proposed for the homologous NS3 protein of BTV [5]. A novel feature of AHSV NS3 was that its expression appeared to have a cytotoxic effect on the host cells [17]. This raises the question of the involvement of NS3 in other virus attributes, such as pathogenicity and virulence (see O'Hara et al., this publication). Our long term aim in studying AHSV NS3, is the elucidation of its structure/function relationship, and specifically its role in viral infection and morphogenesis. We have previously cloned and sequenced genome segment 10 and expressed NS3 of AHSV-3 [16, 17]. The strategy currently being employed is to construct different mutant forms of NS3, express them via the baculovirus expression system, then evaluate and compare their characteristics (such as cytotoxicity) with those of the wild type NS3 protein.

The nucleotide sequences of genome segment 10 of isolates from five AHSV serotypes have been determined and compared to each other, as well as to the cognate genes of other orbiviruses ([1, 4, 6, 11, 16], see also Martin et al., this publication). Based on these comparisons five different domains, which were thought to be possible structural or catalytic sites of the NS3 protein, were identified as potential targets for mutational analyses (Fig. 2). These include: (1 and 2) two hydrophobic domains (TM1 and TM2: aa 116–137 and aa 154–176) that are

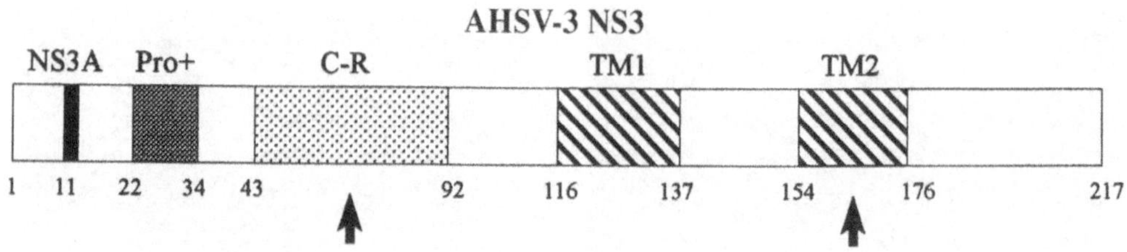

Fig. 2. Regions of the NS3 protein targeted for site-specific mutations. *NS3A* Truncated version of NS3 starting at the second in-phase ATG initiation codon; *Pro+* proline rich region; *C-R* amino acid region showing very high conservation between different NS3 proteins; *TM1*, *TM2* two regions predicted to form transmembrane helices. The arrows indicate mutants detailed in this report. Numbers refer to amino acids present in AHSV-3 NS3

predicted to form transmembrane helical domains; (3) a cluster of conserved proline residues (aa 22–34); (4) an amino acid region (aa 43–92) showing very strong conservation between all AHSV NS3 proteins, as well as among other orbiviruses; and (5) the amino terminus of the NS3 protein (to force expression of only NS3A, by selective deletion of the 5' end of the gene). This report details construction of two of these mutants, i.e. mutants 2 and 4, and characterization of their respective cytotoxic effects when expressed in insect cells.

In constructing the mutated forms of NS3, only specific amino acids were targeted and substituted. This was thought to minimize the chance that the altered protein would be misfolded or inactivated, thereby allowing the contribution of the modified region to the overall properties of the protein to be evaluated. In "mutant 2", the second of the predicted transmembrane regions (TM2: aa 154–176) was modified, from a hydrophobic region containing non-polar amino acids, to a stretch of charged amino acids (aa 165–168). This abolished the membrane-spanning potential of this domain (predicted using the Memsat computer program) [7]. For the construction of "mutant 4" a stretch of six highly conserved amino acids (aa 76–81) were targeted, each of which was substituted with a similar amino acid (e.g. Arg with Lys).

The strategy for generating the mutations was to modify the cDNA copy of AHSV-3 genome segment 10, previously tailored and cloned into the *Bam*H1 site of pUC18, by oligonucleotide-directed mutagenesis using the polymerase chain reaction (PCR). For every mutation a set of two overlapping primers was designed, which contained the specific nucleotide substitutions to encode the desired amino acid changes, as well as a restriction enzyme recognition site for cloning of the PCR generated fragments (Table 1). These primers were used, in conjunction with M13 universal forward/reverse primers, to amplify the segment 10 cDNA as two fragments. This was followed by digestion of the two fragments with the restriction enzyme specific for the primer set, (*Aat*II or *Pst*I) and ligation. The ligated fragment (± 800 bp) was then *Bam*H1 digested to remove the flanking pUC sequences resulting from the PCR, and cloned into the *Bam*H1 site of pBS. The sequences of all mutant recombinant plasmids were verified by sequencing in pBS before expression of the modified NS3 protein.

Table 1. Construction of amino acid substitution mutations of AHSV-3 NS3

	BacNS3-M2	AcNS3-M4
Primer sequences[a]	2A: 5' gggaCgtcCcgCtTAGTGGTACACAG 3' 2B: 5' ccgacGtcGCTGGAAAAATATCAGAT 3'	4A: 5' gaactgcagaTCCaCCATCGCACCAA 3' 4B: 5' ggactgcagaagGAgCCAGAACCGATAC 3'
Nucleotide sequences[b]		
WT Segment 10	512 ATG TTG TTA GCG 523	245 GCG GAA GCA TTA CGT GAT 262
Mutant S10	AAG CGG GAC GTC	GTG GAT CTG CAG AAG GAG
Amino acid sequences		
WT NS3	165 M L L A 168	76 A E A L R D 81
Mutant NS3	K R D V	V D L Q K E

[a] Mispaired bases are shown in lowercase, restriction sites underlined are *Aat* II for 2A/2B, *Pst* I for 4A/4B
[b] Nucleotide substitutions are underlined

Wild type NS3 was previously expressed in the baculovirus system, via the recombinant baculovirus AcAHSV3.10, constructed using the pAcDZ1 transfer vector [17]. The NS3 mutant 4 gene was expressed using this same system, and the resulting baculovirus recombinant was designated AcNS3-M4. The current system of choice in our laboratory for baculovirus expression is the BAC-to-BAC (Life Technologies) system, in which all manipulations and selections are done in bacterial cells. Subsequently the wild type NS3 and NS3 mutant 2 genes were cloned into transfer vector pFastBac1, transposed into the bacmid genome, and expressed in Sf9 insect cells via two recombinant baculoviruses identified as BacNS3 and BacNS3-M2 respectively. As the NS3 protein is not expressed to high levels, the presence of the protein in infected cells was confirmed by immunoblot (Fig. 3). In all cases a single protein band of approximately 24K was present.

As an initial step in characterizing the effect of the site specific mutations in the NS3 protein, an investigation was carried out into the possible cytotoxic effect of the mutant proteins on the host cells. This was achieved by infecting spin cultures of Sf9 cells with the different recombinant baculoviruses, taking samples of cells at different times post infection, staining with trypan blue, and counting the number of stained vs. unstained cells (Fig. 4). Trypan blue is a vital exclusion dye, where cells stained blue are non-viable, probably due to an increased permeability of the cell membrane. For wild type NS3, a direct correlation was observed between the time of expression of the NS3 protein and an increase in cell death.

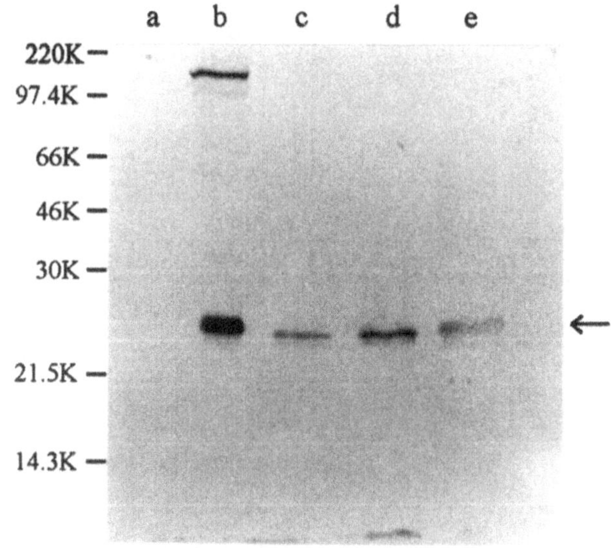

Fig. 3. Expression of AHSV-3 NS3 wildtype and mutant proteins as analysed by Western blot. Monolayers of Sf9 cells were mock infected (*a*), or infected with recombinant baculoviruses AcAHSV3.10 (*b*), BacNS3 (*c*), BacNS3-M2 (*d*) and AcNS3-M4 (*e*) at a multiplicity of 5 pfu per cell, and harvested at 72 h pi. Samples were subject to 15% SDS-PAGE, electroblotted onto a Hybond-C membrane (Amersham), and reacted with monospecific serum to AHSV-3 NS3. The sizes of the molecular weight standards are as shown with the arrow indicating the position of NS3

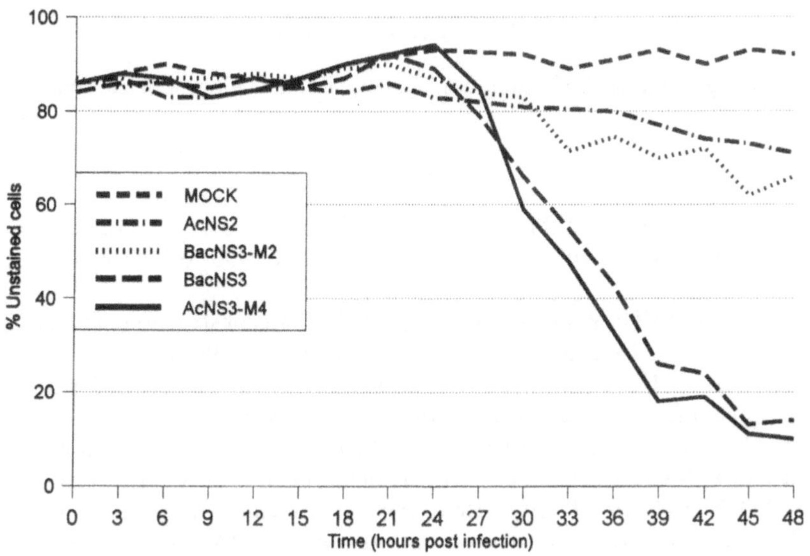

Fig. 4. Representation of the effect of different recombinant baculoviruses expressing wildtype or mutant AHSV NS3 proteins on viability of infected cells. Spin cultures of Sf9 cells were infected at a multiplicity of 5 pfu per cell, samples taken at different times post infection, stained with 0.4% trypan blue in PBS, and the number of impermeable (unstained) cells determined

Observation of the plasma membranes of AcAHSV3.10 infected cells by electron microscopy, showed them to be totally disrupted [17]. The results in Fig. 3 show that the effect of wild type NS3 on cell viability is independent of the specific baculovirus system used to express the protein. The bacmid expressed NS3 (BacNS3), also resulted in only about 10% of cells remaining viable at 48 h pi (as previously found with AcAHSV3.10 expressed NS3).

However, the two mutant NS3 proteins differed radically from each other with respect to their effect on the cells. BacNS3-M4 expression resulted in 86% cell death at 48h pi, similar to what was observed for wild type NS3. The substitution of similar amino acids in the highly conserved region of NS3 in mutant 4 therefore did not affect the cytotoxic properties of the expressed protein. It remains to be determined whether these changes result in changes to other functional or structural properties of the protein. In contrast, BacNS3-M2 expression seemed to have much less of a detrimental effect on the cells and approximately 70% of cells were still viable at 48h pi. This corresponds to the effect of another nonstructural AHSV protein, NS2, which is known not to interact with membranes and could be representative of the general effect of baculovirus infection and protein synthesis in these cells. Based on these observations we postulate that substitution of four amino acids in the second transmembrane region in mutant 2 was sufficient to abolish or diminish the cytotoxic effect of BacNS3-M2. It is thought likely that this was due to an alteration in the structure of the protein which rendered it incapable of interacting with cellular membrane components.

Acknowledgements

The authors are grateful to the Foundation for Research Development for financial support.

References

1. De Sá RO, Zellner M, Grubman MJ (1994) Phylogenetic analysis of segment 10 from African horse sickness virus and cognate genes from other orbiviruses. Virus Res 33: 157–165
2. Hewat EA, Booth TF, Wade RH, Roy P (1992) 3-D Reconstruction of bluetongue virus tubules using cryoelectron microscopy. J Struct Biol 108: 35–48
3. Huismans H, Els HJ (1979) Characterization of the tubules associated with the replication of three different orbiviruses. Virology 92: 397–406
4. Hwang G, Yang Y, Chiou J, Li JK (1992) Sequence conservation among the cognate nonstructural NS3/3A protein genes of six bluetongue viruses. Virus Res 23: 151–161
5. Hyatt AD, Zhao Y, Roy P (1993) Release of bluetongue virus-like particles from insect cells is mediated by BTV nonstructural protein NS3/NS3A. Virology 193: 592–603
6. Jensen MJ, Cheney IW, Thompson LH, Mecham JO, Wilson WC, Yamakawa M, Roy P, Gorman BM (1994) The smallest gene of the orbivirus, epizootic hemorrhagic disease, is expressed in virus-infected cells as two proteins and the expression differs from that of the cognate gene of bluetongue virus. Virus Res 32: 353–364
7. Jones DT, Taylor WR, Thornton JJ (1994) A model recognition approach to the prediction of all-helical membrane protein structure and topology. Biochemistry 33: 3038–3048
8. Maree FF, Huismans H (1997) Characterization of tubular structures composed of nonstructural protein NS1 of African horse sickness virus expressed in insect cells. J Gen Virol 78: 1077–1081
9. Marshall JJA, Fayard B, Roy P (1990) Biophysical studies on the morphology of baculovirus-expressed bluetongue virus tubules. J Gen Virol 71: 1839–1844
10. Mertens PPC, Brown F, Sangar DV (1984) Assignment of the genome segments of bluetongue virus type 1 to the proteins which they encode. Virology 135: 207–217
11. Moss SR, Jones LD, Nuttall PA (1992) Comparison of the nonstructural protein, NS3, of tick-borne and insect-borne orbiviruses. Virology 187: 841–844
12. Nel LH, Huismans H (1991) Synthesis of the virus-specified tubules of epizootic haemorrhagic disease virus using a baculovirus expression system. Virus Res 19: 139–152
13. Roy P (1992) Bluetongue virus proteins. J Gen Virol 73: 3051–3064
14. Stoltz MA, Van der Merwe CF, Coetzee J, Huismans H (1996) Subcellular localization of the nonstructural protein NS3 of African horse sickness virus. Onderstepoort J Vet Res 63: 57–61
15. Urakawa T, Roy P (1989) Bluetongue virus tubules made in insect cells by recombinant baculoviruses: expression of the NS1 gene of bluetongue virus serotype 10. J Virol 62: 3919–3927
16. Van Staden V, Huismans H (1991) A comparison of the genes which encode nonstructural protein NS3 of different orbiviruses. J Gen Virol 72: 1073–1079
17. Van Staden V, Stoltz MA, Huismans H (1995) Expression of nonstructural protein NS3 of African horse sickness virus (AHSV): evidence for a cytotoxic effect of NS3 in insect cells, and characterization of the gene products in AHSV-infected VERO cells. Arch Virol 140: 289–306

Authors' address: Prof. H. Huismans, Department of Genetics, University of Pretoria, Pretoria, 0002 South Africa.

Development of a mouse model system, coding assignments and identification of the genome segments controlling virulence of African horse sickness virus serotypes 3 and 8

R. S. O'Hara*, A. J. Meyer, J. N. Burroughs, L. Pullen, L.-A. Martin,** and **P. P. C. Mertens**

Institute for Animal Health, Pirbright Laboratory, Woking, U.K.

Summary. Attenuated (^{att}) and wild type (^{wt}) strains of the nine AHSV serotypes were evaluated for virulence in adult Balb C mice. Although most were avirulent in this system, isolates of AHSV 1^{att}, 3^{wt}, 3^{att}, 4^{wt}, 4^{att}, 5^{att}, 7^{att} and 8^{att} caused some mortality when administered via an intranasal route. After plaque cloning, only the attenuated vaccine strain of AHSV 7^{att} caused any mortality via an intravenous route. AHSV 3^{att} and AHSV 8^{wt} were virulent (V) and avirulent (AV) (respectively) in the mouse model and were selected as parental strains for production of genome segment reassortants. These progeny virus strains were plaque cloned, then characterised to identify the genome segments that influence virulence of AHSV in the mouse model.

Three virulence phenotypes were observed: fully virulent (V); fully avirulent (A); and a novel intermediate virulence (N) not expressed by either parental strain. Genome segment 2 (encoding outer capsid protein VP2) from the avirulent parent appeared to have a controlling influence in production of the A phenotype. Reassortants with the V phenotype all contained segment 2 from the virulent parent, however in each case they also contained genome segments 5 and 10, also from AHSV 3 (V). Genome segments 5 and 10 encode the smaller outer capsid protein VP5 and the non structural proteins NS3/NS3a, respectively. A combination of genome segments 2, 5 and 6 from the avirulent parent and segment 10 from the virulent parent were found in each of the virus strains with the N phenotype. However, comparison of two reassortants (A79 and A790), which differ only in a single segment, showed that replacement of genome segment 10 from the avirulent parent with that from the virulent parent, conferred the N phenotype on A790.

Present addresses: * Surrey Wildlife Trust, Pirbright, Woking Surrey, U.K.; ** ICRF Oncology Unit, RPMS Hammersmith Hospital, London, U.K.

Introduction

African horse sickness virus (AHSV) is one of nineteen species of viruses classified within the genus *Orbivirus,* family *Reoviridae* [4, 41]. AHSV is usually transmitted via the bite of certain species of *Culicoides* (biting midges) [32, 33] and causes economically important disease in horses (African horse sickness: AHS) [34]. There is evidence that AHSV can also infect other equids, dogs, elephants, camels, cattle, sheep, goats, predatory carnivores and (in special circumstances) humans [34, 41, 49].

AHSV is relatively closely related to bluetongue virus (BTV: the prototype *Orbivirus* species). The capsid of BTV and AHSV has icosahedral symmetry (Fig. 1) and is composed of an inner and an outer capsid shell (Fig. 2) [3, 13, 15, 35, 36, 50]. The inner capsid, which forms the virus core, has two distinct layers of proteins [13, 49]. The inner layer or "subcore" is formed of 120 copies of VP3 (arranged with T = 2 symmetry [13, 48]) and determines both the size and overall symmetry of the entire capsid structure. The VP3(T2) "sub core" is the only shell of the capsid which can self assemble or retain its structure in the absence of the other viral proteins [21, 28, 43]. Formation of the sub core and its association with viral RNA, within the "viral inclusion bodies" (VIB) that are formed in the cytoplasm of infected cells, appears to represent an early stage in the assembly of progeny virus particles [10]. The VP3(T2) capsid shell surrounds the ten genome segments, as well as

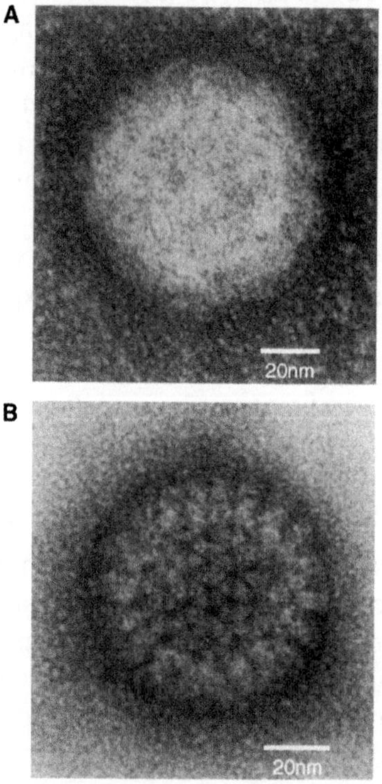

Fig. 1. AHSV virus and core particles. Electron micrographs of particles of AHSV serotype 9 purified using CsCl gradient centrifugation, as described by Burroughs et al. [3] and stained with 2% aqueous uranyl acetate. **A** Virus particles, **B** core particles

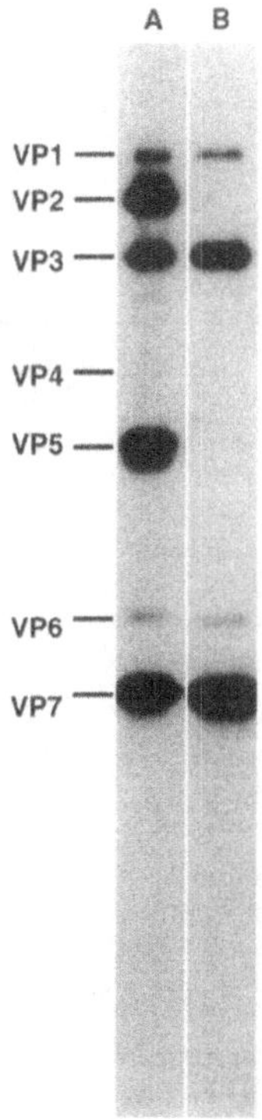

Fig. 2. AHSV structural proteins. Fluorograph of 11% polyacrylamide gel used for SDS page analysis [26] of the structural proteins of purified virus and core particles of AHSV 9, labelled with [^{35}S] methionine as described by Burroughs et al. [3]. **A** Virus particles, **B** core particles

minor virus structural proteins that are the core associated enzymes (VP1, VP4 and VP6) [12, 13, 48]. VP3(T2) makes up the inner surface of a compartment in which the dsRNA segments can function as templates for several rounds of mRNA synthesis by the core associated transcriptase and capping enzymes [7, 12, 35, 37, 51]. The outer surface of the core particle is composed of 780 copies of VP7(T13) (arranged as a lattice with T=13 *l* symmetry), which gives the core structure greater stability and rigidity [13, 21, 28]. VP7(T13) is a highly conserved protein representing the major serogroup (species) specific antigen in both BTV and AHSV. VP7(T13) of AHSV can assemble into large crystalline arrays within the cytoplasm of infected cells (Fig. 3) [3]. These crystals are formed of flat sheets of hexagonal rings, similar to the rings of the VP7(T13) trimers that are observed of the surface of the core particles (Fig. 1) [3, 13, 48]. VP7(T13) on the core surface can mediate cell attachment and penetration (particularly of insect vector cells), resulting in infection by core particles, even in the absence of outer capsid proteins [36]. The outer capsid layer, which is also composed of two major proteins (VP2 and VP5 [3]), contains the major serotype specific, neutralisation antigens and is in centrally involved in cell attachment and penetration, particularly in mammalian systems [2, 18, 19, 36, 47].

BTV and AHSV encode three distinct nonstructural proteins [38] NS1, NS2 and NS3. NS1 forms tubular structures with a distinctive morphology [16, 17], which have an unknown function but are a characteristic feature of orbivirus replication

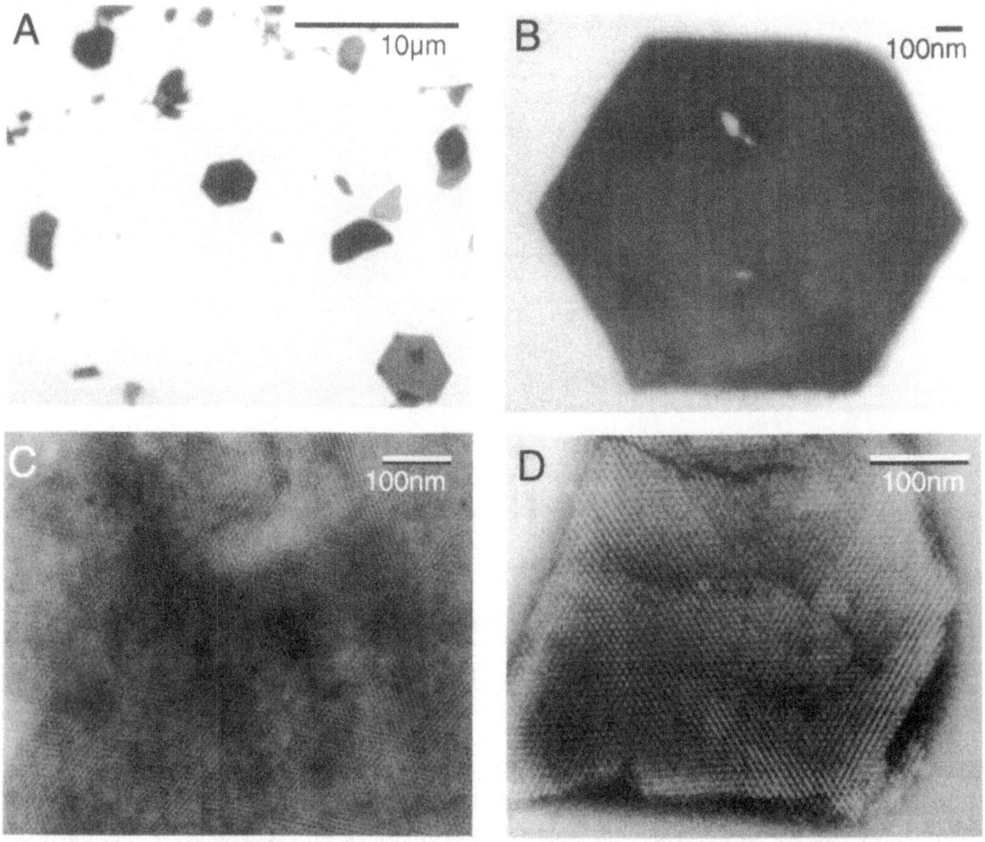

Fig. 3. VP7(T13) crystals. Electron micrographs of VP7(T13) crystals purified from BHK-21 cells infected with AHSV 9, as described by Burroughs et al. [3] and then stained with 2% methylamine tungstate. **A** Typical preparation, **B** Crystal with intact hexagonal outline, **C** Lattice detail from hexagonal crystal, **D** Fragment showing edge detail and hexagonal rings of VP7(T13) trimers

[1, 41]. NS2 is a phosphoprotein with ssRNA binding activity [20, 50]. It forms a major component of the matrix of the VIB, which are the sites of virus replication and assembly in the infected cell cytoplasm [1, 11]. NS3 is a small membrane protein that exists as two forms (NS3 and NS3a) translated from two in frame initiation codons near the 5' end of the segment 10 mRNA [22, 27, 30, 53]. NS3 can be glycosylated [56] and is involved in release of virus particles from infected cells within a few hours of infection [8, 23, 24, 30]. This release occurs prior to visible cytopathic effect or cell lysis. Unlike the conserved NS3/3a proteins of BTV, these small nonstructural proteins of AHSV show significant variations between isolates [6, 22, 27, 42, 46, 52, 57, 58] and can be separated into three distinct groups (α, β and γ) on the basis of their RNA/ protein sequence variation [30].

The genome segments of dsRNA viruses can be separated by electrophoresis in either agarose or polyacrylamide gels, forming patterns of segment migration that are often characteristic of the virus species (agarose gels), or which can be used to distinguish even quite closely related viruses (polyacrylamide gels) [45].

The individual genome segments can be assigned to the proteins for which they code, using a technique pioneered with orthoreoviruses by McCrae and Joklik [31]. This method involves denaturation of the isolated dsRNA genome segments and direct translation of the +ve RNA strand into proteins, using an in vitro assay system and has subsequently been used for BTV and AHSV [14, 38, 45].

A characteristic of the multi segmented RNA viruses, is their ability to exchange genome segments when two genetically compatible but distinct viruses infect the same cell, a process termed "reassortment". The limited number of studies that have been published indicate that viruses within a single *Orbivirus* serogroup (species) can reassort genome segments but viruses in different species cannot [5, 34, 39–41]. This property now forms one of the parameters used to define virus species within each of the genera of the family *Reoviridae* [4, 41].

For each of the serotypes of AHSV, vaccines strains have been produced by multiple passages in mouse brain, resulting in attenuation for their more usual mammalian host (horses). We have evaluated both the wild type (wt) and vaccine strains (att) of each AHSV serotype for virulence, during development of a mouse model system. We subsequently used genome segment reassortants of AHSV 3^{att} and 8^{wt}, to identify the genes that control virulence in an adult mouse model. The proteins encoded by each genome segment were determined in a genome segment assignment study carried out with AHSV 8^{wt}, allowing us to identify which of the viral proteins are also involved in virulence. Their involvement is discussed with respect to protein function.

Materials and methods

Viruses

The viruses used in this study were all originally isolated from infected horses in South Africa with the exception of AHSV 4 (Ricarda), which was isolated from an infected horse in Spain in 1987 (Table 1). Virus stocks were prepared by passage in BHK-21 cells and their titre determined by assay in microtitre plates, as previously described [9, 36]. Purified virus was prepared using methods described by Burroughs et al. [3].

Genome segment coding assignments

The genome segments of AHSV 8 ^{wt} were separated by polyacrylamide gel electrophoresis, recovered individually from the gel and translated, using a commercially available in vitro translation assay system (Amersham), into the proteins which they encode. These experiments were carried out as described by Mertens et al. [38, 45], using a modification of the technique pioneered by McCrae and Joklik [31]. The coding assignments were "generalised" as far as possible for other members of the AHSV species, by using the uniform migration pattern of AHSV genome segments during agarose gel electrophoresis (as described previously for BTV [45]).

Production and characterisation of AHSV reassortants

Reassortant virus strains were produced by co-infection of BHK cells with plaque cloned strains of AHSV 3^{att} (virulent parent, in the mouse model system) and AHSV 8^{wt} (avirulent parent, in the mouse model system), as described by O'Hara [41]. Reassortant progeny

Table 1. African horse sickness viruses

Virus[a]	Isolate designation		Passage history[b]
AHSV 1^{wt}	29/63	18/11/87	MB3-BHK8
AHSV 2^{wt}	82/61	26/08/68	MB3-BHK6
AHSV 3^{wt}	13/63	04/12/87	MB3-BHK6
AHSV 4^{wt}	RICARDA	21/04/89	MB1-BHK6
AHSV 5^{wt}	30/62	26/08/68	MB3-BHK6
AHSV 6^{wt}	39/63	23/07/75	MB3-BHK5
AHSV 7^{wt}	31/62	09/09/75	MB4-BHK6
AHSV 8^{wt}	10/62	25/11/63	MB3-BHK6
AHSV 9^{wt}	90/61	02/05/73	MB3-BHK6
AHSV 1^{att}	A501	11/07/73	+ BHK2
AHSV 2^{att}	OD	21/05/75	+ BHK3
AHSV 3^{att}	L	29/06/60	+ BHK3
AHSV 4^{att}	VRYHEID	29/06/60	+ BHK3
AHSV 5^{att}	VH	29/06/60	+ BHK3
AHSV 6^{att}	114	29/06/60	+ BHK3
AHSV 7^{att}	KAREN	29/06/60	+ BHK3
AHSV 8^{att}	18/60	27/04/76	+ BHK4
AHSV 9^{att}	90/61	02/05/73	+ BHK4

[a] Viruses designated with the suffix 'wt' are wild type strains that have been passaged in mouse brain and BHK cells a small number of times during isolation. Viruses designated with the suffix 'att' are attenuated vaccine strains that have undergone multiple passage (100 + times) in neonatal mouse brain

[b] Passage history in mouse brain and BHK-21 cell cultures is indicated. The passage history of the vaccine strains is only indicated after they had been attenuated

viruses were also plaque cloned at least three times in BHK 21 cells. The parental origin of each of the genome segments (except segment 9) was then determined by electrophoresis of genomic dsRNAs in 9% polyacrylamide gels. Genome segment 9 of the two parental strains were found to comigrate during SDS PAGE, across a wide range of gel concentrations. The parental origins of genome segment 9 from each of the reassortants was determined by sequence analysis and comparison to the sequence of segment 9 from the two parental strains (Meyer, unpubl. data). AHSV 8^{wt} and AHSV 3^{att} have 96% nucleotide sequence homology in genome segment 9.

DNA manipulation and sequencing

Double stranded viral RNA was purified as described by Martin and Mertens [29]. cDNAs of segment 9 from the parental AHSV strains and reassortants were produced by reverse transcription and amplified in a polymerase chain reaction (RT-PCR). The oligonucleotide primers used were upstream (5'GTTAAATAAGTTGTTTCATG...3') and downstream (5'GTAAGTTTTAAGTTGCCCTC...3'). The amplified DNA was subsequently purified by LMP agarose gel electrophoresis. The required DNA band was excised and purified using phenol: chloroform. This was then cloned into pGEM-T and used to transform competent

DH5α *E. coli*. The bacteria were grown on ampicillin IPTG + XGAL agar plates. Positive (white) bacterial colonies were "mini prep'd" and checked for presence of the inserted DNA by restriction endonuclease analysis. Sequencing was carried out with the ALFexpress Autoread Sequencing Kit (from Pharmacia), using CY5-labelled M13 primers. M13 Reverse Sequence Primer 5'-d [CAGGAAACAGCTATGAC]-3' and M13 Universal Sequencing Primer 5'-d[GTAAAACGACGGCCAGT]-3'.

Results

Genome segment coding assignments

As an initial part of this study, the genome segment coding assignments of AHSV were determined using a plaque cloned isolate of AHSV 8^{wt} (A). The dsRNA genome segments were separated by SDS PAGE in 10% gels, recovered and analysed by both PAGE and agarose gel electrophoresis (AGE) to check their purity and migration order (Fig. 4). The genome segment migration pattern was found to be similar for each of the AHSV strains, when analysed by AGE (data not shown). Genome segments 5 and 6 were found to migrate in the reverse order

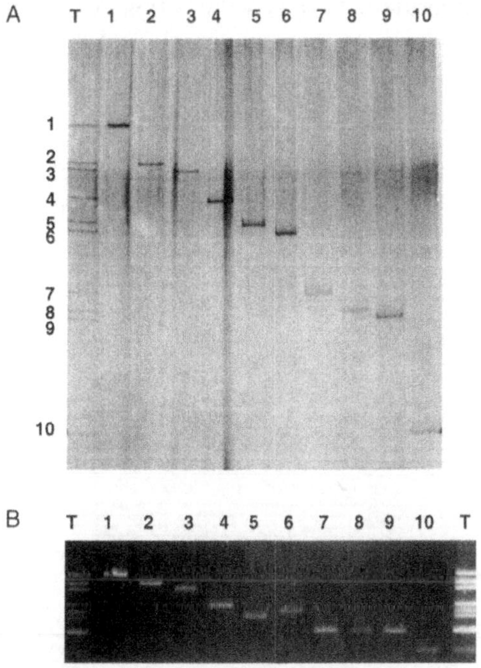

Fig. 4. Separation of AHSV 8 genome segments. The individual genome segments of AHSV 8^{wt} were separated by 10% SDS PAGE [26], stained with ethidium bromide and visualised under UV illumination. The bands were excised and the RNA, representing individual genome segments recovered [38]. The separated segments were checked for purity by 10% SDS PAGE (**A** silver staining) and by AGE (**B** ethidium bromide staining/UV). *T* contains total unfractionated genomic dsRNA. *1–10* contain separated genome segments 1–10 respectively. Genome segment nomenclature is based on migration order during 10% SDS PAGE. The order of migration of genome segments 5 and 6 is reversed during AGE

during AGE, as compared to PAGE (Fig. 4). Similar observations have been reported for some BTV isolates [45]. Unlike BTV, genome 7, 8 and 9 of AHSV isolates comigrate during AGE.

Each of the separated genome segments of AHSV 8^{wt} was denatured with 20 mM methyl mercuric hydroxide and translated in a rabbit reticulocyte lysate translation assay system (Amersham). The protein products, which were labelled by the incorporation of [^{35}S] methionine (Fig. 5), were compared to proteins present in purified virus particles, [^{35}S] methionine labelled viral proteins synthesised in infected cells [3] and molecular weight markers (data not shown). In this way the identity of the genome segment encoding each of the viral proteins was determined (Fig. 6). Using the migration order of the genome segments during AGE, it is possible to generate a coding assignment which is valid for all AHSV isolates, although because genome segments 7, 8 and 9 comigrate in this gel system, we have based the genome segment nomenclature used here ("genome segments 1 to 10") on their order of migration during 10% SDS PAGE.

The coding assignment (Fig. 6) is similar to that published for BTV using similar methods [38, 45] but differs significantly from that previously published for AHSV [14]. The major differences appear to involve the products of genome segment 9, which we have identified as VP6/ VP6a/ VP6b/ VP6c, using a similar

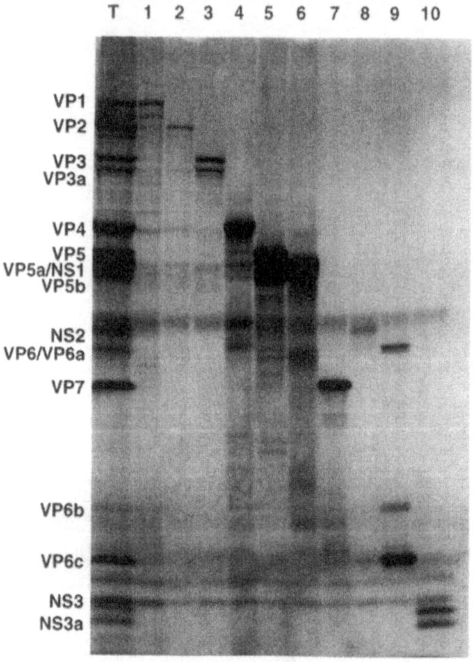

Fig. 5. AHSV 8 RNA translation products. The separated genome segments of AHSV 8^{wt} (see Fig. 4) were denatured in 10 mM methyl mercuric hydroxide and translated individually in a commercially available rabbit reticulocyte lysate (Amersham). The [^{35}S] methionine labelled proteins were analysed by SDS PAGE [26] and detected by fluorography. *T* contains the translation products of total unfractionated genomic RNA. *1–10* contain the translation products of the separated genome segments 1 to 10 respectively

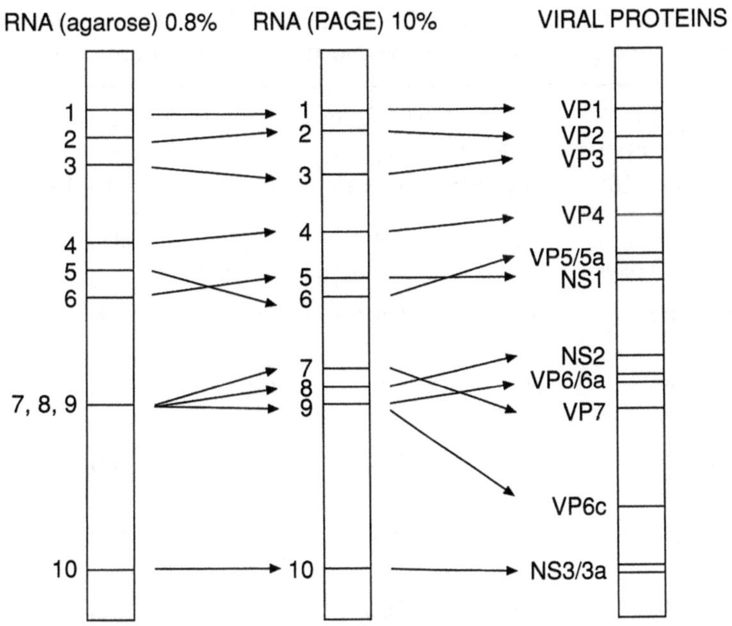

Fig. 6. Genome segment coding assignment for AHSV 8^{wt}

(and therefore we believe correct) nomenclature to that adopted for the homologous proteins (VP6/ VP6a) of BTV (the prototype orbivirus). This nomenclature is not the same as that previously used by Grubman and Lewis [14].

Development of a mouse model

In order to establish a mouse model system, which would distinguish between strains with different virulence characteristics, groups of five adult or ten neonate Balb C mice were inoculated with different strains of AHSV via a number of different routes (Table 2). Each of the viruses tested caused some clinical signs and some mortality in neonate mice (data not shown) and although some variation in virulence was observed, the differences were not sufficiently clear cut to allow a study of virulence characteristics by genome segment reassortment. Studies in neonates were therefore discontinued. In contrast initial studies in adult mice showed that some viruses (AHSV 1^{att}, AHSV 3^{wt} and AHSV 4^{att}) caused a significant level of mortality, while others produced no apparent effect. Of the two serotypes (1 and 4) where both wild type and vaccine strains were compared in adult mice, the viruses which had been attenuated by multiple passage in mouse brain were significantly more virulent. In each case the highest mortality (up to 100%) was caused by intranasal inoculation, although some mice also showed clinical signs and died after conjunctival (eye drops) or intravenous inoculation. The clinical signs observed after conjunctival inoculation may reflect the physical connection between the eye and nasal passages. Further studies used intravenous, or intranasal inoculation only.

Table 2. Evaluation of different routes of inoculation and their effect on AHSV virulence in adult Balb C mice, showing percentage mortality and mean time before death

Route of inoculation	Virus isolate and titre ($TCID_{50}$/ml)					
	$AHSV1^{att}$ (1×10^8)	$AHSV1^{wt}$ (7×10^7)	$AHSV3^{wt}$ (2×10^9)	$AHSV4^{att}$ (4×10^8)	$AHSV4^{wt}$ (1×10^8)	$AHSV9^{att}$ (2×10^8)
Intranasal (20 µl)	100% 4.0 days	0% –	80% 7.0 days	100% 5.8 days	0% –	0% –
Subcutaneous (250 µl)	0% –	0% –	0% –	0% –	0% –	0% –
Intramuscular (250 µl)	0% –	0% –	0% –	0% –	0% –	0% –
Intraperitoneal (250 µl)	0% –	0% –	0% –	0% –	0% –	0% –
Intradermal (50 µl)	0% –	0% –	0% –	0% –	0% –	0% –
Conjunctival (20 µl)	20% 5.0 days	0% –	20% 8.0 days	0% –	0% –	0% –
Intravenous (20 µl)	40% 5.0 days	0% –	0% –	20% 8.0 days	0% –	0% –

Selection of parental strains for production of reassortants

Attenuated and wild type isolates of the different AHSV serotypes were assayed for virulence in adult mice by inoculation of a fixed amount of virus (1×10^5 $TCID_{50}$) via an intranasal or intravenous route (Table 3). None of the wild type strains produced any detectable clinical signs via either route. Several of the attenuated strains (serotypes 2, 4, 6 and 9) also failed to produce any clinical signs when administered in these amounts by either route. However, attenuated viruses of serotypes 1, 3, 5, 7 and 8 did cause a significant level of mortality when administered via an intranasal route. Only AHSV 7^{att} caused any clinical signs or mortality via an intravenous route.

On the basis of these initial results (with uncloned virus isolates) attenuated and wild type strains of AHSV serotypes 3, 4, 7 and 8 were selected as potential parents for reassortment (to provide a pair of viruses that were either virulent or avirulent in the adult mouse model). Each of these viruses was plaque cloned three times and reassessed for virulence at higher concentrations in adult mice (Table 4). From the results obtained, AHSV 8^{wt} or 7^{wt} were selected as potential "avirulent" (A) parent strains, while the vaccine strains $AHSV3^{att}$, 4^{att}, 7^{att} and 8^{att} were selected as potential "virulent" (V) parents. The genome segments of each possible combination of these A and V strains were compared by SDS PAGE, in order to find one pair whose genome segments could be distinguished on the basis of their electrophoretic migration (data not shown). The combination which gave the highest number of distinguishable genome segments, by these techniques, was

Table 3. Virulence of attenuated (^{att}) and wild type (^{wt})[a] strains of different AHSV serotypes, in adult Balb C mice when 1 x 10^5 $TCID_{50}$ (total) was administered via an intranasal or intravenous route

	Route of inoculation[c]	
	Intranasal (20 µl)	Intravenous (20 µl)
AHSV strain[b]	Percentage mortality, MTBD[d]	
AHSV 1^{att}	40% 6 days	0% –
AHSV 2^{att}	0% –	0% –
AHSV 3^{att}	60% 8 days	0% –
AHSV 4^{att}	0% –	0% –
AHSV 5^{att}	80% 8.4 days	0% –
AHSV 6^{att}	0% –	0% –
AHSV 7^{att}	100% 5.4 days	20% 10.0 days
AHSV 8^{att}	100% 10.4 days	0% –
AHSV 9^{att}	0% –	0% –

[a] Directly comparable experiments were conducted with uncloned wild type strains of each serotype but no mortality was observed by either route (data not shown)

[b] All isolates were uncloned

[c] Isolates were diluted to the same starting titre (5 x 10^6 $TCID_{50}$/ml)

[d] MTBD Mean time before death (in days) for those animals which died or were euthanased

AHSV3^{att} (V) and AHSV 8^{wt} (A) (Fig. 7). However even with this combination of viruses, only nine of the genome segments could be resolved simply by electrophoretic analyses. Genome segment 9 of AHSV3^{att} and 8^{wt} comigrated over a wide range of polyacrylamide gel concentrations and could be distinguished only by sequence analysis (Meyer, unpubl. data).

The dose response of the two parental strains of AHSV were evaluated over a wide concentration range (Table 5). Although AHSV 8^{wt} (A) caused no detectable clinical signs or mortality over the entire concentration range tested (1 x 10^3 to 1 x 10^8), AHSV3^{att} (V) caused significant mortality (20%) at a titre of 1 x 10^5 $TCID_{50}$/ml, which rose to 100% at titres ≥ 1 x 10^7 $TCID_{50}$/ml. All analyses of the virulence of reassortant viruses were therefore carried out using virus stocks having titres ≥ 1 x 10^7 $TCID_{50}$/ml (titrated in BHK-21 cells).

Table 4. Virulence of attenuated (^{att}) and wild type (^{wt}) strains of AHSV serotypes 3, 4, 7 and 8 (after plaque cloning), in adult Balb C mice when administered via an intranasal route[c]

AHSV strain[a]	[Titre: $TCID_{50}$/ml]	Percentage mortality	MTBD[b]
AHSV 3^{wt} **(PC)**	[1 x 10^9 $TCID_{50}$/ml]	80%	9.8 days
AHSV 3^{att} **(PC)**	[3 x 10^8 $TCID_{50}$/ml]	100%	8.2 days
AHSV 4^{wt} **(PC)**	[1 x 10^7 $TCID_{50}$/ml]	20%	12.0 days
AHSV 4^{att} **(PC)**	[1 x 10^9 $TCID_{50}$/ml]	100%	6.0 days
AHSV 7^{wt} **(PC)**	[5 x 10^9 $TCID_{50}$/ml]	0%	–
AHSV 7^{att} **(PC)**	[6 x 10^8 $TCID_{50}$/ml]	100%	5.2 days
AHSV 8^{wt} **(PC)**	[1 x 10^8 $TCID_{50}$/ml]	0%	–
AHSV 8^{att} **(PC)**	[3 x 10^8 $TCID_{50}$/ml]	0%	–

[a] Strains identified as **(PC)**, were all plaque cloned (three times)

[b] *MTBD* Mean time before death (in days) for those animals which died or were euthanased

[c] Each inoculation contained 20 µl of virus stock

Table 5. Mouse dose response for plaque cloned parental strains of AHSV 3^{att} (V) and AHSV 8^{wt} (A)[a] administered via an intranasal route (20 µl)

Virus strain	Titre ($TCID_{50}$/ml)	[numbers of mice which died each day] Days post infection 1 2 3 4 5 6 7 8 9 10 11 12 15 16 17 18	Mortality	MTBD[b]
AHSV 3^{att}	1 x 10^8	[3][2]	100%	7.4
	1 x 10^7	[1][2][2]	100%	8.2
	1 x 10^6	[1][3]	80%	7.8
	1 x 10^5	[1]	40%	10.0
	1 x 10^4		0%	–
	1 x 10^3		0%	–

[a] Similar experiments were carried out with the plaque cloned parental strain of AHSV 8^{wt} (A), however even at the highest titre, no clinical signs were detected and the total mortality was 0%

[b] *MTBD* Mean time before death in days for those animals which died or were euthanased

Production and characterisation of reassortants

Reassortant AHSV strains were produced by coinfection of BHK-21 cells with AHSV3^{att} (V) and AHSV 8^{wt} (A). The progeny virus strains were subsequently plaque purified a minimum of three times and characterised with regard to the parental origins of their genome segments by SDS PAGE (Fig. 8). The parental origins of segment 9 were determined in each case by sequence analysis and comparison to the sequence of the two parental strains (Meyer, unpubl. data).

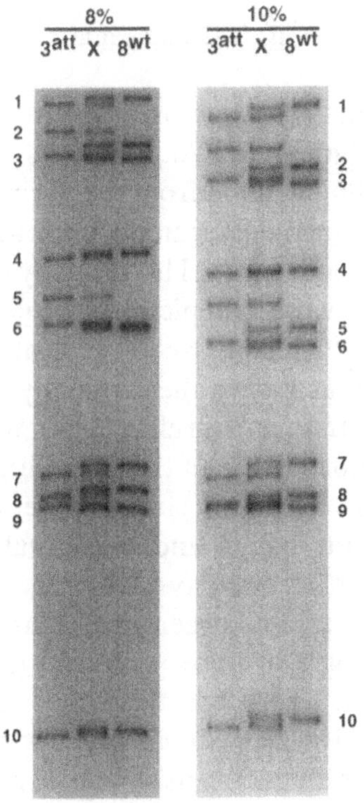

Fig. 7. Electrophoretic comparison of the RNAs of parental strains of AHSV. Comparisons of the migration patterns of AHSV 8^{wt} (A) and AHSV 3^{att} (V) by 8% and 10% SDS PAGE. The RNA was labelled with [^{32}P] pCp and RNA ligase [38]. The bands were detected by autoradiography of the dried gel. The lanes contain the dsRNA of the two viruses as indicated. The lanes marked "X" contained a mixture of RNA from both viruses

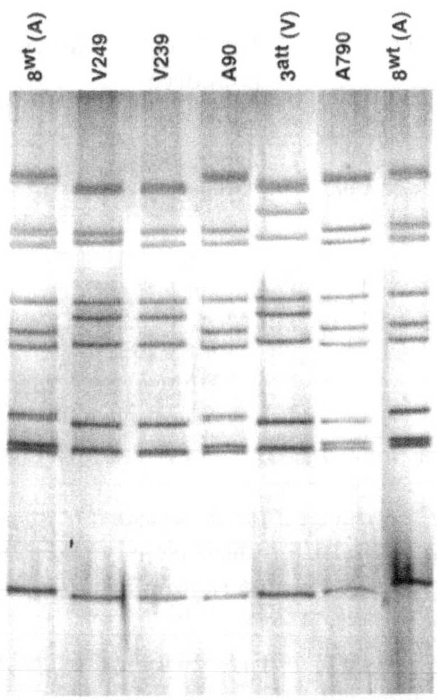

Fig. 8. Electrophoretic analysis of the genomic dsRNAs of reassortant viruses. Electrophoretic analysis of the genomic dsRNAs of reassortant viruses and comparison to parental strains by 9% SDS PAGE. The tracks are loaded with the RNA of the plaque purified reassortants and parental strains as indicated. Bands were detected by silver staining

Each of the reassortants was then assayed for virulence by inoculation of groups of 5 adult BALB C mice with 20 μl of virus (> 1 x 10^7 $TCID_{50}$/ml) via an intranasal route. The results obtained from these virulence assays, as well as the parental derivation of the genome segments present in each of the reassortant viruses is summarised in Tables 6–9. The progeny virus strains fell into three distinct groups: those that were completely avirulent (A) and resembled the avirulent parent (AHSV 8^{wt}) (Table 6); those that were virulent (V), causing similar clinical signs and mortality to the virulent parent (AHSV3^{att}) (Table 7); and those with a novel intermediate virulence (N), which caused distinct clinical signs and lower levels of mortality than the fully virulent strains (Table 8).

The mice infected with "virulent" strains were all characterised by a rapid onset of clinical signs (within 6 days) and they usually died, or were severely ill and were culled, within a further two days. All of the mice infected with the other reassortant strains had produced no observable signs of infection at this time post infection. Some of the groups of mice produced no clinical signs of infection throughout the experiment. These had been inoculated with strains that were designated avirulent (A). However after day eight some of the groups of mice gradually started to produce clinical signs, which were different to those observed with the virulent strains. They lost condition until about 13 days post infection, showed no secretions from their orifices and were not nervous or agitated. Severely affected individuals were culled during this period. From 14 to 18 days

Table 6. Derivation of genome segments in six AHSV reassortants with "avirulent" (A) characteristics

Virus strain[a]	Derivation of genome segments[b]										Virulence[c] phenotype	Mortality	Morbidity	MTBD[d] (days)
	1	2	3	4	5	6	7	8	9	10				
P-3^{att} (V)	v	v	v	v	v	v	v	v	v	v	V	100%	100%	7.2
P-8^{wt} (A)	a	**a**	a	a	a	a	a	a	a	a	A	0%	0%	–
R-V2	v	**a**	v	v	v	v	v	v	v	v	A	0%	0%	–
R-V24	v	**a**	v	a	v	v	v	v	v	v	A	0%	0%	–
R-A79	a	**a**	a	a	a	a	v	a	v	a	A	0%	0%	–
R-V126	a	**a**	v	v	v	a	v	v	v	v	A	0%	0%	–
R-V239	v	**a**	a	v	v	v	v	v	a	v	A	0%	0%	–
R-A3580	a	**a**	v	a	v	a	a	v	a	v	A	0%	0%	–

[a] Strain nomenclature: *P* parental strain, *R* reassortant strain. For the reassortants, "A" or "V" indicates which parent the strain most closely resembles, followed by numbers to indicate which genome segments (genome segments 1–9 = 1–9, and segment 10 = 0) are derived from the other parental strain

[b] Derivation of those genome segments shown in bold and boxed in was identical for all of the avirulent (A) strains analysed

[c] Virulence phenotype: *A* avirulent phenotype, *V* virulent phenotype, *N* novel virulence phenotype

[d] *MTBD* Mean time before death in days for those animals which died or were euthanased

Table 7. Derivation of genome segments in nine AHSV reassortants with "virulent" (V) virulence characteristics

Virus strain[a]	Derivation of genome segments[b] 1 2 3 4 5 6 7 8 9 10	Virulence[c] phenotype	Mortality	Morbidity	MTBD[d] (days)
P-3[att] (V)	v **v** v v **v** v v v v **v**	V	100%	100%	7.2
P-8[wt] (A)	a a a a a a a a a a	A	0%	0%	–
R-V1	a **v** v v **v** v v v v **v**	V	80%	80%	7
R-V3	v **v** a v **v** v v v v **v**	V	80%	100%	9
R-V6	v **v** v v **v** a v v v **v**	V	100%	100%	7
R-V34	v **v** a a **v** v v v v **v**	V	100%	100%	9
R-V39	v **v** a v **v** v v v a **v**	V	100%	100%	7
R-V49	v **v** v a **v** v v v a **v**	V	100%	100%	8
R-V78	v **v** v v **v** v a a v **v**	V	100%	100%	10
R-V368	v **v** a v **v** a v a v **v**	V	100%	100%	7.4
R-V3469	v **v** a a **v** a v v a **v**	V	80%	100%	9

[a] Strain nomenclature: *P* parental strain, *R* reassortant strain. For the reassortants, "A" or "V" indicates which parent the strain most closely resembles, followed by numbers to indicate which genome segments (genome segments 1–9 = 1–9, and segment 10 = 0) are derived from the other parental strain

[b] Derivation of those genome segments shown in bold and boxed in was identical for all of the avirulent (V) strains analysed

[c] Virulence phenotype: *A* avirulent phenotype, *V* virulent phenotype, *N* novel virulence phenotype

[d] *MTBD* Mean time before death in days for those animals which died or were euthanased

post infection, the condition of the remaining mice improved, resulting in an apparently full recovery. This group of virus strains are designated as having "novel virulence" (N) (Table 8).

Discussion

By means of genome segment reassortment we have been able to analyse the involvement of individual genome segments of AHSV 3[att] (V) and AHSV 8[wt] (A) in determination of virulence, using an adult mouse model system. This system is clearly different from the "normal" host for AHSV (ie. horses), since the vaccine strains were in most cases more virulent for mice that the wild type strains. This is perhaps unsurprising since these viruses were attenuated for horses by multiple (> 100) passages in mouse brain. The most effective route for AHSV inoculation of adult mice (for production of clinical signs) proved to be the intranasal route. This may reflect the accessibility of nerve tissues in the nasal cavity and good connections to the brain. Interestingly, the very limited number of cases of disease in humans caused by AHSV, have been linked to inhalation of freeze dried material derived from neurotropic vaccine strains [49].

Table 8. Derivation of genome segments in seven AHSV reassortants with "novel" (N) intermediate virulence characteristics

Virus strain[a]	Derivation of genome segments[b] 1	2	3	4	5	6	7	8	9	10	Virulence[c] phenotype	Mortality	Morbidity	MTBD[d] (days)
P-3[att] (V)	v	v	v	v	v	v	v	v	v	**v**	V	100%	100%	7.2
P-8[wt] (A)	a	**a**	a	a	**a**	**a**	a	a	a	a	A	0%	0%	–
R-A90	a	**a**	a	a	**a**	**a**	a	a	v	**v**	N	20%	80%	11
R-A790	a	**a**	a	a	**a**	**a**	v	a	v	**v**	N	40%	80%	14
R-A80	a	**a**	a	a	**a**	**a**	a	v	a	**v**	N	0%	60%	–
R-A170	v	**a**	a	a	**a**	**a**	v	a	a	**v**	N	80%	100%	11.8
R-A480	a	**a**	a	v	**a**	**a**	a	v	v	**v**	N	20%	100%	13
R-V1256	a	**a**	v	v	**a**	**a**	v	v	v	**v**	N	40%	80%	13
R-A7890	a	**a**	a	a	**a**	**a**	v	v	v	**v**	N	80%	80%	115

[a] Strain nomenclature: *P* parental strain, *R* reassortant strain. For the reassortants, "A" or "V" indicates which parent the strain most closely resembles, followed by numbers to indicate which genome segments (genome segments 1–9 = 1–9, and segment 10 = 0) are derived from the other parental strain

[b] Derivation of those genome segments shown in bold and boxed in was identical for all of the avirulent (N) strains analysed

[c] Virulence phenotype: *A* avirulent phenotype, *V* virulent phenotype, *N* novel virulence phenotype

[d] *MTBD* Mean time before death in days for those animals which died or were euthanased

Table 9. Combinations of genome segments present in parental and reassortant strains of AHSV 3[att] (V) and AHSV 8[wt] (A)

AHSV strain or phenotype	Parental origin of genome segments 2	5	6	10
AHSV 3[att] (V)	v	v	v	v
AHSV 8[wt] (A)	a	a	a	a
Virulent (V)	v	v	a or v	v
Avirulent (A)	a	a or v	a or v	a or v
Novel, (intermediate) virulence	a	a	a	v

Genome segment 2 from the virulent parent was found in each of the V phenotype reassortants but not in any of the A or N group viruses. Segment 2 codes for the outer capsid protein VP2, which is implicated in cell attachment and penetration, and in virus neutralisation by antibodies. It therefore appears likely that VP2 and it role in cell entry may have an important influence in determination of virulence characteristics in mammals, possibly by affecting tissue or cell tropism. Each of the virulent strains also contains genome segment 5 and segment 10 from the virulent parent. However, both of these segments also appear in the other groups of viruses and do not therefore appear to have a full controlling influence on the virulent phenotype. Genome segments 5 and 10 from the virulent parent could be required for expression of the virulent phenotype, even if it is controlled primarily by segment 2. It is also possible that there is some "linkage" between these segments during infection and replication, possibly reflecting functional interaction of the proteins for which they code. Genome segment 5 codes for outer capsid protein VP5, which is known to interact with VP2 in the outer protein shell of the intact virus particle. Genome segment 10 codes for the small nonstructural proteins NS3 and NS3a, which may interact with the outer capsid proteins as a consequence of their role in release of virus from the infected cell [24, 30].

The avirulent (A) reassortant viruses all contain genome segments 2 from the avirulent parent (AHSV 8^{wt}) but each of the other segments can apparently be derived from either parental strain. One reassortant, which is identified as "V2", only differs from the virulent parent (AHSV 3^{att}) in genome segment 2. This confirms the role of segment 2 and VP2 in control of the virulence phenotype. Although all of the reassortant strains with the "novel" virulence characteristics also contained genome segment 2 from the avirulent parent (like the A strains) they each contain both genome segments 5 and 6 from the avirulent parent and segment 10 from the virulent parent. One particularly interesting pair of reassortants is A79 which is avirulent and A790 which has N type virulence. This demonstrates that the presence of the segment 10 from the virulent parent has significantly increased the virulence of reassortant strain A790, conferring the N phenotype.

The possible combinations of genome segments found in each of the groups of reassortants are summarised in Table 9. All of the reassortant strains, except one (A79), contain genome segment 10 from the virulent parent. This suggests that segment 10 from AHSV 3^{att} (V) has some selective advantage over that from AHSV 8^{wt} (A) and may correlate with earlier release from infected cells [30]. Reassortant A79 was isolated by deliberately selecting small plaques during its purification. A rapid cell release mechanism and therefore presumably more rapid spread of the virus within infected mice, may also explain the increased virulence conferred by the "v" form of genome segment 10. However, most of the avirulent reassortant strains also contain the "v" form of segment 10 but not in combination with the "a" form of genome segment 5 (see reassortants A3580 and V126: Table 6). This suggests that segment 5, which encodes the smaller outer capsid outer capsid protein VP5, may play a role in expression of the N phenotype and therefore in control of virulence. As already discussed for the "V" group strains, this appears to correlate with the role of the outer capsid components in cell attachment and penetration.

The genome segments that control virulence of these strains of AHSV in adult BALB C mice, encode proteins that are primarily involved in cell entry or cell exit. It appears likely that they exercise this control by affecting the infection of cells, tissue tropism and overall spread of the virus within the infected animals. Viral proteins that are involved in the replication processes of the virus within the successfully infected cell, although essential, may have relatively less influence on variations in its virulence characteristics. These data correlate with results of other reassortant studies with AHSV that have used horses and which also implicate segment 10 in the control of virulence [25]. However, we feel that some caution must be exercised in extrapolation of the results from our mouse model to other host species, particularly horses that are infected via the bite of insect vectors.

The mouse model has been further characterised by O'Hara [44] and (particularly in combination with the plaque cloned vaccine strain of AHSV 7^{att} that was virulent via an intravenous route) has proved to be a useful model for studies of the immune response to AHSV [54, 55].

Although we have used vaccine strains of AHSV that overall were significantly more virulent than their wild type counterparts, sequencing studies have shown few base changes in genome segment 10 between the wild type and attenuated strains of the same serotypes (only one amino acid change was detected between AHSV 3^{wt} and AHSV 3^{att} [30]). In view of the apparently dominant role of the outer capsid proteins in control of virulence, the differences caused by attenuation may be reflected in sequence alterations in segment 2 and possibly to a lesser extent segment 5. However the two parental strains used in the studies presented here have segment 10 belonging to the β and γ sequence groups respectively (as identified by Martin et al. [30]), having approximately 35% amino acid and nucleotide sequence variation in segment 10/NS3. Sequence data also suggest that NS3 of AHSV3 does not have the potential glycosylation sites [6, 30]. It therefore appears more likely that the differences in virulence phenotype caused by exchange of genome segment 10, reflect the relatively large differences that exist between the α, β and γ groups, rather than the affects of attenuation.

Acknowledgements

The authors wish to thank Stephen Archibald and Bernard Clark for preparation of figures. These studies were supported by the Ministry of Agriculture Fisheries and Food (UK) and by EC contract No. 8001-CT91-0211.

References

1. Brookes SM, Hyatt AD, Eaton BT (1993) Characterization of virus inclusion bodies in bluetongue virus-infected cells. J Gen Virol 74: 525–530
2. Burrage TG, Trevejo R, Stonemarschat M, Laegreid WW (1993) Neutralizing epitopes of African horse sickness virus serotype-4 are located on VP2. Virology 196: 799–803
3. Burroughs JN, O'Hara RS, Smale CJ, Hamblin C, Walton A, Armstrong R, Mertens PPC (1994) Purification and properties of virus particles, infectious sub viral particles, cores and VP7 crystals of African horse sickness virus serotype 9. J Gen Virol 75: 1849–1857
4. Calisher CH, Mertens PPC (1998) Taxonomy of African horse sickness viruses. In: Mellor PS, Baylis M, Hamblin C, Calisher CH, Mertens PPC (eds) African horse sickness. Springer, Wien New York, pp 3–11 (Arch Virol [Suppl] 14)
5. Cowley JA, Gorman BM (1989) Cross-neutralization of genetic reassortants of bluetongue virus serotypes 20 and 21. Vet Microbiol 19: 37–51

6. de Sa RO, Zellner M, Grubman MJ (1994) Phylogenetic analysis of segment 10 from African horse sickness virus and cognate genes from other orbiviruses. Virus Res 33: 157–165
7. Devi R, Burroughs JN, Mertens PPC, Roy P (1998) Transmethylase 1 and Transmethylase 2 activities associated with the Cap protein (VP4) of BTV, in native cores, core like particles and as a purified, baculovirus expressed protein. (submitted)
8. Eaton BT, Hyatt AD, Brookes SM (1990)The replication of bluetongue virus. Curr Top Microbiol Immunol 162: 89–118
9. Finney DJ (1964) Statistical method in biological assay, 2nd ed. Griffin, London
10. Fu H (1996) Mechanisms controlling the infection of *Culicoides* biting midges with bluetongue virus. Ph.D. Thesis, University of Hertfordshire
11. Fu H, Leake CJ, Mertens PPC, Mellor PS, The barriers to bluetongue virus infection dissemination and transmission in the vector Culicoides variipennis (Diptera: Ceratopogonidae) (submitted)
12. Gouet P, Grimes JM, Diprose JM, Malby R, Burroughs JN, Zeintara S, Stuart, DI, Mertens PPC (1998) Crystallographic studies show ordered double-stranded RNA in bluetongue virus (submitted)
13. Grimes JM, Burroughs JN, Gouet P, Diprose JM, Malby R, Zeintara S, Mertens PPC, Stuart DI (1998) The atomic structure of the bluetongue virus core. Nature (in press)
14. Grubman MJ, Lewis SA (1992) Identification and characterization of the structural and non structural proteins of AHSV and determination of the genome coding assignments. Virology 186: 444–451
15. Hewat EA, Booth TF, Roy P (1992) Structure of bluetongue virus particles by cryoelectron microscopy. J Struct Biol 109: 61–69
16. Hewat EA, Booth TF, Wade RH, Roy P (1992) 3-D reconstruction of bluetongue virus tubules using cryo-electron microscopy. J Struct Biol 108: 35–38
17. Huismans H, Els HJ (1979) Characterization of tubules associated with the replication of the different orbiviruses. Virology 92: 397–406
18. Huismans H, Van Der Walt NT, Cloete M, Erasmus BJ (1987) Isolation of a capsid protein of bluetongue virus that induces a protective immune response in sheep. Virology 61: 3589–3595
19. Huismans H, Van Dijk AA (1990) Bluetongue virus structural components. Curr Top Microbiol Immunol 162: 21–42
20. Huismans H, Van Dijk AA, Bauskin AR (1987) In vitro phosphorylation and purification of nonstructural protein of bluetongue virus with affinity for single stranded RNA. J Virol 61: 3589–3595
21. Huismans H, Van Dijk AA, Els HJ (1987) Uncoating of parental bluetongue virus to cores and subcore particles in infected L cells. Virology 157: 180–188
22. Hwang G, Yang Y, Chiou J, Li JK (1992) Sequence conservation among the cognate nonstructural NS3/3A protein genes of six bluetongue viruses. Virus Res 23: 151–161
23. Hyatt AD, Gould AR, Coupar B, Eaton BT (1991) Localization of the non-structural protein NS3 in bluetongue virus-infected cells. J Virol 72: 2263–2267
24. Hyatt AD, Zhao Y, Roy P (1993) release of bluetongue virus-like particles from insect cells is mediated by BTV non-structural protein NS3/3a. Virology 193: 592–603
25. Laegried WW, Burrage TG, Stone-Marschat MA, Skrowronek AJ, Hussein A (1995) The S10 gene of African horse sickness virus is a major virulence determinant. Abstracts, Fifth International Symposium on double-stranded RNA viruses (P46)
26. Laemmli UK (1970) Cleavage of structural proteins during assembly of the head of bacteriophage T4. Nature 227: 680–685
27. Lee JW, Roy P (1986) Nucleotide sequence of a cDNA clone of RNA segment 10 of bluetongue virus (serotype 10). J Gen Virol 67: 2833–2837

28. Loudon PT, Roy P (1991) Assembly of five bluetongue virus proteins expressed by recombinant baculoviruses: inclusion of the largest protein VP1 in the core and virus like particles. Virology 180: 798–802
29. Martin L-A, Mertens PPC (1998) Purification and characterisation of viral ds RNA genome profiles by cross hybridisation. Methods Mol Biol (in press)
30. Martin L-A, Meyer AJ, O'Hara RS, Fu H, Mellor PS, Knowles NJ, Mertens PPC (1988) Phylogenetic analysis of African horse sickness virus segment 10: sequence variation, virulence characteristics and cell exit. In: Mellor PS, Baylis M, Hamblin C, Calisher CH, Mertens PPC (eds) African horse sickness. Springer, Wien New York, pp 281–293 (Arch Virol [Suppl] 14)
31. McCrae MA, Joklik WK (1978) The nature of the polypeptides encoded by each of thee 10 double stranded RNA segments of reovirus type 3. Virology 89: 578–593
32. Meiswinkel R (1998) The imicola group (subgenus *Avaritia*): Short notes on the taxonomy, biology and geographic distribution of 10 species. In: Mellor PS, Baylis M, Hamblin C, Calisher CH, Mertens PPC (eds) African horse sickness. Springer, Wien New York, pp 69–83 (Arch Virol [Suppl] 14)
33. Mellor PS, Rawlings P, Baylis M, Welby MP (1998) Effect of temperature on African horse sickness virus infection in Culicoides. In: Mellor PS, Baylis M, Hamblin C, Calisher CH, Mertens PPC (eds) African horse sickness. Springer, Wien New York, pp 155–163 (Arch Virol [Suppl] 14)
34. Mertens PPC (1994) Orbiviruses and coltiviruses – general features. In: Webster RG, Granoff A (eds) Encyclopedia of virology, vol 2. Academic Press, London, pp 941–956
35. Mertens PPC, Burroughs JN, Anderson J (1987) Purification and properties of virus particles, infectious subviral particles, and cores of bluetongue virus serotype 1 and 4. Virology 157: 375–386
36. Mertens PPC, Burroughs JN, Walton A, Wellby MP, Fu H, O'Hara RS, Brookes SM, Mellor PS (1996) Enhanced infectivity of modified bluetongue virus particles for two insect cell lines and for two *Culicoides* vector species. Virology 217: 582–593
37. Mertens PPC, Burroughs JN, Wade-Evans AM, Le Blois H, Oldfield S, Basak A, Loudon P, Roy P (1992) Analysis of guanylyltransferase and transmethylase activities associated with bluetongue virus cores and recombinant baculovirus-expressed core-like particles. In: Walton TE, Osburn BI (eds) Bluetongue, African horse sickness and related orbiviruses Second International Symposium on bluetongue, African horse sickness and related orbiviruses. CRC Press, Boca Raton, pp 404–415
38. Mertens PPC, Brown F, Sangar DV (1984) Assignment of the genome segments of bluetongue virus type 1 to the proteins which they encode. Virology 135: 207–217
39. Mertens PPC, Pedley S, Cowley JA, Corteyn AH, Jeggo MH, Jennings DM, Gorman BM (1989) Analysis of the roles of bluetongue virus outer capsid proteins VP2 and VP5 in determination of virus serotype. Virology 170: 561–565
40. Mertens PPC Sangar DV (1985) Analysis of the terminal sequences of the genome segments of four orbiviruses. Virology 140: 55–67
41. Mertens PPC, et al (in preparation) Family *Reoviridae*. In: Seventh Report of the International Committee on Taxonomy of Viruses
42. Moss SR, Jones LD, Nuttall PA (1992) Comparison of the nonstructural protein, NS3, of tick-borne and insect-borne orbiviruses. Virology 187: 841–844
43. Moss SR, Nuttall PA (1994) Subcore and core-like particles of Broadhaven virus (BRDV), a tick bourne orbivirus, synthesised from baculovirus expressed VP2 and VP7, the major core proteins of BRDV. Virus Res 32: 401–407
44. O'Hara RS (1994) Identification of the genome segments and proteins controlling the virulence of African horse sickness virus. Ph.D. Thesis, University of Reading

45. Pedley S, Mohamed MEH, Mertens PPC (1988) Analysis of the genome segments of six different bluetongue virus isolates using RNA-RNA hybridisation: a generalised coding assignment for bluetongue viruses. Virus Res 10: 381–390
46. Sailleau C, Moulay S, Zientara S (1997) Nucleotide sequence comparison of the segments S10 of the nine African horse sickness virus serotypes. Arch Virol 142: 965–978
47. Stonemarschat MA, Moss SR, Burrage TG, Barber ML, Roy P, Laegreid WW (1996) Immunization with VP2 is sufficient for protection against lethal challenge with African horse sickness virus type-4. Virology 220: 219–222
48. Stuart DI, Gouet P, Grimes J, Malby R, Diprose J, Zientara S, Burroughs JN, Mertens PPC (1998) Orbivirus particle structure. In: Mellor PS, Baylis M, Hamblin C, Calisher CH, Mertens PPC (eds) African horse sickness. Springer, Wien New York, pp 235–250 (Arch Virol [Suppl] 14)
49. Swanepoel R, Erasmus BJ, Williams R, Taylor MB (1992) Encephalitis and chorioretinitis associated with neurotropic African horse sickness virus infection in laboratory workers. Part III. Virological and serological investigations. S Afr Med J 81: 458–461
50. Thomas CP, Booth TF, Roy P (1990) Synthesis of bluetongue virus encoded phosphoprotein and formation of inclusion bodies by recombinant baculovirus in insect cells: it binds the single stranded RNA species. J Gen Virol 71: 2073–2083
51. Van Dijk AA, Huismans H (1988) The in vitro activation and further characterisation of the bluetongue virus associated transcriptase. Virology 104: 347–356
52. van Staden V, Huismans H (1991) A comparison of the genes which encode nonstructural protein NS3 of different orbiviruses. J Gen Virol 72: 1073–1079
53. van Staden V, Stoltz MA, Huismans H (1995) Expression of nonstructural protein NS3 of African horse sickness virus (AHSV): evidence for a cytotoxic effect of NS3 in insect cells, and characterization of the gene products in AHSV infected Vero cells. Arch Virol 140: 289–306
54. Wade-Evans AM, Pullen L, Hamblin C, O'Hara RS, Burroughs JN, Mertens PPC (1997) African horse sickness virus, VP7, subunit vaccine protects mice against a lethal heterologous serotype challenge. J Gen Virol 78: 1611–1616
55. Wade-Evans AM, Pullen L, Hamblin C, O'Hara RS, Burroughs JN, Mertens PPC (1998) VP7 from African horse sickness virus serotype 9 protects mice against a lethal, heterologous serotype challenge. In: Mellor PS, Baylis M, Hamblin C, Calisher CH, Mertens PPC (eds) African horse sickness. Springer, Wien New York, pp 211–219 (Arch Virol [Suppl] 14)
56. Wu X, Chen S-Y, Iwata H, Compans RW, Roy P (1992) Multiple glycoproteins synthesized by the smallest RNA segment (S10) of bluetongue virus. J Virol 66: 7104–7112
57. Zientara S, Sailleau C, Moulay S, Crucière C (1995) Differentiation of African horse sickness viruses by polymerase reaction and segment 10 restriction patterns. Vet Microbiol 47: 365–375
58. Zientara S, Sailleau C, Plateau E, Moulay S, Mertens PPC, Crucière C (1998) Molecular epidemiology of African horse sickness virus based on analyses and comparisons of genome segments 7 and 10. In: Mellor PS, Baylis M, Hamblin C, Calisher CH, Mertens PPC (eds) African horse sickness. Springer, Wien New York, pp 221–234 (Arch Virol [Suppl] 14)

Authors' address: Dr. P. P. C. Mertens, Institute for Animal Health, Pirbright Laboratory, Ash Road, Pirbright, Woking, Surrey, GU24 ONF, U.K.

Phylogenetic analysis of African horse sickness virus segment 10: sequence variation, virulence characteristics and cell exit

L.-A. Martin*, **A. J. Meyer**, **R. S. O'Hara**, **H. Fu**, **P. S. Mellor**, **N. J. Knowles**, and **P. P. C. Mertens**

Institute for Animal Health, Pirbright Laboratory,
Woking, Surrey, U.K.

Summary. African horse sickness virus (AHSV) genome segment 10 encodes the non-structural proteins NS3/NS3a, which is involved in release of virus from cells. Full length segment 10 cDNAs were amplified by reverse transcription-polymerase chain reaction, from isolates of AHSV serotypes 2, 3, 4, 5, 7, 8 and 9. These cDNAs were cloned, sequenced and their phylogenetic relationships analysed. High levels of sequence homology were detected in segment 10 from some isolates of different serotypes, confirming that they could be grouped on this basis (serotypes 4, 5, 6 and 9 (group α); serotypes 3 and 7 (group β); serotypes 1, 2, and 8 (group γ)). However, data from bluetongue virus (the prototype orbivirus) indicate that the AHSV serotype is determined exclusively by the structural outer coat proteins VP2 and VP5, encoded by genome segments 2 and 5 respectively. Therefore, as a direct consequence of genome segment reassortment between AHSV strains from different serotypes, the differences observed in segment 10 do not give a reliable indication of virus serotype. Segment 10 of AHSV 3 (virulent) and AHSV 3^{att} (attenuated) were also analysed. These strains, together with AHSV 8, have been used to study of the genetic basis of virulence using reassortment (O'Hara et al., this publication). Virus release studies, using *Culicoides* cell cultures, indicate that differences in segment 10 of AHSV 3att and 8 can influence the timing of virus release from the infected cell.

Introduction

African horse sickness virus (AHSV) is a member of the genus *Orbivirus* in the family *Reoviridae* [2]. It is the causative agent of African horse sickness (AHS), an economically important disease in *Equidae*, which is infectious but non-contagious and is transmitted by species of *Culicoides* (biting midge). The virus genome

---

* Present address: ICRF Oncology Unit Royal Post-graduate Medical School, Hammersmith Hospital, London, U.K.

consists of 10 segments of double-stranded RNA, each encoding a different viral protein [21]. There are seven structural proteins, which combine to form a double-shelled virus particle [1] and three distinct non-structural proteins, NS1, NS2 and NS3/3a. The exact function of the non-structural proteins remains unclear. The smallest of these proteins NS3/3a are encoded by genome segment 10. Sequence analysis of this gene from several bluetongue virus (BTV) serotypes, has indicated the presence of two hydrophobic domains, as well as the presence of two potential glycosylation sites within the NS3 protein [8, 15]. More recently these motifs have also been noted in other orbivirus species including AHSV [3, 28, 29]. Further sequence analysis has revealed that the sequence of segment 10 from different AHSV strains of different serotypes is not so highly conserved as in BTV. However, in view of the non structural nature of NS3/NS3a (the proteins encoded by segment 10), analysis of these variations in sequence cannot be used as a reliable means of identifying AHSV serotype [32, 33].

The function of NS3 during AHSV virus replication remains uncertain. However, immuno-localisation studies, carried out on BTV infected cells, have demonstrated that NS3 is associated with the plasma membrane and golgi apparatus during virus replication. Virus like particles, synthesised in *Spodoptera frugiperda* cells by recombinant baculoviruses expressing BTV structural proteins, were only released if the cells were also infected with baculoviruses expressing NS3/NS3a [9]. These data strongly suggest that NS3 may be involved in virus release. Studies of AHSV virulence characteristics have been carried out using a mouse model system and genome segment reassortants between virulent and avirulent strains [21]. Exchange of genome segment 10, and therefore variations in the protein sequence of NS3, were shown to influence the virulence characteristics of the progeny virus strain. Here we describe virus release assays, which demonstrate that the virulent and avirulent parental strains of AHSV, that were used in the earlier studies [21], are released from the cells of a *Culicoides* cell line (KC cells) at different times post infection. The role of genome segment 10 and its effect on the timing of virus release from KC cells, were also analysed using genome segment reassortment.

Materials and methods

Tissue culture cells and viruses

BHK-21 cells were grown in Eagles medium containing 5% foetal calf serum at 37°. *C. variipennis* cells (KC cells) were propagated in Schneiders insect medium (Sigma) containing 5% foetal calf serum at room temperature, ca. 22°.

The AHSV serotypes used were originally obtained from the Veterinary Research Institute, Onderstepoort, South Africa as low passage mouse brain extracts and unless otherwise stated are regarded as 'wild type' (wt) strains [21]. They were passaged in BHK-21 cells prior to at least three rounds of plaque purification. AHSV # A90 is a reassortant virus generated from parental strains AHSV 3^{att} (virulent in mice) and AHSV 8^{wt} (avirulent in mice), the genotype of this virus is shown in Table 1 (for further details see O'Hara et al. [21]). Propagated virus was purified as previously described by Burroughs et al. [1, 18]. The designation, origin and database accession numbers of all virus sequences examined in this study are shown in Table 2.

Table 1. Genome segment derivation of AHSV reassortant #A90 [21]

	1	2	3	4	5	6	7	8	9	10	Type
3^{att} (V)	v	v	v	v	v	v	v	v	v	v	V
8^{wt} (A)	a	a	a	a	a	a	a	a	a	a	A
#A90	a	a	a	a	a	a	a	a	v	v	V

DNA manipulation and sequencing

Double stranded viral RNA was purified as described by Martin and Mertens [16]. cDNAs of segment 10 from each of the AHSV serotypes (2, 3, 3^{att}, 4, 5, 7, 8 and 9) were produced by reverse transcription followed by amplification by polymerase chain reaction (RT-PCR). The oligonucleotide primers used were upstream (5'GTTTAAATTATCCCTTGTC...3') and downstream (5'GTAAGTCGTTATCCCGGCT...3') based on the sequences described by van Staden and Huismans [28]. The cDNA's were ligated into pGemT (Promega) and sequenced by the dideoxy chain-termination method [25].

Sequence analysis

The complete segment 10 nucleotide sequences were aligned using CLUSTALW [27] and "fine-tuned" by hand using the multiple sequence alignment editor GeneDoc [20]. Pairwise comparisons were made of nucleotide and predicted amino acid sequences, using a program written by one of the authors (N. J. Kowles, unpubl.) and tables of percentage differences were assembled. Unrooted trees were constructed from the distance matrices, using the neighbour-joining algorithm of Saitou and Nei [23] as implemented in the PHYLIP package program NEIGHBOUR [5]. The trees were drawn using the program TREEVIEW [22].

Infectivity/release assay for C. variipennis *(KC) cells*

Ten fold serial dilutions of AHSV 3^{att}, AHSV 8^{wt} and AHSV #A90 were prepared and inoculated onto microtitre plates containing confluent KC cells, the plates were then incubated at room temperature for seven days. As viral CPE is not observed in KC cells [7, 30], virus replication was detected by removing 20 µl of the contents from each well, at 7 days pi and transferring them in duplicate to microtitre plates containing confluent monolayers of BHK-21 cells. The development of CPE was then monitored after a further 7 days incubation at 37° (according to the method of Mertens et al. [17]). The titration end point titres were calculated using the method of Spearmann and Karber [6].

Results

Comparison of the sequence of genome segment 10 of AHSV isolates from different serotypes

Segment 10 of AHSV strains from serotypes 2, 3, 3^{att}, 4, 5, 7, 8 and 9 were amplified by RT-PCR. Fragments of the correct size were gel purified and ligated into the T-tailed vector pGemT. Clones from three separate amplifications were sequenced in each case. No differences were observed in the % base composition of the segment 10 cDNAs from different strains. However, the number of nucleotides in the full length segment 10 did vary. Genome segment 10: from strains of

AHSV 2, and 8 is 764 bp in length; from the strains of AHSV 3, 3att, 4, 5, and 7 is 758 bp; and from strains of AHSV 9 is 756 bp in length. Each segment 10 gene has a large open reading frame with two in frame initiation codons, which are near to the 5' end (Fig. 1) but the position of which varies between the strains from different serotypes. The first initiation codon, which is generally accepted as the translation start [12], lies at position 20–22 in strains of AHSV 2, 3, 3^{att}, 4, 5, 7 and 8; whilst in AHSV 9 it lies at position 19–21. The location of the second initiation codon also varies: between nucleotides 50–52 for strains of serotypes 2, 3, 3^{att}, 7 and 8; position 53–55 for strains of AHSV 4 and 5; and at nucleotides 52–54 for AHSV 9.

```
                    10        20        30        40        50        60
Consensus   MnLAaIaenYmmHNgnqraIVPYvPPPYafanAPtlggqasqMEsmSLGILNQAMSSttG  60
AHSV2-P     ----S-SQS--S--E-E-S----I----H.PT--A-AVS-----T------------SA-  59
AHSV8-P     ----S-SQS--S--E-E-S----I----H.PT--A-AVS-----T------------SA-  59
AHSV3-P     -S--T------------------------Y----------GE------------------  60
AHSV3-PA    -S--T------------------------Y----------GE------------------  60
AHSV7-P     -S--T------------------------Y----------GE------------------  60
AHSV4-P     -------K--S----ESG----------N--S---FSQRT-----V--------------  60
AHSV5-P     -------K--S----ESGT---------N--S---FSQRT-----V--------------  60
AHSV9-P     ----------S----ESG----------N--S---FSQRT-----V--------------  60

                    70        80        90       100       110       120
Consensus   ASgALKDEKAAFGAmAEaLRDPEPIRqIKkqVGirTLKhLKiELasMRRkkaiLkivifi  120
AHSV2-P     --------------------------K--R----QX--A--V--SG----RL----IM--  119
AHSV8-P     --------------V-----------K--R----Q---T--V--SG-----L----IM--  119
AHSV3-P     --R---------------------------H--L---------------RY---RV---M  120
AHSV3-PA    --R---------------------------H--L---------------RY---RV--LM  120
AHSV7-P     ------------------------------H--L---------------RY---RV---M  120
AHSV4-P     -----------------R--------------------N--M---T-----SA---T-L-  120
AHSV5-P     --------------------------------------N--M---T-----SA---M---  120
AHSV9-P     --------------------------------------N--M---T-----SA---M---  120

                   130       140       150       160       170       180
Consensus   sgcVTmATSmvGGlsIvDneIAKHledlkkddWlsKtiHGLNLLCTTmLLAAgKISdKiq  180
AHSV2-P     CAN------L---I----ED-----AFDG-G--V---V--------------N---E-VR  179
AHSV8-P     CAN------L---M----ED-----AFDG-G--V---V--------------N---E-VR  179
AHSV3-P     ----------A---T-I----...Y---SG-G----------------------------  177
AHSV3-PA    ----------A---T-I----...Y---SG-G----------------------------  177
AHSV7-P     ----------A---T-I-K--...YD--SG-G----------------------------  177
AHSV4-P     -----L---------------...F--Y--N---M-A----------V------------  177
AHSV5-P     -----L------------D--...-G-Y-NN---M------------V------------  177
AHSV9-P     -----L------------DQ-...-D-Y--N---M------------V------------  177

                   190       200       210       220
Consensus   EEIsRTKRDIAKReSYVsAAsMSWnGDtevlLkevkyGdS  220
AHSV2-P     ---A---------Q------T---D--SVTP-RD-----.  218
AHSV8-P     ---A---------Q------T---D--SVT--RD--C--.  218
AHSV3-P     ------------------------S---S-----------  217
AHSV3-PA    ------------------------S---S-----------  217
AHSV7-P     -----------------P------S---S------NI---  217
AHSV4-P     --------------------------------QGI-----  217
AHSV5-P     -----------------------------MS-QGI---E-  217
AHSV9-P     -----------------------------M--QGT---E-  217
```

Fig. 1. Alignment of the predicted amino acid sequences of NS3/3a of the AHS viruses studied. Dots have been used to represent deletions; dashes denote consensus sequence

Alignment of the amino acid sequence for NS3 (derived from the nucleotide sequences) revealed variations in the length of the mature protein (Fig. 1). NS3 from strains of serotypes 3, 3att, 4, 7 and 9 are 217 amino acids, whilst NS3 from strains of serotypes 2, 5 and 8 are composed of 218 amino acids. Several motifs are also evident from the protein sequence. Two conserved hydrophobic regions are present, spanning nucleotides 116 to 137 and 154 to 170, which are thought to represent transmembrane domains [28, 29]. Potential glycosylation sites also exist, at positions 9-11 in strains of AHSV 4 and AHSV 9, and residues 122–124 in strains of AHSV 8. Interestingly strains of AHSV3 and 3att have no potential glycosylation sites [3]. Other motifs include a cluster of prolines residues between amino acids 22–34, together with a series of conserved cysteines within the hydrophobic domains. Further comparison of the sequences from AHSV 3 and 3^{att} show a phenylalanine to lysize change at residue 119. This lies within the first hydrophobic domain, but as both are polar moities it is considered unlikely that this single change would be significant.

Although there are major areas of conservation between NS3 from the different AHSV strains, the degree of sequence variation is significant when compared to NS3 from BTV, which is more highly conserved. Segment 10 from all strains of BTV serotypes so far sequenced is 822 nucleotides in length [8, 15]. Similarly the glycosylation sites in NS3 of BTV are conserved, as are the positions of the cysteine residues and hydrophobic domains [8, 15].

Phylogenetic analysis of the segment 10 sequences showed that strains of the nine AHSV serotypes fall into three groups (Fig. 2). Group α contains serotypes 4, 5, 6 and 9; group β contains serotype 3, 3att and 7; and group γ is formed by serotypes 1, 2 and 8. Different strains within each serotype do not cluster together. There is approximately 35% variation between the members of either the α or β groups, when compared with the γ group and approximately 25% difference between α and β group members, in both nucleotide or amino acid sequences (segment 10/NS3). The previously published sequences of AHSV 2-US, AHSV 3-US, AHSV 4, AHSV 5-US, AHSV 8-US and AHSV 9-US (Table 2) are essentially similar to those determined in this work and are therefore not shown in Fig. 2. Similarly AHSV3-Pa shows only one nucleotide difference to AHSV3-P and is also not shown in the phylogenetic comparison.

Comparison of the release characteristics of reassortant virus #A90 and its parental strains

KC cell cultures were infected at a concentration of 10^6 particles/ml with either AHSV 3^{att} (virulent in the mouse model), AHSV 8^{wt} (avirulent in the mouse model), or with the reassortant AHSV # A90 (identical to the AHSV 8^{wt} parent, except that it contains segments 9 and 10 from the AHSV 3^{att} and has intermediate virulence [21]). The time course of virus production and its release from the cells was analysed by titration of tissue culture supernatants, collected over a 48 h period, in BHK-21 cells (Fig. 3). After an initial eclipse period of 4–7 h, the level of infectious virus produced by the KC cells infected with either AHSV 3^{att}, AHSV # A90 in-

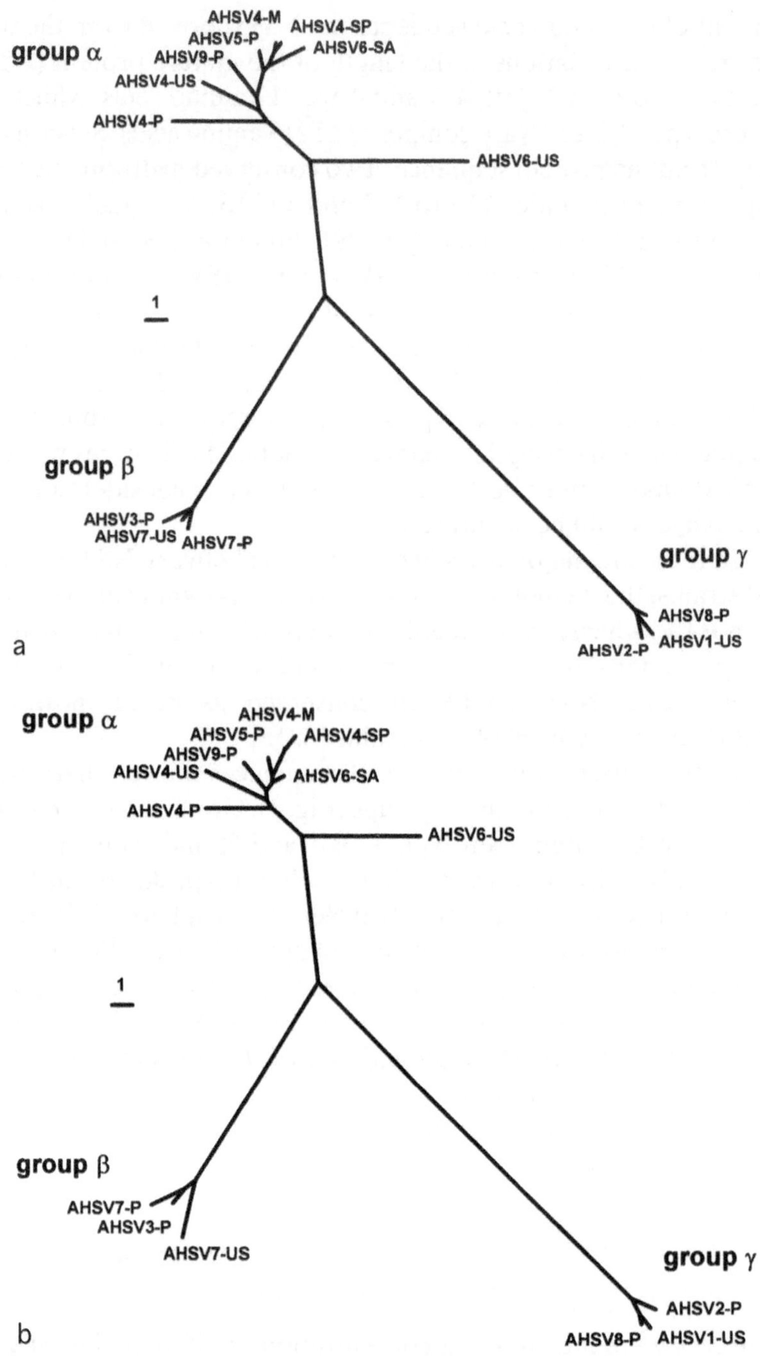

Fig. 2. Unrooted neighbour-joining trees depicting groupings for segment 10 from isolates of various African horse sickness virus serotypes. **a** Nucleotide differences; **b** amino acid differences. Bar marker represents 1% difference in nucleotide sequence, or deduced amino acid sequence

Table 2. Designation and origin of the African horse sickness virus segment 10 sequences examined

Serotype	Designation	Strain	Origin	Length	Accession no.	Reference	Reference
1	AHSV 1-US	USDA	South Africa	764	U02711	[3]	De Sa et al., 1994
2	AHSV 2-US	USDA	South Africa	764	U59279	[24]	Sailleau et al., 1997
2	AHSV 2-P	Pirbright	South Africa	764		This work	This work
3	AHSV 3-US	USDA	South Africa	758	D12479; D01131	[28]	Van Staden and Huismans, 1991
3	AHSV 3-P	Pirbright	South Africa	758	AJ007303	This work	This work
3	AHSV 3-Pa	Pirbright-att	South Africa	758	AJ007304	This work	This work
4	AHSV 4-US	USDA	South Africa	753	Z48734	[24, 31]	Zientara, 1995; Sailleau et al., 1997
4	AHSV 4	SA	South Africa	758	U02712	[3]	De Sa et al., 1994
4	AHSV 4-SP	SP	Spain, 1988	667	L27347	[11]	Laviada et al., 1995
4	AHSV 4-M	M90	Morocco, 1990	747	Z48735	[24, 31]	Zientara, 1995; Sailleau et al., 1997
4	AHSV 4-P	Pirbright	South Africa	758	AJ007305	This work	This work
5	AHSV 5-US	USDA	South Africa	746	U60188	[24]	Sailleau et al., 1997
5	AHSV 5-P	Pirbright	South Africa	746	AJ007309	This work	This work
6	AHSV 6-US	USDA	South Africa	755	U60189	[24]	Sailleau et al., 1997
6	AHSV 6-SA	UP	South Africa	755	U26171	[26]	Stoltz et al., unpub.
7	AHSV 7-US	USDA	South Africa	758	U60190	[24]	Sailleau et al., 1997
7	AHSV 7-P	Pirbright	South Africa	758	AJ007306	This work	This work
8	AHSV 8-US	USDA	Iran	764	U02713	[3]	De Sa et al., 1994
8	AHSV 8-P	Pirbright	South Africa	764	AJ007307	This work	This work
9	AHSV 9-US	USDA	Iran	756	D12480; D01132	[28]	Van Staden and Huismans, 1991
9	AHSV 9-P	Pirbright	South Africa	756	AJ007308	This work	This work

UP University of Pretoria

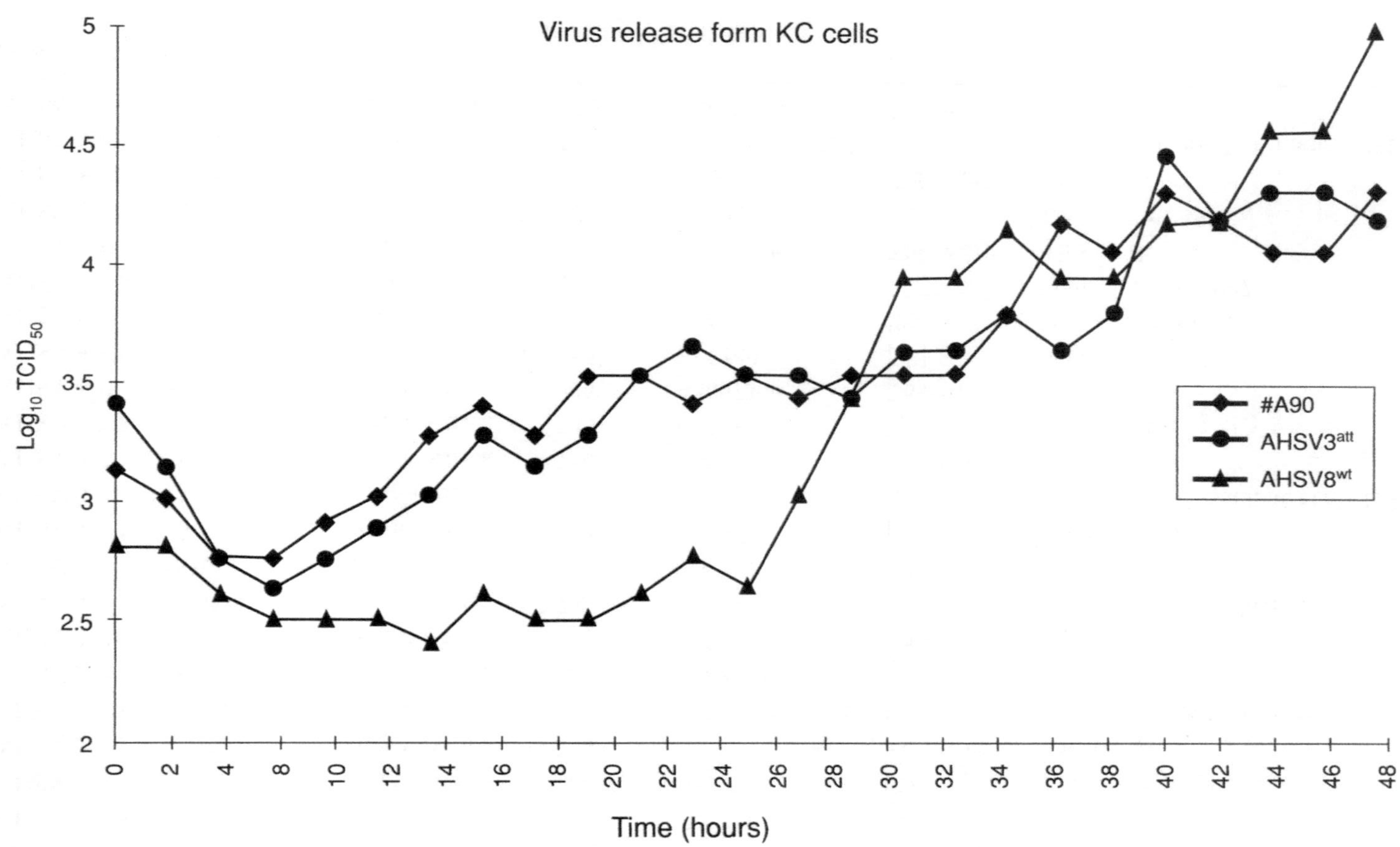

Fig. 3. Time course for release of infectious virus particles from KC cells infected with reassortant virus #A90, or its parental strains AHSV 3^{att} or AHSV 8^{wt}

creased steadily, up to the termination of the experiment at 48 h. In contrast KC cells infected with the avirulent parental strain AHSV 8^{wt}, released much less virus between 6 and 24 h pi (ca. 10^3 BHK $TCID_{50}$/ml). However, after 24 h there was a sharp increase in the amount of virus released into the tissue culture medium (ca. 10^4 BHK $TCID_{50}$/ml). Although no CPE was detected in the KC cells infected with either the reassortant virus #A90 or the parental strains, the final virus titre for AHSV 3^{att} and #A90 at 48 h pi was approximately $10^{4.5}$ BHK $TCID_{50}$/ml. At 48h pi the cells infected with AHSV 8^{wt} had released a slightly higher titre of virus into the cell culture supernatant (approximately 10^5 BHK $TCID_{50}$/ml).

Discussion

As discussed previously, the function of NS3/3a during AHSV replication is still uncertain. Evidence for the role of NS3/3a in BTV replication tends to suggest that it mediates virus exit [9, 10] and this also appears to be true for NS3 of AHSV. One of the most interesting features of segment 10 from AHSV is the variability in the sequence compared to the cognate gene from BTV, Palyam virus and EHDV [11, 19]. It has been suggested that these variations in the sequence of segment 10 from AHSV may be used to diagnose serotype [32]. However, these differences in sequence observed between strains from different serotypes can be deceptive. Firstly virus serotype is controlled solely by the outer capsid proteins VP2 and VP5, to which neutralising antibodies are raised. Secondly our studies and those of Laegried [13] have shown that reassortment and exchange of genome segment 10 between viruses of different serotypes is possible. Variations in AHSV NS3 sequence therefore do not correlate with virus serotype [33], this is clearly demonstrated in Fig. 1. Each of the groups of viruses (α, β and γ) contain several different serotypes and the reassortant study described by O'Hara et al. [21] generated a virus #A90, in which segments 9 and 10 of AHSV 3^{att} has been included in a virus which is otherwise identical to AHSV 8^{wt}. Since none of the capsid proteins have been altered the virus serotype remains as type 8 but as a result of changes in its NS3 sequence the #A90 virus has moved from group α to group β.

O'Hara et al. [21] demonstrated with a reassortment study that segment 10 can influence virulence. It has been suggested that NS3 mediates this via an effect on the rate/timing of virus release from the cell [13]. Previous studies have indicated that NS3 from BTV can mediate cell exit by budding through the cell membrane [4, 7, 9, 10, 13]. Although orbiviruses can exit from mammalian cells by two processes, extrusion or budding (Fig. 4), exit from KC cells was only observed by budding [7].

Analysis of the time course of virus release from KC cells of the two parental strains of reassortant #A90, showed that AHSV 3^{att} (virulent) started after 6 hours pi, while a significant increase in the titre of AHSV 8^{wt} (avirulent) was observed only after 24 h pi. Reassortant #A90 which is identical to AHSV 8^{wt} in all segments except 9 and 10 was released from cells with a similar pattern to that for AHSV 3^{att}. Since genome segment 9 encodes a small internal virus core protein (VP6), these data strongly indicate that changes in segment 10 (which #A90 had

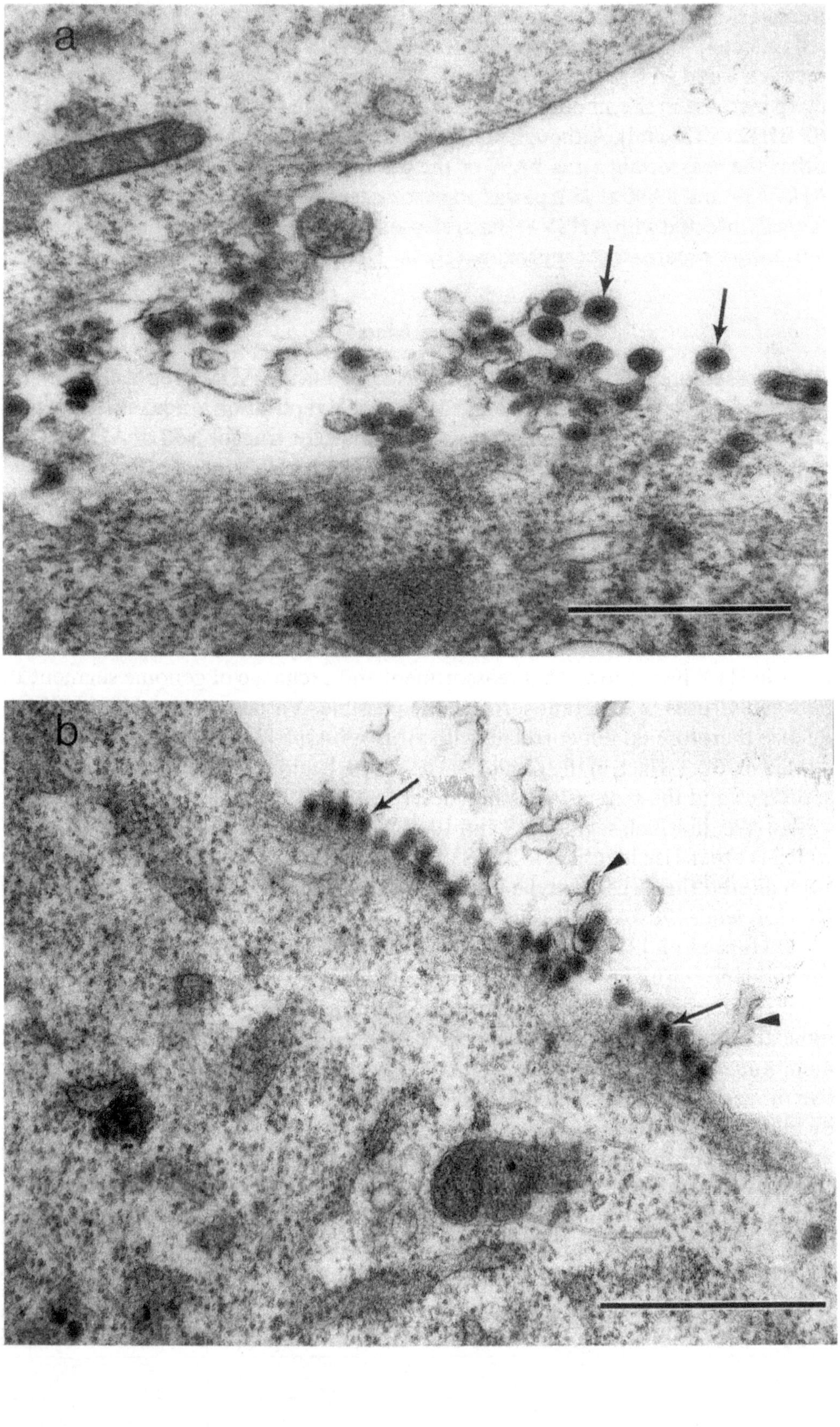
a
b

inherited from AHSV 3^{att}) had altered the timing of virus release. These data confirm the role for NS3 in virus release from cells and support the hypothesis that segment 10 (NS3/3a) may influence virulence phenotypes by altering the timing of virus release. Further studies to compare reassortants #A79 and #A790 [21] are currently in progress.

The AHSV 3 and AHSV 3^{att} strains showed only a single amino acid change in NS3, which is thought unlikely to significantly affect protein function. The virulence changes in these strains of AHSV 3 are therefore thought more likely to be dependent on changes in the outer capsid proteins (particularly VP2), which were also implicated in the determination of virulence [21].

However a larger number of amino acid differences (81 out of 218 in total = 37% amino acid sequence variation) were also observed between the two parental strains of AHSV 3^{att} and AHSV 8^{wt} used in reassortant studies [21], reflecting the position of their NS3 proteins in the β and γ groups (Fig. 2). The differences in virus release characteristics, or in virulence, which were caused by exchange or variation in genome segment 10, may therefore reflect differences overall in the NS3 proteins from these α, β and γ groups (although this will only be confirmed by further experimentation).

Important barriers to dissemination of orbivirus infection in the vector insect (and therefore transmission of the virus during feeding) may involve the release of virus from infected cells [7]. Any virus or cell dependent factor which can influence the timing or efficiency of virus particle release from the infected cells, could therefore affect the vector status of the insect. Viruses from these three groups (α, β and γ) may also show some differences in the timing of their release from cells within the adult insect vector. The relevance of these three groups of NS3 in specific insect vectors and their affect on vector status in the field, remains to be determined.

Acknowledgements

KC cells were kindly provided by Sally Wechsler, USDA Laramie. This work was supported by EC contract No. 8001-CT91-0211, by MAFF (UK) and BBSRC (UK).

References

1. Burroughs JN, O'Hara RS, Smale CJ, Hamblin C, Walton A, Armstrong R, Mertens PPC (1994) Purification and properties of virus particles, infectious sub viral particles, cores and VP7 crystals of African Horse sickness virus serotype 9. J Gen Virol 75: 1849–1857

Fig. 4. Transmission electron micrograph of BTV 1 infection of BHK-21 cells (20 h pi). **a** Mature double shelled virus particles were released by budding from an infected cell, acquiring an envelope derived from the cell membrane in the process. The resulting membrane enveloped virus particles (MEVP) were present outside the cell (arrows). **b** BTV virions were released by extrusion from the infected cell surface, without acquiring an envelope from the cell membrane (arrows). Cell debris was associated with the released particles (arrowheads). Bar: 500 nm

2. Calisher CH, Mertens PPC (1998) Taxonomy of African horse sickness viruses. In: Mellor PS, Baylis M, Hamblin C, Calisher CH, Mertens PPC (eds) African horse sickness. Springer, Wien New York, pp 3–11 (Arch Virol [Suppl] 14)
3. de Sa RO, Zellner M, Grubman MJ (1994) Phylogenetic analysis of segment 10 from African horse sickness virus and cognate genes from other orbiviruses. Virus Res 33: 157–165
4. Eaton BT, Hyatt AD, Brookes SM (1990) The replication of bluetongue virus. Curr Top Microbiol Immunol 162: 89–118
5. Felsenstein J (1993) PHYLIP (Phylogeny Inference Package) version 3.5c. Distributed by the author. Department of Genetics, University of Washington, Seattle, 1993
6. Finney DJ (1964) Statistical method in biological assay, 2nd ed. Griffin, London
7. Fu H, Leake CJ, Mertens PPC, Mellor PS (1998) The barriers to bluetongue virus infection dissemination and transmission in the vector *Culicoides variipennis* (Diptera: Ceratopogonidae) (in preparation)
8. Hwang G, Yang Y, Chiou J, Li JK (1992) Sequence conservation among the cognate nonstructural NS3/3A protein genes of six bluetongue viruses. Virus Res 23: 11–161
9. Hyatt AD, Gould AR, Coupar B, Eaton BT (1991) Localization of the non-structural protein NS3 in bluetongue virus-infected cells. J Virol 72: 2263–2267
10. Hyatt AD, Zhao Y, Roy P (1993) Release of bluetongue virus-like particles from insect cells is mediated by BTV non-structural protein NS3/NS3A. Virology 193: 592–603
11. Jensen MJ, Cheney IW, Thompson LH, Mecham JO, Wilson WC, Yamakawa M, Roy P, Gorman BM (1994) The smallest gene of the orbivirus, epizootic hemorrhagic disease is expressed in virus-infected cells as two proteins and the expression differs from that of the cognate gene of bluetongue virus. Virus Res 32: 353–364
12. Kozak M (1991) Structural features in eukaryotic mRNAs that modulate the initiation of translation. J Biol Chem 266: 19867–19870
13. Laegried WW, Burrage TG, Stone-Marschat MA, Skrowronek AJ, Hussein A (1995) The S10 gene of African horse sickness virus is a major virulence determinat. Abstracts, Fifth International Symposium on double-stranded RNA viruses (P46)
14. Laviada MD, Roy P, Sanchez-Vizcaino JM, Casal JI (1995) The use of African horse sickness virus NS3 protein, expressed in bacteria, as a marker to differentiate infected from vaccinated horses. Virus Res 38: 205–218
15. Lee JW, Roy P (1986) Nucleotide sequence of a cDNA clone of RNA segment 10 of bluetongue virus (serotype 10). J Gen Virol 67: 2833–2837
16. Martin L-A, Mertens PPC (1998) Purification and characterisation of RNA. Methods Mol Biol (in press)
17. Mertens PPC, Burroughs JN, Walton A, Wellby MP, Fu H, O'Hara RS, Brookes SM, Mellor PS (1996) Enhanced infectivity of modified bluetongue virus particles for two insect cell lines and for two *Culicoides* vector species. Virology 217: 582–593
18. Mertens PPC, Burroughs JN, Anderson J (1987) Purification and properties of virus particles, infectious sub-viral particles and cores of bluetongue virus serotypes 1 and 4. Virology 170: 561–565
19. Moss SR, Jones LD, Nuttall PA (1992) Comparison of the nonstructural protein, NS3, of tick-borne and insect-borne orbiviruses. Virology 187: 841–844
20. Nicholas KB, Nicholas HB Jr (1996) GeneDoc: a tool for editing and annotating multiple sequence alignments. Distributed by the author
21. O'Hara RS, Meyer AJ, Pullen L, Martin L-A, Mertens PPC (1998) Development of a mouse model system and identification of the individual genome segments of African horse sickness virus serotypes 3 and 8 involved in determination of virulence. In: Mellor PS, Mertens PPC, Baylis M, Hamblin C (eds) African horse sickness. Springer, Wien New York, pp 259–279 (Arch Virol [Suppl] 14)

22. Page RDM (1996) TREEVIEW: An application to display phylogenetic trees on personal computers. CABIOS 12: 357–358
23. Saitou N, Nei M (1987) The neighbour-joining method: a new method for reconstructing phylogenetic trees. Mol Biol Evol 4: 406–425
24. Sailleau C, Moulay S, Zientara S (1997) Nucleotide sequence comparison of the segments S10 of the nine African horse sickness virus serotypes. Arch Virol 142: 965–978
25. Sanger F, Nicklen S, Coulson AR (1977) DNA sequencing with chain-terminating inhibitors. Proc Natl Acad Sci USA 74: 5463–5467
26. Stoltz MA, van Staden V, Napier GB, Huismans H (1997) Unpublished data
27. Thompson JD, Higgins DG, Gibson TJ (1994) CLUSTAL W: improving the sensitivity of progressive multiple sequence alignment through sequence weighting, positions-specific gap penalties and weight matrix choice. Nucleic Acids Res 22: 4673–4680
28. van Staden V, Huismans H (1991) A comparison of the genes which encode non-structural protein NS3 of different orbiviruses. J Gen Virol 72: 1073–1079
29. van Staden V, Stoltz MA, Huismans H (1995) Expression of nonstructural protein NS3 of African horse sickness virus (AHSV): evidence for a cytotoxic effect of NS3 in insect cells, and characterization of the gene products in AHSV infected Vero cells. Arch Virol 140: 289–306
30. Wechsler SJ, McHolland LE, Tabachnick WJ (1989) Cell-lines from *Culicoides-variipennis* (Diptera, Ceratopogonidae), support replication of bluetongue virus. J Invertebr Pathol 54: 385–393
31. Zientara S (1995) Thesis, Laboratoire de Virologie Equine, CNEVA
32. Zientara S, Sailleau C, Moulay S, Crucière C (1995) Differentiation of African horse sickness viruses by polymerase reaction and segment 10 restriction patterns Vet Microbiol 47: 365–375
33. Zientara S, Sailleau C, Plateau E, Moulay S, Mertens PPC, Crucière C (1998) Molecular epidemiology of African horse sickness virus based on analyses and comparisons of genome segments 7 and 10. In: Mellor PS, Baylis M, Hamblin C, Calisher CH, Mertens PPC (eds) African horse sickness. Springer, Wien New York, pp 317–327 (Arch Virol [Suppl] 14)

Authors' address: Dr. P. P. C. Mertens, Institute for Animal Health, Pirbright Laboratory, Ash Road, Pirbright, Woking, Surrey, GU24 ONF, U.K.

[illegible]
[illegible] models and [illegible]
[illegible]

[illegible]

Vaccines and diagnosis

Future international management of African horse sickness vaccines

J. A. House

USDA, Animal and Plant Health Inspection Service, Veterinary Services, National Veterinary Services Laboratory, Foreign Animal Disease Diagnostic Laboratory, Greenport, New York, U.S.A.

Summary. Three types of African horse sickness (AHS) vaccine, namely adult mouse brain, modified live vaccine and inactivated viral vaccine (IVV) are reviewed. The results of efficacy trials carried out with each vaccine type highlight the advantages of the IVV. Vaccination with African horse sickness virus serotype 4 IVV, given as 2 separate doses, provided full protection against subsequent, homologous challenge. The absence of any detectable viraemia after challenge would also prevent infection of insect vectors. The advantages of establishing international vaccine banks for AHS are discussed.

Introduction

African horse sickness viruses (AHSV) are arthropod borne viruses belonging to the family *Reoviridae*, genus *Orbivirus*. The viruses cause a severe disease, African horse sickness (AHS), in horses and mortality in susceptible animals can exceed 90%. Low but significant levels of cross-neutralisation have been reported between some of the 9 serotypes [5] viz 1 and 2; 3 and 7; 5 and 8; 6 and 9. Vaccination against AHS has been used successfully for many years and remains an effective and essential part of any control strategy. This paper reviews three major types of vaccine which have been used to protect horses against AHS and gives consideration to future vaccine strategies. Although an effective experimental subunit vaccine against AHSV has been described [30], no commercial preparations are yet available. The value and use of these genetically engineered vaccines are discussed elsewhere [30, 33].

Adult mouse brain vaccines (AMBV)

The neurotropic strains of AHSVs, long used as vaccines, were prepared following approximately 100 passages in adult mice [1]. These vaccines were used extensively in South Africa until 1988 although it is not clear if they are still being produced and used in other countries. There are several reports that some of these vaccines caused encephalitis in horses [27, 29, 31] and in some cases failed to induce protection [9, 17]. Encephalitis and choriorentinitis have also been reported in laboratory workers exposed to these AMBVs [32]. Offcials deciding on the use of AMBVs must, therefore, take into consideration the safety and efficacy problems associated with these products. The immune response generated after vaccination with AMBVs cannot be distinguished from those elicited by natural infection.

The efficacy of the AMBV seed viruses has been evaluated as a potential source of vaccine for the USA [13]. Individual ponies were each inoculated with one of the 9 monovalent vaccine viruses, except for AHSV 5 and 6 where two ponies were inoculated against each serotype. A febrile response occurred in several animals following vaccination but only those vaccinated against AHSV 3 and 9 developed detectable neutralising antibodies. Each vaccinated pony was challenged with the virulent, homologous strain of AHSV. Ponies vaccinated with the AMBVs for serotypes 1, 4, 5, 6 and 7 developed a primary immune response after challenge. The two ponies vaccinated against AHSV 1 and 4 and one of the ponies vaccinated against AHSV 5 died with typical signs of AHS on days 26, 21 and 9 after challenge with virulent virus of the homologous serotype. Ponies vaccinated against AHSV 2 and 8 responded anamnestically after challenge and both survived. The remaining two ponies which were inoculated with AMBV for serotypes 3 and 9 showed no increases in neutralising antibody titre after challenge and also survived. Control animals inoculated with each of the virulent AHSV serotypes died showing typical signs of AHS.

In a second study 4 ponies were vaccinated with AMBV for AHSV 4. No neutralising antibodies were detected in any of the animals 21 days post vaccination (DPV). Three of the four vaccinates and the non-vaccinated control pony died with typical signs of AHS following challenge with homologous virus. The highly variable results of these small efficacy trials suggest that these vaccines would not be suitable for use in the USA.

Modified live vaccine (MLV)

Attenuation of AHSVs by adaption to growth in monkey kidney cell lines and their use as polyvalent MLVs has been reported [8, 11]. These vaccines were based on the finding that those viruses producing large plagues were avirulent [10]. The Onderstepoort Biological Products (OBP) facility in South Africa is the major world supplier of MLVs although Biopharma in Morocco produced a similarly derived monovalent MLV for use in Morocco during the recent epizootic of AHSV 4 in that country (1989–1991).

The OBP quadrivalent MLVs are designed for use in South Africa where multiple AHSV serotypes occur. According to the package insert, these MLVs consist of 2 quadrivalent products, one containing serotypes 1, 3, 4 and 5 and the second containing serotypes 2, 6, 7 and 8. The 2 injections are administered 3 weeks apart. Annual re-vaccination is recommended and full protection against all serotypes is usually expected after 3 to 4 years. Literature on these products is limited but like the AMBVs the immune response generated after vaccination with MLVs cannot be distinguished from those elicited by natural infection.

A monovalent AHSV 4 MLV prepared by OBP was used in Spain during the 1987–1990 epizootic. However, studies in Spain showed that approximately 10% of a group of equidae were antibody negative 1 to 3 months after primary vaccination [12]. These authors also reported anamnestic antibody responses following the re-vaccination of 2 ponies that were recorded sero-negative after primary vaccination suggesting that some sero-negative vaccinated animals might be protected. At the request of Spanish officials three ponies were also vaccinated at the Plum Island Animal Disease Center in the USA with the AHSV 4 MLV. The ponies remained clinically healthy. All animals seroconverted but no viraemia was detected after vaccination. The ponies were challenged on day 28 with virulent AHSV 4. The non-vaccinated control pony died 15 days post challenge (DPC). One vaccinated pony remained clinically normal and had no viraemia. A second pony experienced a viraemia of $10^{2.8}$ mouse lethal dose 50% (MLD_{50}) of AHSV 4/ml of blood on 3 DPC. The third pony was febrile for 9 days and had a viraemia which persisted from 3 to 11 DPC, ranging from $10^{2.8-4.8}$ MLD_{50} of AHSV 4/ml of blood [14]. Slight congestion of the oral and nasal membranes were also observed with the 2 viraemic ponies. In the field these signs would probably have gone unnoticed. For bluetongue viruses (BTVs), the prototype orbivirus, it is estimated that a viraemia greater than $10^{4.0}$ infectious particles per ml of blood is suffcient to infect *Culicoides* [21]. If this value is similar for AHSVs then one of the three ponies in this study would have been a potential source of AHSV for insect vectors. In conclusion, the AHSV 4 MLV afforded protection against severe clinical signs and death following challenge but did not protect 2 of 3 ponies against viraemia. The vaccine seed used for the production of AHSV 4 MLV has since been replaced (Van Dijk, pers. comm.).

Inactivated viral vaccine (IVV)

Experimental inactivated vaccines were reported as early as 1966 [27] but the process was never commercialised. However, during the 1987–1990 AHSV epizootic in Spain, a commercial IVV for AHSV 4 was developed and used experimentally in Spain and Morocco [7].

The efficacy of the AHSV 4 IVV as a single dose vaccine was evaluated following the vaccination of 9 ponies [15]. All ponies remained clinically healthy and developed AHSV specific neutralising antibodies by 28 DPV. The 9 vaccinated and 3 non-vaccinated control ponies were inoculated with virulent AHSV 4. The 3 non-vaccinated control ponies died 7 DPC exhibiting clinical signs characteristic of AHS. Of the vaccinated ponies, three ponies exhibited a febrile response one had congestion of the oral and nasal mucosa and another had a viraemia where viral titres ranged $10^{2.6-3.1}$ MLD_{50}/ml of blood. The remaining 6 vaccinated animals remained clinically healthy. An anamnestic response in antibodies was detected in 7 of the 9 vaccinated animals, indicating virus replication had occurred. From these results it was concluded that a single dose of IVV would protect horses from severe clinical signs and death but not necessarily from the development of a viraemia.

The effects of administering two doses of AHSV 4 IVV has since been evaluated [16]. Five ponies were given a single dose of the vaccine on day 0 and then re-vaccinated 56 DPV. Neutralising antibody titres against AHSV 4 reached peak values 13 days after the second vaccination. The 5 vaccinated and 2 non-vaccinated control ponies were challenged 20 days after the second vaccination with virulent AHSV 4. The 2 non-vaccinated control ponies died 7 and 8 DPC exhibiting clinical signs characteristic of AHS whereas the 5 vaccinated animals were fully protected and did not show clinical signs of disease or viraemia. Furthermore, no boost in neutralising antibody titre was recorded after challenge. Consequently it was concluded that vaccination using the 2-dose IVV regime would not only give full protection but also it would prevent the infection of insect vectors which is a critical consideration for eradicating AHSV from non-enzootic countries.

The AHSV non-structural proteins (NS1, NS2 and NS3) can be removed during the purification of the virus particles [4]. Therefore, provided that an effective purification method is used in the preparation of IVV, assays which can detect antibodies to these non-structural proteins, can be used to differentiate vaccinated from naturally infected animals during an epizootic [18]. It is therefore important that the purity of the virus preparation used to make the IVV is adequately monitored.

One important limitation of such tests for antibodies to non-structural proteins, results from the variable nature of the AHSV NS3 protein [20]. Not all NS3 proteins from different strains of AHSV will serologically cross react. This is in direct contrast to BTV where NS3 is a highly conserved protein. The use of such tests to detect antibodies to AHSV NS3 and reliance on them to distinguish vaccinated from infected animals, can therefore only be valid if it has been shown that antibodies to NS3 from all of the different viruses that are circulating in the field can be detected.

Alternative tests for antibodies to the other more conserved non-structural proteins (NS1 and NS2) could be developed and used, as described for BTV [2]. With further development, validation and implementation of such tests, a region or country could safely undertake a prophylactic vaccination campaign with a 2-dose regime of IVV with the added advantage of being able to differentiate between vaccinated and infected animals if a risk of an epizootic was anticipated.

Vaccine standards

Internationally, vaccine standards are becoming more sophisticated and as such good manufacturing practices, quality control and the use of the master seedlot principle are now being applied in the development and production of most new vaccines. The retrospective application of these standards to existing products, however, has lagged.

Consideration should also be given to the type of vaccines available and the consequence of their use. For instance, all vaccines containing attenuated live virus should be examined to determine the potential for that virus to revert to virulence, particularly if a viraemia occurs as a result of vaccination and where there would be a potential for vector insects to become infected. The backpassage test is a relatively simple method of determining if reversion occurs and involves inoculating blood from vaccinated animals into naive recipients and then looking for clinical signs and viraemia. Backpassage data are not available for AHS, AMBVs and MLVs. Ideally the vaccine should also prevent viraemia on subsequent exposure of the vaccinated animal to field viruses thereby eliminating the possibility of it acting as a source of virus, "silently" perpetuating the epizootic. Furthermore, if live vaccines, particularly polyvalent, are given to animals that are incubating field viruses or that are naturally infected soon after vaccination, there is the potential for reassortment to occur between the vaccine and field viruses and thereafter the emergence of new strains. Studies of genome segment reassortment between wild type and vaccine strains of AHSVs [19], BTVs [6, 22] and Kemerovo serogroup orbiviruses [23–26] have demonstrated the production of progeny virus strains which can have novel serological properties, or altered virulence characteristics. The virulence of the progeny viruses can even be greater than that of either parental strain. The generation of such potentially dangerous reassortant viruses and their subsequent survival and spread, may be relatively rare events. However, reassortment has been shown to occur between related orbivirus strains in infected tissue cultures as well as in their mammalian hosts and insect vectors. Live vaccines of these multi-segmented RNA viruses must therefore represent a risk potentially leading to the development of novel and more dangerous strains of virus in the field.

The future role of AHSV vaccines

In the enzootic regions of Africa, vaccination has been used essentially as a method of preventing animal losses, infection of vector insects being a secondary consideration. The AMBVs may continue to serve in enzootic regions because of the absence of an easily available alternative but the safety and efficacy studies reported in this paper highlight the problems inherent in the continual use of these vaccines. Although cell culture derived MLVs appear effective and safe in terms of animal health, sufficient firm data is not yet available to confirm this assertion. In light of the recent changes in vaccine regulations and because of the problems associated with the continuous production and quality control of the AMBVs and MLVs, considerable interest is now being shown in the production of IVVs.

Because of the disease and economic implications of AHS in non-enzootic regions, the official policy has always been to eradicate the virus as quickly as possible. However, a stamping out policy is often difficult to enforce due to the great value of equines and because of animal welfare concerns. Vaccination is therefore always a major component in any AHSV control and eradication programme. The vaccine of choice must be safe, efficacious and should adequately prevent the infection of insect vectors. The 2-dose IVV regime appears to be the only type of vaccine reviewed here that satisfies all of these requirements.

Vaccine banking

In non-enzootic regions, regulatory officials may need AHS vaccines on an emergency basis. Vaccines could be obtained from countries routinely producing them but in many cases the production site may not be certified and live agents may have been used to produce vaccines that cause diseases exotic to the inquiring country. Also in the case of AHS the market is small and spare capacity may not be available to readily supply vaccines at short notice.

In the case of other vaccines, for example foot-and-mouth disease, vaccine banking is being practised by several countries and consortiums to overcome such problems and assure availability. Banking of vaccine stocks or concentrates is also practised by certain vaccine manufacturers giving the ability to respond rapidly to vaccine requests. In such cases, banked vaccines are pretested and certified to meet prescribed safety and efficacy requirements. However, efficiently managing a vaccine bank requires accurate estimations of the number of doses and the serotypes likely to be required. These types of data can be obtained via international exchange of information through the OIE [3], an organisation which is at the forefront of animal disease control and eradication.

References

1. Alexander RA, Du Toit PJ (1934) The immunization of horses and mules against horsesickness by means of the neurotropic virus of mice and guinea pigs. Onderstepoort J Vet Sci 2: 375–391
2. Anderson J, Mertens PPC, Herniman KAJ (1993) A competitive ELISA for the detection of anti-tubule antibodies using monoclonal antibody against bluetongue virus non-structural protein NS 1. J Virol Methods 43: 167–176
3. Anonymous (1992) African horse sickness. In: Manual of standards for diagnostic tests and vaccines. Office International Des Epizooties, Paris pp 91–100
4. Burroughs JN, O'Hara R, Smale CJ, Hamblin C, Walton A, Armstrong R, Mertens PPC (1994) Purification and properties of virus particles, infectious sub-viral particles, cores and VP7 crystals of African horse sickness virus serotype 9. J Gen Virol 75: 1849–1857
5. Coetzer JAW, Erasmus BJ (1994) African horsesickness. In: Coetzer JAW, Thomson GR, Tustin RC (eds) Infectious diseases of livestock, vol 1. Oxford University Press, Oxford, pp 460–475
6. Cowley JA, Gorman BM (1989) Cross-neutralisation of genetic reassortants of bluetongue virus serotypes 20 and 21. Vet Microbiol 19: 37–51
7. Dubourget P, Preaud JM, Detraz N, Lacoste F, Fabry AC, Erasmus B, Lombard M (1992) Development, production and quality control of an industrial inactivated vaccine against AHS type IV. In: Walton TE, Osburn BI (eds) Bluetongue, Africa horse sickness and related orbiviruses. Proc 2nd Int Symp. CRC Press, Boca Raton, pp 874–886
8. Erasmus BJ (1965) The attenuation of viscerotropic horsesickness virus in tissue culture. Bull Off Int Epiz 64: 697–702

9. Erasmus BJ (1966) The attenuation of African horsesickness virus: problems associated with the use of different host systems. In: Proc 1st Int Conf Equine Inf Dis, Stresa 1966, pp 208–213
10. Erasmus BJ (1972) The pathogenesis of African horse sickness. In: Proc 3rd Int Conf Equine Inf Dis, Paris, pp 1–11
11. Erasmus BJ (1976) A new approach to polyvalent immunization against African horsesickness. In: Bryans JT, Gerber H (eds) Equine infectious diseases. Veterinary Publications, Princeton, pp 401–403
12. Hamblin C, Mellor PS, Graham SD, Hooghuis H, Montejano RC, Cubillo MA, Boned J (1991) Antibodies in horses, mules and donkeys following monovalent vaccination against African horse sickness. Epidemiol Infect 106: 365–371
13. House JA, House C (1989) Immunization studies on African horse sickness. Proc 93rd Ann Mtg US Animal Health Assoc, pp 357–365
14. House JA, House C, Mebus CA (1992) A review of African horse sickness with emphasis on selected vaccines. Ann NY Acad Sci 653: 228–232
15. House JA, Lombard M, House C, Dubourget P, Mebus CA (1992) Efficacy of an inactivated vaccine for African horse sickness virus serotype 4. In: Walton TE, Osburn BI (eds) Proc 2nd Int Symp bluetongue, African horse sickness and related orbiviruses. CRC Press, Boca Raton, pp 891–895
16. House JA, Lombard M, Dubourget P, House C, Mebus CA (1994) Further studies on the efficacy of an inactivated African horse sickness serotype vaccine. Vaccine 12: 142–144
17. Howell PG (1963) Observations on the occurrence of African horsesickness amongst immunised horses. Onderstepoort J Vet Res 30: 3–10
18. Laviada MD, Roy P, Vizcaino-Sanchez JM, Casal JI (1995) The use of African horse sickness virus NS3 protein, expressed in bacteria, as a marker to differentiate infected from vaccinated horses. Vir Res 38: 205–218
19. O'Hara RS, Meyer AJ, Pullen L, Martin LA, Mertens PPC (1998) Development of a mouse model system and identification of the individual genome segments of African horse sickness virus serotypes 3 and 8 involved in determination of virulence. In: Mellor PS, Baylis M, Hamblin C, Calisher CH, Mertens PPC (eds) African horse sickness. Springer, Wien New York, pp 259–279 (Arch Virol [Suppl] 14)
20. Martin LA, Meyer AJ, O'Hara RS, Fu H, Knowles NJ, Mertens PPC (1998) Phylogenetic analysis of African horse sickness virus genome segment 10: Does sequence variation influence virulence characteristics? In: Mellor PS, Baylis M, Hamblin C, Calisher CH, Mertens PPC (eds) African horse sickness. Springer, Wien New York, pp 281–293 (Arch Virol [Suppl] 14)
21. Mellor PS (1990) The replication of bluetongue virus in *Culicoides* vectors. Curr Top Microbiol Immunol 162: 143–161
22. Mertens PPC, Pedley S, Cowley J, Burroughs JN, Corteyn AH, Jeggo MH, Jennings DM, Gorman BM (1989) Analysis of the roles of bluetongue outer capsid proteins, VP2 and VP5, in determination of virus serotype. Virology 170: 561–565
23. Moss SR, Ayres CM, Nuttall PA (1987) Assignment of the genome segment coding for the neutralising epitope(s) of Orbiviruses in the Great Island serogroup (Kemerovo serogroup). Virology 157: 137–144
24. Nuttall PA, Moss SR, Jones LD, Carey D (1989) Identification of the major genetic determinant for neurovirulence of tick-borne orbiviruses. Virology 172: 428–434
25. Nuttall PA, Moss SR, Carey D, Jones LD, Jacobs SC (1990) Genetic determinants modulating the pathogenic phenotype of tick-borne orbiviruses. Virology 174: 430–435
26. Nuttall PA, Jacobs SC, Jones LD, Carey D, Moss SR (1992) Enhanced neurovirulence of tick-borne orbiviruses resulting from genetic modulation. Virology 187: 407–412

27. Nobel TA, Neumann F (1961) Vaccination against African horse sickness and postvaccination reactions in Israel. OAH Vaccine Readings, pp 168–173
28. Ozawa Y, Bahrami S (1966) African horse-sickness killed-virus tissue culture vaccine. Can J Comp Med Vet Sci 30: 311–314
29. Pavri KM, Anderson CR (1963) Isolation of a vaccine strain of African horse-sickness virus from brains of two horses given polyvalent vaccine. Ind J Vet Sci 33: 215–219
30. Roy P (1998) New generation of African horse sickness vaccines based on structural and molecular studies of the virus particles. In: Mellor PS, Baylis M, Hamblin C, Calisher CH, Mertens PPC (eds) African horse sickness. Springer, Wien New York, pp 177–202 (Arch Virol [Suppl] 14)
31. Shah KV, Chinoy DN, Gokhale TB (1963) Investigations of African horse-sickness in India. II. Reactions in non-immune horses after vaccination with the polyvalent African horse-sickness vaccine. Ind J Vet Sci 34: 2: 75–83
32. Swanepoel R, Erasmus BJ, Williams R, Taylor MB (1991) Encephalitis and chorioretinitis associated with neurotropic African horsesickness virus infection in laboratory workers. III. Virological and serological investigations. S Afr Med J 81: 458–461
33. Wade-Evans AM, Pullen L, Hamblin C, O'Hara RS, Burroughs JN, Mertens PPC (1998) VP7, from African horse sickness virus serotype 9, protects mice against a lethal heterologous serotype challenge. In: Mellor PS, Baylis M, Hamblin C, Calisher CH, Mertens PPC (eds) African horse sickness. Springer, Wien New York, pp 211–219 (Arch Virol [Suppl] 14)

Author's address: Dr. J. A. House, United States Department of Agriculture, Animal and Plant Health Inspection Service, Veterinary Services, National Veterinary Services Laboratory, Foreign Animal Diseases Diagnostic Laboratory, P.O. Box 848, Greenport, Plum Island, New York 11944, U.S.A.

Application of an indirect fluorescent antibody assay for the detection of African horse sickness virus antibodies

H. El Hasnaoui[1], **M. El Harrak**[2], **A. Tber**[3], **A. Fikri**[3], **K. Laghzaoui**[3], and **M. H. Bikour**[4]

[1] Laboratoire Régional d'Analyses et de Recherches Vétérinaires de Casablanca, Casablanca, Morocco

[2] Laboratoire Biopharma, Rabat-Akkari, Morocco

[3] Direction de l'Elevage, Ministère de l'Agriculture et de la Mise en Valeur Agricole, Rabat, Morocco

[4] Laboratoire Régional d'Analyses et de Recherches Vétérinaires de Tanger, Tanger, Morocco

Summary. An indirect fluorescent antibody (IFA) technique was used to screen and quantify antibodies against African horse sickness virus (AHSV) in equine sera. Results obtained with the IFA assay were compared directly with those obtained with standard complement fixation (CF) and virus neutralisation (VN) tests using horse sera from experimental studies and samples from the field. Positive fluorescent antibody titres were detected from as early as 7 days after primary vaccination and persisted for at least six months. The IFA technique offers a clear advantage over CF tests, where the antibodies are often of shorter duration and where sera from donkeys and mules are frequently anticomplementary. The sensitivity and specificity of the IFA test compared with the VN test were 98% and 83.3%, respectively. The IFA test is rapid, relatively easy to perform and inexpensive, and can be recommended as an alternative assay for screening different species of equidae in AHSV control and surveillance programmes.

Introduction

African horse sickness (AHS) is an infectious but non-contagious viral disease of equidae. The causal agent, AHS virus (AHSV), is a member of the family *Reoviridae*, genus *Orbivirus* [6] and is transmitted by arthropods of the genus *Culicoides* [8]. There are 9 known virus serotypes, which are differentiated by virus neutralisation (VN) tests [6].

AHSV is usually restricted to sub-Saharan Africa, however, incursions into countries beyond this area do occur occasionally, the most recent being a 5 year epizootic involving Spain, Portugal and Morocco (1987–1991) [9]. The appearance of AHSV serotype 4 (AHSV4) in these countries was of major economic importance. In such situations the early identification of AHSV is essential to ensure that control and eradication measures are implemented without delay. Continual serological surveillance is also vital to determine the extent and rate of spread of the virus, particularly in non-horse equines, which rarely exhibit clinical signs of infection, and also to assess the effectiveness of vaccination and other control strategies. Traditionally the VN and complement fixation (CF) tests have been used in such epidemiological studies but both have inherent disadvantages. The VN test is expensive and time consuming, while with the CF test, AHSV antibodies can be short lived and donkey and mule sera are often anticomplementary [4, 7].

This paper describes the use of a rapid, sensitive, inexpensive and specific, indirect fluorescence antibody (IFA) technique for the qualitative and quantitative assessment of antibodies against AHSV in Morocco.

Materials and methods

Experimental and field sera

Twenty-two horses confirmed as being AHSV sero-negative by the CF test were housed in insect proof accommodation at the Laboratoire Officiale de Control de Medicaments et Serum Veterinaire, Rabat, Morocco. Each horse was then vaccinated twice, 21 days apart with the recommended dose of inactivated, purified vaccine (Equipest, batch No. 00D04101, Rhône-Mérieux, Lyon, France) [3]. Serum samples were obtained from each animal on 2 occasions prior to vaccination (44 negative control sera) and at 10 intervals thereafter until 180 days post vaccination (dpv).

Serum samples were also available from 179 horses, 95 mules and 260 donkeys, collected in 1990 from Tanger, Tetouan and Larache provinces. These samples had been collected 1 to 5 months after vaccination with an attenuated AHSV 4 vaccine (Horse Lyopest, Lot No. CO3, Biopharma, Rabat, Morocco).

All sera had been heat inactivated at 56°C for 30 min and then stored at –20°C before use.

Indirect fluorescent antibody (IFA) test

Monolayer cultures of BHK-21 cells (ATCC) were grown in Eagles MEM containing 5% foetal calf serum (FCS), 100 i.u. penicillin, 100 µg streptomycin and 2.5 µg fungizone per ml. Each 175 cm^3 flask was inoculated with 2.0 ml of AHSV 4 containing $10^{2.0}$ $TCID_{50}$/ml. The virus was adsorbed for 30 min at 37°C and then overlayed with Eagles MEM containing 2% FCS. Non-infected BHK-21 cultures were included as negative controls. The cells were harvested from the infected and non-infected cultures 3 days after infection when the infected cultures showed 20% cytopathic effect (CPE). Cells were washed twice and resuspended in phosphate buffered saline (PBS) pH 7.2. Fifty microlitre drops of infected and non-infected cells, each containing approximately 2×10^4 cells, were dispensed in single rows onto microscope slides. The slides were air dried and fixed in acetone for 10 min. Sera were tested either at a single dilution (1:5) or in a two-fold dilution series prepared in PBS. Twenty microlitre amounts of each serum dilution were then added to the infected and non-infected, acetone fixed cells and incubated in a humidity chamber for 1 hour at 37°C. The microscope slides were washed 3 times in PBS and then rinsed once in distilled water. Twenty microlitre amounts of fluorescein isothiocyanate conjugated to rabbit anti-horse IgG (Sigma Chemical Co., St Louis, Mo., USA) were then added to each cell spot. Conjugate was used at a dilution giving optimal intensity of fluorescence with a minimum background staining. Slides were incubated for 1 h at 37°C and washed as before. The cells were examined for cytoplasmic staining using a fluorescent microscope. Sera giving end point titres greater than reciprocal $\log_{10}$ 0.7 were recorded positive.

Virus neutralisation (VN) test

Test sera were assayed using a modification of the VN test as described previously [5]. Briefly, a two-fold dilution series of each test serum, prepared in Eagles MEM, was incubated with an equal volume of AHSV 4 at a final concentration of 100 $TCID_{50}$ of virus per 100 µl. After incubation for 1 hour at 37°C the virus-serum mixtures were inoculated onto established monolayers of BHK-21 cells in 96-well microtitre plates (100 µl/well). The serum/virus mixtures were discarded after adsorption for 1 h at 37°C. Plates were then washed with PBS and overlaid with 200 µl of Eagles MEM containing 2% FCS and antibiotics. The plates were incubated at 37°C for 5 days and examined daily for CPE. The titre of the test serum was taken as the highest dilution giving complete inhibition of CPE. Serum titres above reciprocal $\log_{10}$ 0.47 were considered positive.

Complement fixation (CF) test

The CF tests were carried out in 96 well round-bottomed microtitre plates [1] where all dilutions were made in cold veronal buffer containing 0.1% gelatin. Briefly, antigen was prepared by acetone ether extraction from homogenates of suckling mouse brain infected with AHSV 4 [2]. Test sera were diluted in a two-fold dilution series (50 µl/well). Fifty microlitres each of antigen (2 units) and guinea-pig complement (5 units) were added to each serum dilution. Plates were incubated for 18 h at 4°C followed by 30 min at 37°C. Sensitised erythrocytes, at a concentration of 2.8%, were added (50 µl/well) and the plates incubated for a further 30 min at 37°C. Finally the plates were centrifuged at 1000 rpm for 5 min. End-point titres were recorded as the highest dilution giving 50% or greater, fixation. Positive and negative antigen and serum controls, haemolysin, veronal buffer controls and a titration of the complement used were included for each test. Complement fixation titres equal to or greater than reciprocal $\log_{10}$ 0.7 were considered positive.

Results

No fluorescence was detected with the 44 negative horse sera at a dilution above reciprocal $\log_{10}$ 0.6.

Figure 1 shows the mean antibody responses recorded by IFA, VN and CF tests for sera collected at intervals from 22 horses following primary and secondary vaccination. Antibodies were detected by IFA and CF tests in 75% and 23% of sera, respectively, collected at 7 dpv and in 27% of sera collected at 14 dpv by VN tests (Table 1). Ability to detect CF antibodies declined dramatically between 60 and 90 dpv, and by 150 dpv all sera were recorded as negative. However, high levels of AHSV specific antibodies were detected by the IFA test in 95% of the vaccinated horses until at least 150 dpv. Furthermore, positive titres of antibody were recorded by both IFA and VN tests in 50 and 100% of sera, respectively, collected at 180 dpv (Fig. 1, Table 1).

The results of screening 534 horse, mule and donkey sera by IFA, VN and CF tests are given in Table 2. There was a good agreement between the results obtained with the IFA and VN tests. However, anticomplementary effects pre-

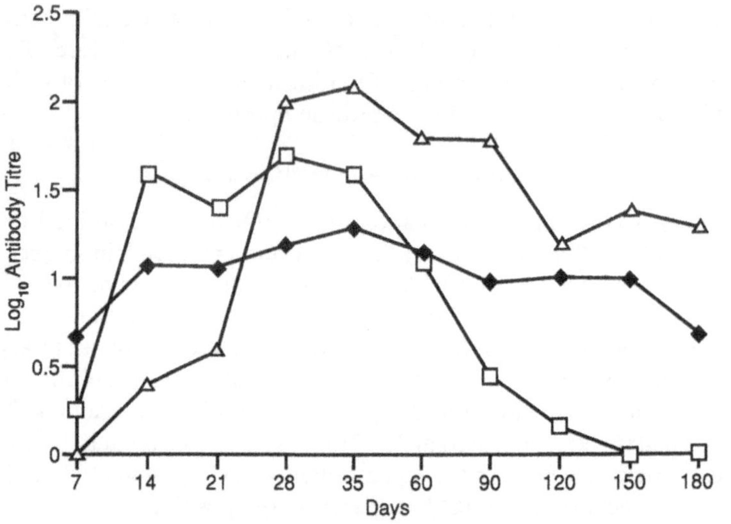

Fig. 1. Antibody responses in horses following immunisation with an inactivated AHSV 4 vaccine. Titres are given as arithmetic means from 22 horses for IFA (◆), VN (△), CF (□) tests

Table 1. Detection of AHSV antibodies by IFA, VN and CF tests in sera from 22 horses vaccinated with an attenuated AHSV vaccine on days 0 and 21

Days after the primary vaccination	IFA	CF	VN
7	75[a]	23[a]	0[a]
14	100	100	27
21	100	100	86
28	100	100	100
35	100	100	100
60	100	95	100
90	95	50	100
120	91	18	100
150	95	0	100
180	50	0	100

IFA Indirect fluorescent antibody; *CF* complement fixation, *VN* virus neutralisation
[a] Percentage positive

vented confirmation of either a positive or a negative result in 33.7% of the mule and 81.9% of the donkey sera tested by CF. The numbers of horse sera recorded positive by CF (55.9%) were also significantly lower than those recorded positive by IFA (95%) and VN (97.2%). Overall, positive antibody titres were detected in 94.3, 95.5 and 38% of sera tested by IFA, VN and CF tests, respectively. Table 3 shows the calculated relative sensitivity (98%) and specificity (83.3%) of the IFA test in comparison to the VN test results. Overall, there was a good agreement between the two tests (97.3%).

Table 2. Detection of AHSV 4 antibodies by IFA, VN and CF tests in field sera from vaccinated equidae

Species	Number of sera tested	IFA		VN		CF		
		P	N	P	N	AC	P	N
Horse	179	170	9	174	5	0	100	79
Mule	95	90	5	90	5	32	63	0
Donkey	260	244	16	246	14	213	40	7
Totals	534	504	30	510	24	245	203	86

IFA Indirect fluorescent antibody; *VN* virus neutralisation; *CF* complement fixation, *P* positive; *N* negative; *AC* anticomplementary

Table 3. The relationship between VN and IFA tests for the detection of antibodies to AHSV 4 in 534 sera collected from vaccinated equidae

	Number of sera		Virus neutralization test		
	P	N	Sensitivity[a] (%)	Specificity[b] (%)	Overall agreement[c]
IFA					
Positive	500	4	98%		
Negative	10	20		83.3%	97.3%

[a] Relative sensitivity: (IFA and VN positives / VN positives) x 100

[b] Relative specificity: (IFA and VN negatives / VN negatives) x 100

[c] Overall agreement: (IFA and VN positives) + (IFA and VN negatives) / (VN positives + VN negatives)

IFA Indirect fluorescent anitibody; *P* positive; *N* negative

Discussion

Rapid and accurate sero-monitoring of equidae during an epizootic of AHS can provide valuable information on the spread of disease and the effectiveness of vaccination. The CF test has been used for many years for the detection of AHSV group specific antibodies and is an OIE designated test method. However, the assay is not ideal for testing sera from horses as the complement fixing antibodies are often of short duration [4, 7]. Furthermore, because mule and donkey sera are notoriously anticomplementary, the CF test does not provide accurate or comprehensive epidemiological data when mixed populations of equidae are being tested. The results presented here confirm the shorter duration of antibody detection by the CF test with sera from AHSV vaccinated horses, and the high percentage of mule and donkey sera that are anticomplementary.

The VN test is used for the detection of AHSV serotype specific antibodies and can provide discriminative data in countries where more than one AHSV serotype occurs. The assay is not usually considered for large scale sero-surveillance, particularly in epizootic areas where disease is caused by a single AHSV serotype. This is because VN tests are expensive in terms of reagents, are labour intensive and require incubation for 5 days before results can be reported. However, the VN test is sensitive and specific and is generally considered to provide a standard against which other assay methods are judged.

The analysis of the results presented in this paper for 543 field sera collected from vaccinated equidae shows a good agreement between the IFA and VN tests (97.3%). Differences in sensitivity, however, were observed when testing serial samples taken from vaccinated horses. The IFA technique appeared to be more sensitive than the VN test in detecting AHSV antibodies produced during the early stages of the humoral response (7–12 dpv), but was less sensitive in the detection of AHSV antibodies in sera collected more than 3 to 5 months after vaccination. Despite this the IFA test as described here is economic, rapid and relatively easy to perform and being more sensitive and specific than the CF test may prove useful as an alternative to this test for the detection of antibodies against AHSV in sera from different species of equidae.

References

1. Bernard G (1975) Adaptation de la microtechnique de fixation de complément au diagnostic de la peste équine. Rev Elev Med Vet Pays Trop 28: 451–457
2. Casals J (1949) Acetone ether extracted antigens for complement fixation with certain neurotropic viruses. Proc Soc Exp Biol Med 70: 339–343
3. Dubourget P, Preaud JM, Detraz N, Lacoste F, Fabry AC, Erasmus B, Lombard M (1992) Development, production and quality control of an industrial inactivated vaccine against AHS type IV. In: Walton TE, Osburn BI (eds) Bluetongue, African horse sickness and related orbiviruses. Proc 2nd Int Symp. CRC Press, Boca Raton, pp 874–886
4. Hazrati A, Mirchamsy H, Bahrami S (1972) Comparative studies on the serological responses of horses to African horse-sickness virus. In: Bryans JT, Gerber H (eds) Equine infectious diseases III. Proc 3rd Int Con Equine Inf Dis. Karger, Basel, pp 69–80
5. Hermiman KAJ, Gumm ID, Owen L, Taylor WP, Sellers RF (1980) Distribution of bluetongue virus and antibodies in some countries of the eastern hemisphere. Bull Off Int Epiz 92: 581–586
6. Holmes HI, Boccardo G, Estes MK, Furuichi MK, Hoshino Y, Joklik WK, Mc Crae M, Mertens PPC, Milne RG, Samal KSK, Shikata E, Winton JR, Uyeda I, Nuss DL (1995) Family Reoviridae. In: Murphy FA, Fauquet CM, Bishop DHL, Ghabrial SA, Jarvis AW, Martelli GP, Mayo MA, Summers MD (eds) Virus Taxonomy. Classification and Nomenclature of Viruses. Sixth Report of the International Committee on Taxonomy of Viruses. Springer, Wien New York, pp 208–239 (Arch Virol [Suppl] 10)
7. Howell PG (1963) African horse sickness. In: Emerging diseases of animals. Food and Agric Org, Rome No 61, 73–108
8. Mellor PS, Boned J, Hamblin C, Graham S (1990) Isolations of African horse sickness virus from vector insects made during the 1988 epizootic in Spain. Epidemiol Infect 105: 447–454
9. Mellor PS (1993) African horse sickness: transmission and epidemiology. Vet Res 24: 199–212

Authors' address: Dr. M. H. Bikour, Laboratoire Régional d'Analyses et de Recherches Vétérinaires de Tanger, Rue Ibnou Toumart, B.P. 3, Tanger, Morocco.

Validation of ELISA for the detection of African horse sickness virus antigens and antibodies

C. Rubio[1], **M. A. Cubillo**[1], **H. Hooghuis**[1], **J. M. Sanchez-Vizcaino**[2], **M. Diaz-Laviada**[2], **E. Plateau**[3], **S. Zientara**[3], **C. Crucière**[3], and **C. Hamblin**[4]

[1] Laboratorio de Sanidad y Producción Animal de Algete, Algete, Madrid, Spain
[2] Centro de Investigacion en Sanidad Instituto Nacional de Investigacion y Tecnologia, Agraria y Alimentaria, Valdeolmos, Madrid, Spain
[3] Centre National d'Etudes Veterinaires et Alimentaires Laboratoire Central de Recherches Veterinaires, Maisons-Alfort, France
[4] Institute for Animal Health, Pirbright Laboratory, Woking, Surrey, U.K.

Summary. The mortality rate in susceptible populations of horses during an epizootic of African horse sickness (AHS) may be in excess of 90%. Rapid and reliable assays are therefore essential for the confirmation of clinical diagnoses and to enable control strategies to be implemented without undue delay. One of the major objectives of a recent European Union funded project was the validation of newly developed diagnostic assays which are rapid, sensitive, highly reproducible and inexpensive, for the detection of African horse sickness virus (AHSV) antigens and antibodies. The Laboratorio de Sanidad y Produccion Animal (LSPA) in Algete, Spain was charged with the responsibility of co-ordinating and supplying samples of viruses and antisera to the participating laboratories in Spain, France and the United Kingdom. The panels comprised 76 antigen samples for assay by indirect sandwich ELISAs and 53 serum samples for antibody detection by either indirect or competitive ELISAs. Results generated by ELISA for each laboratory were analysed in LSPA in terms of their relative sensitivities and specificities. There was a good agreement between the ELISAs used for either antigen or antibody detection. The participating groups agreed that any field sample giving a doubtful result would always be retested by ELISA and an alternative assay.

Introduction

African horse sickness (AHS) is an infectious but non-contagious viral disease affecting all species of equidae. The disease is caused by a dsRNA virus of the genus *Orbivirus*, family *Reoviridae* which is transmitted by biting midges, *Culicoides* species [10]. Nine different serotypes of AHS virus (AHSV) have been identified [3].

The incursion of AHSV serotype 4 (AHSV) into Spain in July 1987 resulted in an epizootic which spanned three countries over a five year period, Spain (1987–1990), Portugal (1989) and Morocco (1989–1991) [1, 9, 12]. As a consequence of the AHSV outbreaks in Spain, a three year international project sponsored by the European Union was carried out to validate rapid, sensitive and specific diagnostic assays for the laboratory confirmation of clinical diagnoses and to determine the epidemiology of the disease in Europe. Although several different AHSV ELISAs have been developed in the Northern hemisphere [4, 5, 7, 8, 15] and validated in the laboratories of their origin, no cross validation between these assays had ever been undertaken. This paper reports an independent, comparative analysis of results obtained for these assays in the different laboratories in terms of their sensitivities and specificities for the detection of AHSV antigens and/or antibodies using encoded panels of samples.

Materials and methods

Antigen samples

A panel of 76 antigen samples, representing field material, laboratory isolates and stock viruses, were used to cross-validate the AHSV antigen detection assays. Antigen samples were submitted to each of the participating laboratories on ice then stored at –70°C until used. The field material included homogenised spleen samples from 14 confirmed cases of AHSV 4 in horses from Spain, homogenised spleen samples from 20 clinically heathy, negative animals, one batch of homogenised *Culicoides variipennis* which had been infected with AHSV 4 and one non-infected batch of *C. variipennis*. The remaining 40 laboratory strains of virus represented the 9 AHSV serotypes, bluetongue virus serotype 1 (BTV 1), 3 serotypes of epizootic haemorrhagic disease of deer virus (EHDV 1, 2 and 3) and Akabane virus. These viruses had all been passaged through either tissue culture cells or suckling mouse brain.

Antibody samples

A panel of 53 serum samples were used to validate the AHSV antibody detection ELISAs. An aliquot of each serum was submitted to each of the participating laboratories on ice. Samples were then stored at –20°C until used. Positive samples included 15 antisera from horses vaccinated against AHSV 4 and 9, and convalescent antisera from both horses and guinea pigs, inoculated with one of the 9 serotypes of AHSV. The negative samples included 13 sera from clinically healthy horses and 2 sera from healthy cattle. Sera raised against other arboviruses were used to check for any cross-reactions. These included a BTV1 positive sheep antiserum, EHDV 1, 2 and 3 cattle antisera and an Akabane virus goat antiserum.

ELISA for antigen detection

Sandwich ELISAs for the detection of AHSV antigens, using either mouse monoclonal antibodies [7] or rabbit polyclonal antibodies [5] for antigen capture, were carried out either in the laboratories where they were developed, Centro de Investigacion en Sanidad Animal, Instituto Nacional de Investigacion y Agrarias (CISA-INIA), Spain [7] and the Institute for Animal Health (IAH), U.K. [5] or in the Laboratorio de Sanidad y Producción Animal (LSPA), the European Reference Centre for AHS in Algete, Spain.

ELISA for antibody detection

Specific antibodies in the test sera were detected using indirect [7] and competitive ELISAs [4, 15]. The assays were performed in the laboratories where they were developed CISA-INIA [7], IAH [4, 15], Centre National d'Etudes Veterinaires et Alimentaires (CNEVA) in France [4] and in LSPA [15].

Analysis

The sensitivity of the ELISAs was determined using the formula *a/a+c* where *a* is the number of samples recorded positive and *c* is the number of false negatives. The specificity of the ELISAs was determined using the formula *d/d+b* where *d* is the number of samples recorded negative and *b* is the number of sera giving a false positive reaction.

Results

The results obtained by each of the participating laboratories for each panel of samples were analysed by LSPA. The numbers of sera used in the analysis varied depending on the format of the assay method used. For example competitive ELISAs which use immune guinea pig antisera as the competitor cannot be used to assay guinea pig sera. Similarly, indirect ELISAs which utilise an anti-horse enzyme conjugate cannot be used to measure responses in species other than equine. As such no results were available from CISA-INIA to exclude the possibility of cross reactions between AHSV and BTV, EHDV or Akabane virus; however, in practice such cross-reactions are considered highly unlikely.

Table 1 shows the sensitivity and specificity of the different ELISAs used for the detection of AHSV antigens. One of the known AHSV positive samples gave a doubtful reaction and one negative horse spleen sample gave a false positive reaction when tested by the CISA-INIA ELISA. The single false positive reaction recorded by CISA-INIA lowered the specificity of this test to 96%. Interpretation of the doubtful result as either positive or negative would result in a sensitivity of either 100 or 98%, respectively.

The results presented in Table 2 show the sensitivity and specificity of the different ELISAs for the detection of AHSV antibodies. One of the known AHSV positive antisera recorded a doubtful reaction by the IAH ELISA. If this doubtful result is considered positive then the sensitivities of all 5 ELISAs were 100%. However, if interpreted as a negative result, the sensitivity of the IAH

Table 1. Sensitivity and specificity of indirect ELISA for the detection of AHSV antigens

Institute Number of samples	CISA-INIA 76	LSPA 76	IAH 76
Detection system	mouse monoclonal	pig polyclonal	guinea pig polyclonal
Sensitivity %	98*–100	100	100
Specificity %	96	100	100
*Number doubtful	1	–	–

Table 2. Sensitivity and specificity of ELISA for the detection of AHSV antibodies

Institute	CISA-INIA	LSPA	CNEVA	IAH	
Number of samples	37	53	43	53	43
Detection system	horse polyclonal	mouse monoclonal	guinea pig polyclonal	mouse monoclonal	guinea pig polyclonal
Sensitivity %	100	100	100	100	96*–100
Specificity %	85*–100	95*–100	100	100	100
*Number doubtful	2	1	–	–	1

ELISA was 96%. Of the samples tested by LSPA and CISA-INIA, 1 and 2 known negatives, respectively, were recorded as doubtful. If these samples were considered negative the specificities would remain at 100% but if they were given as positive the sensitivities would decrease to 95 and 85%, respectively.

Discussion

The purpose of this work was to identify and validate, rapid and reliable diagnostic tests that are both sensitive and specific for the detection of AHSV antigens and antibodies, particularly during the early stages of disease. ELISA was selected because it has distinct advantages over the previously used group-specific assays such as agar gel immunodiffusion and complement fixation, being more specific, sensitive and rapid for the detection of virus antigens and antibodies. Furthermore, at the time that this work was initiated several new AHSV ELISA methodologies had been reported or were under development [4, 5, 7, 8, 15].

Generally, there was a good agreement between the different ELISA methods assessed in this study and analysis of the results obtained suggest that all of these assays would satisfy the basic requirements listed above. The only proviso is that samples giving inconclusive results should be further investigated. For example, when samples from suspect cases give suspicious or negative reactions in the antigen detection ELISAs, attempts should be made to isolate and amplify a virus which can then be identified by ELISA or an alternative test. Similarly, when a serum sample being tested by ELISA for antibodies gives a doubtful or suspect result, it should be either titrated in the ELISA or retested using an alternative test followed by a request for a second sample.

AHSV is exotic to Europe but incursions of virus from enzootic regions do occasionally occur [1, 2, 6, 9, 11–14]. Historically, these epizootics have been caused by a single serotype of AHSV, for example, serotype 9 in 1966 [13, 14] and serotype 4 in 1987 to 1991 [1, 9, 12]. In "emergency" situations, at the start of an outbreak and when the AHSV serotype is unknown, the group-specific ELISA can provide a rapid and reliable confirmation of clinical diagnoses. The option of vaccination with a polyvalent vaccine is then immediately available. Confirmation of the AHSV serotype by virus neutralisation tests will be required at a later stage, whereupon selection of the appropriate monovalent vaccine can be made.

Acknowledgements

We are very grateful for the participation in this evaluation of Dr. B. Erasmus, Veterinary Research Institute, Onderstepoort, South Africa who kindly contributed a number of samples.

References

1. Lhafi A, Tber A, Fikri, A, Laghzaoui K (1992) African horse sickness in Morocco: the epizootics of 1989 and 1990. In: Walton TE, Osburn BI (eds) Bluetongue, African horse sickness and related orbiviruses. Proc 2nd Int Symp. CRC Press, Boca Raton, pp 205–216
2. Diaz Montilla R, Panos Marti P (1967) Epizootologia de la peste equina en Espana. Bull Off Int Epizoot 68: 705–714
3. Gorman BM, Taylor J, Walker J (1983) Orbiviruses. In: Joklik WK, Fraenkel-Conrad H, Wagner RR (eds) The *Reoviridae*. Plenum, London, pp 287–346
4. Hamblin C, Graham SD, Anderson EC, Crowther JR (1990) A competitive ELISA for the detection of group-specific antibodies to African horse sickness virus. Epidemiol Infect 104: 303–312
5. Hamblin C, Mertens PPC, Mellor PS, Burroughs JN, Crowther JR (1991) A serogroup specific enzyme-linked immunosorbent assay (ELISA) for the detection and identification of African horse sickness viruses. J Virol Methods 31: 285–292
6. Howell PG (1960) The 1960 epizootic of African horse sickness in the Middle East and SW Asia. J S Afr Vet Med Assoc 31: 329–334
7. Laviada MD, Babin M, Dominguez J, Sanchez-Vizcaino JM (1992) Detection of African horse sickness virus in infected spleen by sandwich ELISA using two monoclonal antibodies specific for VP7. J Virol Methods 38: 229–242
8. Laviada M, Roy P, Sanchez-Vizcaino JM (1992) Adaption and evaluation of an indirect ELISA and immunoblotting test for African horse sickness antibody detection. In: Walton TE, Osburn BI (eds) Bluetongue, African horse sickness and related orbiviruses. Proc 2nd Int Symp. CRC Press, Boca Raton, pp 646–650
9. Lubroth J (1988) African horse sickness and the epizootic in Spain 1987. Equine Pract 10: 26–33
10. Matthews REF (1982) Classification and nomenclature of viruses. Fourth Report of the International Committee on Taxonomy of Viruses. Intervirology 17: 199
11. Mellor PS, Hamblin C, Graham SD (1990) The occurrence of African horse sickness in Saudi Arabia. Vet Rec 127: 41–42
12. Mellor PS (1991) African horse sickness in Europe. In: Thrusfield MV (ed) Proc Mtg Soc Vet Epidem Prev Med Univ London 17–19 April 1991, pp 27–32
13. Rabah M (1966) La peste equine au Maroc. These Doct Vet, Lyon
14. Sers JL (1967) La peste equine en Algerie. These Doct Vet, Maisons-Alfort
15. Wade-Evans AM, Woolhouse T, O'Hara R, Hamblin C (1993) The use of African horse sickness virus VP7 antigen, synthesised in bacteria, and anti-VP7 monoclonal antibodies in a competitive ELISA. J Virol Methods 45: 179–188

Authors' address: Dr. C. Rubio, Laboratorio de Sanidad y Producción Animal de Algete, Ctra Algete, km 5400, 28110 Algete, Madrid, Spain.

Use of reverse transcriptase-polymerase chain reaction (RT-PCR) and dot-blot hybridisation for the detection and identification of African horse sickness virus nucleic acids

S. Zientara[1], **C. Sailleau**[1], **S. Moulay**[1], **C. Crucière**[1], **M. El-Harrak**[2], **W. W. Laegreid**[3], and **C. Hamblin**[4]

[1] Centre National d'Etudes Veterinaires et Alimentaires, Laboratoire Central de Recherches Veterinaires, Maisons-Alfort, France
[2] Laboratoire Biopharma, Rabat-Akkari, Morocco
[3] Animal Health Systems, USDA, ARS, Meat Animal Research Center, Clay Center, Nebraska, U.S.A.
[4] Institute For Animal Health, Pirbright Laboratory, Woking, Surrey, U.K.

Summary. A coupled reverse transcriptase-polymerase chain reaction assay (RT-PCR) for the detection of African horse sickness virus (AHSV) dsRNA, has been developed using genome segment 7 as the target template for primers. RNA from isolates of all nine AHSV serotypes were readily detected. The potential inhibitory effects of either ethylene diamine tetra acetic acid (EDTA) or heparin on the RT-PCR were eliminated by washing blood samples before lysis of the red blood cells and storage. There was a close agreement in the sensitivity and the specificity of the RT-PCR and an indirect sandwich ELISA. Confirmation of the presence of AHSV using RT-PCR and dot-blot hybridization on blood samples collected from horses experimentally infected with AHSV serotype 4 (AHSV 4) and AHSV serotype 9 (AHSV 9), was achieved within 24 hours, compared to the period of several days required for virus isolation. The RT-PCR and virus isolation methods showed similar levels of sensitivity when used for the detection of AHSV in 3 horses infected with AHSV 4, and in 2 out of 3 horses infected with a less virulent isolate of AHSV 9. Although viraemia was detected in the third horse by virus isolation, from 6 to 14 days after infection, this animal remained consistently negative by RT-PCR. Conversely, AHSV viral RNA was detected by RT-PCR in the blood of 4 donkeys and 4 mules up to 55 days after their experimental infection despite the absence of any detectable infectious virus. RT-PCR is a sensitive and rapid method for detecting AHSV nucleic acids during either the incubation period at the start of an African horse sickness (AHS) epizootic, or for epidemiological investigations in species where clinical signs may be inapparent.

Introduction

African horse sickness (AHS) is an infectious but non-contagious arthropod borne disease of equidae. The disease results in a high mortality in horses, often exceeding 90%. Mules and donkeys are less susceptible, their death rate being less than 50% [7]. The causative agent is African horse sickness virus (AHSV), which is a dsRNA virus classified as a single distinct species, within the genus *Orbivirus*, family *Reoviridae*. AHSV is closely related to the prototype orbivirus species, bluetongue virus (BTV) [3]. Nine antigenically distinct AHSV serotypes have been identified in serum neutralisation assays. The viral genome is composed of 10 segments of dsRNA, which encode a total of 10 distinct proteins. Seven of these are structural proteins, which together with the 10 dsRNA genome segments, form a virus particle with a double-shelled protein capsid. The outer capsid layer is composed of 2 major proteins (VP2 and VP5), which are involved in virus neutralization and antigenic diversity [23]. The inner capsid is composed of 2 major proteins (VP3 and VP7) and 3 minor proteins (VP1, VP4 and VP6/VP6a), all of which are highly conserved between AHSV serotypes [4, 25]. Three distinct non-structural proteins, NS1, NS2 and NS3/NS3A are also encoded by the AHSV genome [8, 10, 14, 19, 31, 32].

AHSV is transmitted primarily by *Culicoides* vector species, mainly *C. imicola* [9], although it may also be transmitted mechanically by other biting insects, e.g. mosquitoes. AHSV has also been isolated from two mixed batches of midges containing *C. pulicaris* and *C. obsoletus* [21]. Although AHS is mainly present in southern, eastern and central Africa, it has also been recorded in the Middle East, parts of Asia and more recently in Spain (1987–1990), Portugal (1989) and Morocco (1989–1991) [22].

The current methods available for confirming the initial clinical diagnosis of AHS are often time consuming (e.g. virus isolation and virus neutralisation assays), or lack sensitivity, particularly during the early stages of disease, for the detection of virus in blood (e.g. complement fixation and ELISA). Due to the rapid course of the disease, death often occurs in horses before the development of detectable levels of antibodies against the AHSV. Therefore, new diagnostic assays based on the molecular biology of AHSV are currently being developed [6, 24, 26, 34–36].

This paper describes the use of a single tube reverse transcriptase-polymerase chain reaction (RT-PCR) for the early detection of AHSV nucleic acids in blood samples and discusses the possible value of this test for determining the presence and duration of viraemia in different species of equidae where clinical signs may be inapparent.

Materials and methods

Test virus samples

A panel of 73 blind samples representing field material, laboratory isolates and stock viruses were provided by Dr. C. Rubio, European AHSV Reference Laboratory, Algete, Madrid. The 34 field samples comprised homogenised spleen tissue from 14 confirmed cases of

AHSV 4 and homogenised spleen tissue from 20 clinically healthy, virus negative animals. Viruses isolated from the field cases were identified and serotyped in virus neutralization (VN) tests [13]. The remaining 38 laboratory strains of virus had previously been isolated and identified, following passage in tissue culture cells, suckling mouse brain or *C. variipennis* midges. One non-infected sample of *C. variipennis* was also included as a negative control. The 73 samples were assayed by indirect sandwich ELISA [11] for the presence of AHSV antigen.

Horse samples

Two groups of 3 clinically healthy adult horses, which were initially sero-negative for AHSV, were housed in the biocontainment laboratories at Plum Island Disease Centre, USA. Each horse of each group was inoculated intravenously with 1.0 ml of either AHSV 4 or AHSV 9 containing 10^4 $TCID_{50}$/ml of infectious virus. The history of these strains has been previously described [17]. Animals were examined daily. Clinical appearance and rectal temperatures were recorded. Ten millilitres each of blood were collected each day in both ethylene diamine tetra acetic acid (EDTA) and heparin. Plasma was removed by aspiration after centrifugation at 500 **g** for 30 min. The blood cells were then washed twice with RPMI 1640 medium (Life Technologies) and lysed in 0.9 volume of double-distilled H_2O for 30 seconds. The lysate was then restored to isotonic conditions by addition of 0.1 volume of 10 x RPMI 1640 medium.

Virus isolations were made in a Vero cell line, (American Tissue Culture Collection (ATCC) CCl 81), grown in 24-well plates with RPMI 1640 medium supplemented with 8% of foetal calf serum, streptomycin (100 µg/ml) and penicillin (100 IU/ml). Plates were incubated at 37°C for 7 days and examined daily for cytopathic effect (CPE). Samples initially negative were freeze thawed and blind passaged. Samples were considered negative if CPE was absent after 5 blind passages. The identity and serotype of any isolate was confirmed by VN tests [13].

Donkey and mule samples

Two groups, of 4 donkeys or 3 mules, were housed in the biocontainment facilities at Biopharma Laboratory in Rabat, Morocco. Initially all 7 animals were clinically healthy and were serologically negative for AHSV. Each animal was inoculated subcutaneously with 1.0 ml containing 7.9×10^4 $TCID_{50}$ of AHSV isolated in 1989 from a dead horse in Larache, Morocco.

Animals were examined daily. Clinical appearance and rectal temperatures were recorded. Ten millilitres each of EDTA and non-EDTA blood were collected on alternate days. The red blood cell (RBC) fractions from the EDTA bloods were separated by centrifugation at 500 **g** for 10 min, washed 3 times in phosphate buffered saline (PBS) and stored in an equal volume of 50/50 v/v glycerol and distilled water, containing 5% phenol and 5% potassium oxalate (OCG). Each sample of lysed RBCs was prepared by diluting them 1 in 10 in PBS. Virus isolation attempts were then made by intravenous inoculation of 20 µl into 5, ten day old embryonated hens eggs, intracerebral inoculation of 20 µl into 5, two day old suckling mice and inoculation of 1.0 ml onto confluent monolayer cultures of both Vero and BHK cells. The RBC fractions were also assayed by indirect sandwich ELISA [11] for AHSV viral antigen.

Sera were decanted from the non-EDTA bloods and assayed by ELISA [20] for the presence of AHSV NS3 antibodies.

Nucleic acid sample preparations

Total RNA was extracted from 500 µl volumes of cell lysates and from RBC samples using a commercial kit "RNA^+" (Bioprobe system, Montreuil, France), an adaptation of the acid guanidinium thiocyanate-phenol chloroform single extraction procedure of Chomczynski and Sacchi [5]. The RNA pellets obtained were dried and resuspended in 10 µl of double distilled water before being assayed in the RT-PCR [28, 36, 37].

Reverse-transcriptase-polymerase chain reaction

The sequences of the primers, P1 and P2, selected for amplification of genome segment 7 were obtained from the published sequence of AHSV 4 segment 7–1179 bp in length [27].

Primer	Sequence (5' 3')	Location (bp)
P1	GTTAAAATTCGGTTAGGATG	1–20
P2 (reverse)	GTAAGTGTATTCGGTATTGA	1179–1159

These oligonucleotides amplify genome segment 7 from all of the nine AHSV serotypes [34]. A 2.5 µl aliquot of dsRNA was denatured with an equal volume of 20 mM methyl mercuric hydroxide, prior to being used as a template for cDNA synthesis [33]. The methyl mercuric hydroxide was then reduced by the addition of 1 µl of 0.7 M β-mercaptoethanol and 2 µl RNAsin (40 IU/µl) for 10 min at room temperature.

All the components required for the reverse transcriptase reaction and PCR were combined in one 0.5 ml Eppendorf tube and both reactions were run on a single thermal cycling programme [30]. Briefly, 4 µl of the denatured dsRNA was added to 100 µl of the following mix: 10 mM Tris, pH 8.8, 50 mM KCl, 400 µM dNTPs, 1.5 mM $MgCl_2$, 0.1 mg/ml gelatine and 10 pmol of each primer. To each tube was also added 10 U AMV-RT (Appligene) and 7 U of *Taq* DNA polymerase (Boerhinger-Mannheim), giving a ratio of RT-*Taq* less than 3:2 [29]. The tubes were overlaid with 100 µl of paraffin oil and then incubated on a thermocycler (PTC 100/60 – Prolabo) as follows: 1 h at 37°C, 5 min at 95°C, then 30 cycles at 55°C for 1 min, 70°C for 2 min and 95°C for 1 min, followed by a terminal extension step of 8 min at 70°C. The nucleic acid products were then analysed by agarose gel electrophoresis.

A 20 µl aliquot of each PCR product, obtained from the 73 virus test samples, or from the blood samples collected from experimental horses, donkeys and mules, were identified by dot-blot hybridisation using internal digoxigenin labelled probes made from AHSV 9 or BTV 1 SA genome segment 7 cDNAs [24]. The dot-blot hybridisation was visualised using the Chemiluminescent Detection Kit (Boerhinger-Mannheim).

Results

Test virus samples

The RT-PCR was evaluated against the indirect sandwich ELISA and virus isolation for its ability to detect AHSV in field samples and laboratory stock viruses (Table 1).

Generally there was good agreement between the results obtained by RT-PCR (using the outer pair of primers P1 and P2 from genome segment 7) and those obtained by ELISA or virus isolation. However, two of the 34 field samples, which were recorded negative by virus isolation and ELISA, were positive by RT-PCR. These spleen samples had been taken from two Andalucian horses, located within the AHSV infected area in Spain, which died suddenly in 1989. Neither of these two animals had developed signs characteristic of AHS. Both were reported to have been vaccinated against AHS but details of their vaccination history were not available.

One of the bluetongue samples, BTV 1, was also recorded positive by RT-PCR. A dot-blot hybridisation assay (using probes synthesised from cDNA

Table 1. Detection of AHSV viral antigen and dsRNA by indirect sandwich ELISA and RT-PCR

Number of samples	Virus	Origin	Virus isolates	Indirect sandwich ELISA	RT-PCR
18	Uninfected	Horse spleen[a]	–	–	–
2	Uninfected	Horse spleen[a]	–	–	+
1	AHSV 1	Horse spleen[a]	+	+	+
11	AHSV 4	Horse spleen[a]	+	+	+
1	AHSV 5	Horse spleen[a]	+	+	+
1	AHSV 7	Horse spleen[a]	+	+	+
1	AHSV 2	Mouse brain	ND	+	+
1	AHSV 3	Mouse brain	ND	+	+
10	AHSV 4	Mouse brain	ND	+	+
2	AHSV 5	Mouse brain	ND	+	+
2	AHSV 6	Mouse brain	ND	+	+
2	AHSV 7	Mouse brain	ND	+	+
2	AHSV 8	Mouse brain	ND	+	+
2	AHSV 9	Mouse brain	ND	+	+
9	AHSV 1–9	Tissue culture	ND	+	+
1	AHSV 4	*C. Variipennis*	ND	+	+
1	Negative	*C. Variipennis*	ND	–	–
1	BTV 1	Tissue culture	ND	–	+
1	BTV 4	Tissue culture	ND	–	–
1	EHDV 1	Tissue culture	ND	–	–
1	EHDV 2	Tissue culture	ND	–	–
1	EHDV 3	Tissue culture	ND	–	–
1	Akabane virus	Tissue culture	ND	–	–

[a] Field samples

AHSV African horse sickness; *BTV* bluetongue virus; *EHDV* epizootic hemorrhagic disease virus; *ND* not done

clones of BTV 1 SA and AHSV 9 genome segment 7) was carried out on the PCR product to determine its viral origin. The amplified DNA product only gave a positive hybridisation signal with the AHSV 9 probe (data not shown), confirming that the PCR product originated from genome segment 7 of AHSV. These results suggest that the BTV 1 sample was contaminated with AHSV.

Horse study

The 3 horses infected with AHSV 4 developed clinical signs characteristic of the pulmonary form of AHS and each died 6 days post infection (dpi). Two of the horses infected with AHSV 9 developed signs and lesions typical of the cardiac and mixed form of AHS and died 11 and 14 dpi. The third horse infected with AHSV 9 developed the fever form of the disease and survived for the duration of the study (14 dpi).

Table 2 shows the RT-PCR results obtained with the samples of horse blood collected in either EDTA or heparin and after washing twice in isotonic buffer. There was no significant difference ($p = 0.05$) between the RT-PCR results obtained on blood samples collected in either anticoagulant.

The comparisons between the results obtained by virus isolation and RT-PCR for the 2 groups of 3 horses infected with either AHSV 4 or AHSV 9 are given in Table 3. Infectious AHSV 4 and viral RNA was detected, from day 1 until death, in all blood samples collected from the 3 horses infected with AHSV 4. Although virus isolations were made over several days from all 3 horses infected with AHSV 9, viral RNA was only detected by RT-PCR in the blood of 2 of these animals.

Table 2. Number of samples positive or negative using heparin and EDTA for blood collection

	Heparin	EDTA
RT-PCR (positive)	26	23
RT-PCR (negative)	34	37

Table 3. Detection of AHSV 4 and AHSV 9 in the blood of six infected horses (1 to 6) by virus isolation in cell culture and RT-PCR

DPI	AHSV 4						AHSV 9					
	1		2		3		4		5		6	
	VI	PCR	VI	PCR	VI	PCR	VI	PCR	VI	PCR	VI	PCR
1	+	+	+	+	+	+	−	−	−	−	−	−
2	+	+	+	+	+	+	−	−	−	−	−	−
3	+	+	+	+	+	+	+	−	−	−	−	−
4	+	+	+	+	+	+	−	−	−	−	−	−
5	+	+	+	+	+	+	−	−	−	−	−	−
6	+	+	+	+	+	+	−	+	+	−	+	−
7							+	+	−	−	+	−
8							+	+	−	+	+	−
9							+	+	+	+	+	−
10							+	+	+	+	+	−
11							+	+	+	+	−	−
12									+	+	+	−
13									+	+	+	−
14									−	+	+	−

DPI Days post infection; *VI* virus isolation

Donkey and mule study

Some of the animals developed a mild febrile response and signs of inappetence from day 1 after infection but this was of short duration and thereafter all animals remained clinically healthy.

No infectious virus or viral antigens were detected in any of the blood samples by virus isolation or ELISA but specific antibodies against NS3 of AHSV 4 were detected in the sera of all animals from 15 dpi. Table 4 shows that AHSV viral RNA was detected by RT-PCR in the blood of mules as early as 5 dpi and persisted in one animal at least (mule 3), until day 55. Initial detection of AHSV viral RNA was later in donkeys (9 to 13 dpi). AHSV-specific nucleic acids were

Table 4. Detection of AHSV 4 in the blood of infected mules and donkeys by RT-PCR and dot-blot hybridisation

	Donkey								Mule					
	1		2		3		4		1		2		3	
DP1	PCR	NS3	PCR	NS3	PCR	NS3	PCR	NS3	PCR	NS3	PCR	NS3	PCR	NS 3
1	–		–		–		–		–		–		–	
3	–		–		–		–		–		–		–	
5	–		–		–		–		–		+		+	
7	–	–	–	–	–	–	–	–	+	–	+	–	+	–
9	–		–		+		–		–		+		+	
11	+		–		–		+		+		+		+	
13	+		+		+		+		+		+		+	
15	+	+	+	+	+	+	+	+	+	+	+	+	+	+
17	+		+		+		+		+		+		+	
19	+		+		+		+		+		+		+	
21	+	+	+	+	+	+	+	+	+	+	+	+	+	+
23	–		+		+		+		+		+		+	
25	+		+		+		+		+		+		+	
27	+		+		+		+		+		+		+	
29	–	+	+	+	+	+	+	+	+	+	–	+	+	+
31	–		–		+		+		+		+		+	
33	+		–		+		–		+		+		–	
35	–	+	+	+	+	+	+	+	+	+	+	+	–	+
37	–		–		–		+		+		+		+	
39	–		–		–		–		+		–		–	
41	–		–		+		–		–		+		–	
43	–	+	+	+	+	+	–	+	+	+	+	+	+	+
45	+		+		+		–		+		–		–	
47	–		–		+		–		+		+		+	
49	–	+	–	+	–	+	–	+	–	+	–	+	–	+
51	–		+		–		–		+		–		–	
53	–		–		–		–		–		–		–	
55	–		–		–		+		–		–		+	
57	–	+	–	+	–	+	–	+	–	+	–	+	–	+

not detected by RT-PCR in the blood from eleven French donkeys used as negative controls (data not shown).

Discussion

A good agreement was obtained, in terms of sensitivity and specificity, between the results generated by RT-PCR and indirect ELISA for the 73 test samples. Three samples were recorded positive by RT-PCR but were AHSV negative by virus isolation and/or indirect sandwich ELISA. Two of these were spleen tissue from Spanish horses that had died suddenly without clinical signs of AHS. Although it is possible that both samples were contaminated with AHSV in the laboratory, the horses were from infected regions of Spain and were reported to have been vaccinated against AHSV 4 sometime earlier. Therefore the AHSV RNA detected could have been due to residual virus from the vaccine. The only definite false positive recorded by RT-PCR in this study was with a sample of BTV 1 but this was later shown to have been due to contamination with AHSV. Laboratory contamination of samples with foreign nucleic acid is well documented [16]. In the absence of other information, samples recorded as positive by RT-PCR should be confirmed by an alternative method, e.g. ELISA or virus isolation [11, 12, 18, 28, 36]. Despite this, the high sensitivity of the PCR and its ability to amplify AHSV RNA in blood from suspect cases which can then be identified by dot-blot hybridisation within 24 hours gives this test a significant temporal advantage over other existing assay methods and emphasises its potential value, particularly in emerging disease situations.

Several factors that can influence and inhibit the performance of PCR have been reported [1]. These include intrinsic factors, such as the porphyrin ring of the haem group, extrinsic factors such as the presence of heparin and chelating agents such as EDTA [15]. Each sample of horse blood tested here was washed twice before RNA extraction. The amplification products obtained from these washed and lysed blood samples confirmed the work of Beutler et al. [2] that 2 washes are sufficient to remove the extrinsic inhibitory factors. Five of the 6 experimentally infected horses showed clinical signs of AHS characteristic of the pulmonary or cardiac/mixed forms of disease. These animals were readily detected as AHSV positive by PCR before and after the onset of clinical signs. The sixth horse only exhibited the fever form of disease and since it survived, its infection could well have gone unnoticed under field conditions. The failure to detect AHSV RNA by RT-PCR in any of the samples collected from this animal is of concern, particularly since infectious virus was consistently detected by virus isolation in tissue culture, albeit only after several blind passages. The failure of RT-PCR in this instance could have been due to the inhibitory effects of intrinsic factors bound to the haemoglobin, which may have been compounded if only a very low level of virus was present. Mortality in mules and donkeys is reported to be less than 50% [7]. In this study no clinical signs were observed in 4 donkeys and 3 mules inoculated with a strain of AHSV that was lethal for horses. Although no infectious virus was detected in the blood of these animals, consistently positive RT-

PCR reactions, up to 55 dpi, did confirm the presence of AHSV viral RNA. The presence of antibodies against AHSV NS3 showed that the virus had replicated in both species.

Taken as a whole, these findings demonstrate the value of the RT-PCR, particularly if coupled with dot-blot hybridisation assays, for the rapid confirmation of AHSV in blood. These methods have the potential to help determine the duration of viraemia in the absence of detectable levels of infectious virus and where clinical signs are absent. Such data could be particularly important in helping to control the spread of disease during an epizootic where non-horse equines predominate and where time is of the essence.

Acknowledgements

We are grateful to Dr. Lenihan, Veterinary Research Laboratory, Abbotstown, Castleknock, Dublin for useful discussion. This work was funded by an EU grant, Contract 8001-CT91-0211.

References

1. Barker RH, Trairat B, Courval JM, Suwonkerd W, Rimwungtragoon K, Wirth DF (1992) A simple method to detect *Plasmodium falciparum* from blood samples using the polymerase chain reaction. Am J Med Hyg 46: 416–426
2. Beutler E, Gelbart T, Kuhl W (1990) Interference of heparin with the polymerase chain reaction. Bio Techniques 9: 166
3. Borden EC, Shope RE, Murphy FA (1971) Physiochemical and morphological relationships of some arthropod-borne viruses to bluetongue virus – a new taxonomic group. Physiochemical and serological studies. J Gen Virol 13: 261–271
4. Bremer CW, Huismans H, Van Dijk AA (1990) Characterization and cloning of the African horse sickness virus genome. J Gen Virol 71: 793–799
5. Chomczynski P, Sacchi N (1987) A single-step method of RNA isolation by acid guanidinium thiocyanate-phenol-chloroform extraction. Anal Biochem 162: 156–159
6. Chuma T, Le Blois H, Sanchez-Vizcaino JM, Diaz-Laviada M, Roy P (1992) Expression of the major core antigen VP7 of the African horse sickness virus by a recombinant baculovirus and its use as a group-specific diagnostic reagent. J Gen Virol 73: 925–931
7. Coetzar JAW, Erasmus BJ (1994) African horse sickness. In: Coetzar JAW, Thomson GR,Tustin RC (eds) Infectious diseases of livestock. Oxford University Press, Oxford, pp 460–475
8. Devaney MA, Kendall J, Grubman KC (1988) Characterisation of a non-structural phosphoprotein of two orbiviruses. Virus Res 11: 151–164
9. Du Toit RM (1944) The transmission of bluetongue and horse sickness by *Culicoides*. Onderstepoort J Vet Res 50: 536–541
10. Grubman MJ, Lewis SA (1992) Identification and characterization of the structural and nonstructural proteins of African horse sickness virus and determination of the genome coding assignments. Virology 186: 444–451
11. Hamblin C, Mertens PPC, Mellor PS, Burroughs JN, Crowther JR (1991) A serogroup specific enzyme linked immunosorbent assay for the detection and identification of African horse sickness viruses. J Virol Methods 31: 285–292
12. Hamblin C, Anderson EC, Mellor PS, Graham SD, Mertens PPC, Burroughs JN (1992) The detection of African horse sickness virus antigens and antibodies in young equidae. Epidemiol Infect 108: 193–201

13. Herniman KAJ, Gumm ID, Owen L, Taylor WP, Sellers RF (1980) Distribution of bluetongue virus and antibodies in some countries of the eastern hemisphere. Bull Off Int Epiz 92: 581–586
14. Huismans H, Els H (1979) Characterization of the tubules associated with replication of three different orbiviruses. Virology 92: 397–406
15. Holodniy M, Kim S, Katzenstein D, Konrad M, Groves E, Merigan TC (1991) Inhibition of human immunodeficiency virus gene amplification by heparin. J Clin Microbiol 29: 676–679
16. Kwok S, Higuchi R (1989) Avoiding false positives with PCR. Nature 339: 237–238
17. Laegreid WW, Skowronek A, Stone-Marschat M, Burrage T (1993) Characterization of virulence variants of African horse sickness virus. Virology 195: 836–839
18. Laviada MD, Babin M, Dominguez J, Sanchez-Vizcaino JM (1992) Detection of African horse sickness virus in infected spleens by a sandwich ELISA using two monoclonal antibodies specific for the VP7. J Virol Methods 38: 229–242
19. Laviada MD, Arias M, Sanchez-Vizcaino JM (1993) Characterization of African horse sickness virus serotype 4-induced polypeptides in Vero cells and reactivity in Western blotting. J Gen Virol 74: 81–87
20. Laviada MD, Roy P, Sanchez-Vizcaino JM, Casal JI (1995) The use of African horse sickness virus NS3 protein, expressed in bacteria, as a marker to differentiate infected from vaccinated horses. Virus Res 38: 205–221
21. Mellor PS, Boned J, Hamblin C, Graham S (1990) Isolation of African horse sickness virus from vector insects made during the 1988 epizootic in Spain. Epidemiol Infect 105: 447–454
22. Mellor PS (1993) African horse sickness: transmission and epidemiology. Vet Res 24: 199–212
23. Mizukoshi N, Sakamoto K, Iwata A, Tsuchiya T, Ueda S, Watanabe T, Kamada M, Fukusho A (1992) The complete sequence of African horse sickness virus serotype 4 (vaccine strain) RNA segment 5 and its predicted polypeptide compared with NS 1 of bluetongue virus. J Gen Virol 73: 2425–2428
24. Moulay S, Zientara S, Sailleau C, Cruciere C (1995) Detection of African horse sickness viruses by dot blot hybridization using digoxigenin labelled probe. Mol Cell Probes 9: 233–237
25. Oellermann RA, Els HJ, Erasmus BJ (1970) Characterization of African horse sickness virus. Arch Ges Virusforsch 29: 163–174
26. Rodriguez M, Hooghuis H, Castano M (1993) Current status of the diagnosis and control of African horse sickness. Vet Res 24: 189–197
27. Roy P, Hirasawa T, Fernandez M, Blinov VM, Sanchez-Vizcaino JM (1991) The complete sequence of the group-specific antigen, VP 7, of African horse sickness disease virus serotype 4 reveals a close relationship to bluetongue virus. J Gen Virol 72: 1237–1241
28. Sailleau C, Moulay S, Crucière C, Laegreid WW, Zientara S (1997) Detection of African horse sickness virus in the blood of experimentally infected horses: comparison of virus isolation and a PCR assay. Res Vet Sci 62: 229–232
29. Sellner LN, Coellen RJ, Mackenzie JS (1992) Reverse-transcriptase inhibits *Taq* polymerase activity. Nucleic Acids Res 20: 1487–1490
30. Sellner LN, Coellen RJ, Mackenzie JS (1992) A one-tube, one manipulation RT-PCR reaction for detection of Ross River virus. J Virol Methods 40: 255–264
31. Van Staden V, Huismans H (1991) A comparison of the genes which encode non-structural protein NS3 of different orbiviruses. J Gen Virol 72: 1073–1079
32. Van Staden V, Theron J, Greyling BJ, Huismans H, Nel LH (1991) A comparison of the nucleotide sequences of cognate genes of three different orbiviruses. Virology 185: 500–504

33. Wade-Evans AM, Mertens PPC, Bostock CJ (1990) Development of the polymerase chain reaction for the detection of bluetongue virus in tissue samples. J Virol Methods 30: 15–24
34. Zientara S, Sailleau C, Moulay S, Crucière C (1993) Diagnosis and molecular epidemiology of the African horse sickness virus by the polymerase chain reaction and restriction patterns. Vet Res 24: 385–395
35. Zientara S, Sailleau C, Moulay S, Crucière C (1994) Diagnosis of the African horse sickness virus serotype 4 by a one tube, one manipulation RT-PCR reaction from infected organs. J Virol Methods 46: 179–188
36. Zientara S, Sailleau C, Moulay S, Wade-Evans A, Crucière C (1995) Application of the polymerase chain reaction to the detection of African horse sickness viruses. J Virol Methods 53: 47–54

Authors' address: Dr. S. Zientara, Centre National d'Etudes Veterinaires et Alimentaires, Laboratoire Central de Recherches Veterinaires, 22 rue Pierre Curie, Boite Postale 67, 94703 Maisons-Alfort Cedex, France.

Western immunoblotting as a method for the detection of African horse sickness virus protein-specific antibodies: Differentiation between infected and vaccinated horses

S. Idrissi Bougrine, O. Fassi Fihri, and **M. M. Fassi Fehri**

Département de Microbiologie Immunologie Maladies Contagieuses, Institut Agronomique et Vétérinaire Hassan II, Rabat, Morocco

Summary. A Western immunoblotting procedure has been developed for the detection of African horse sickness virus (AHSV) protein-specific antibody responses. This assay readily identifies antibodies specific for at least 4 distinct, AHSV proteins, including VP5, NS1, NS2 and NS3/NS3a. By using the AHSV non-structural proteins as 'markers', the Western blotting procedure could be employed to provide a reliable means of discriminating between animals vaccinated with a purified, inactivated AHSV vaccine and those either naturally infected or vaccinated with a live, attenuated AHSV vaccine.

Introduction

African horse sickness (AHS) is an arthropod-borne virus disease of equids caused by an *Orbivirus* belonging to the family *Reoviridae* [7]. The genome of African horse sickness virus (AHSV) is composed of 10 double-stranded RNA segments, each of which encodes at least one polypeptide [5, 19]. There are 7 structural proteins forming a double-shelled virus particle and at least 4 non-structural proteins which are synthesized in infected cells [5, 13].

AHSV is currently enzootic in tropical Africa [7]. The recent epizootics of AHS that occurred in Spain, Portugal and Morocco between 1987 and 1991 [15–17] resulted in considerable economic losses, particularly in terms of equestrian trade and competition.

Live, attenuated AHSV vaccines were used as part of the control and eradication programmes but their use results in virus replication and does not therefore permit discrimination between naturally infected and vaccinated animals. The non-structural proteins of AHSV, viz NS1, NS2 and NS3/3a can be removed during purification of virus particles [3]. Animals that are vaccinated with an inactivated and purified AHSV vaccine therefore should not develop antibodies against these proteins. Based on this premise an ELISA has already been developed, using a recombinant NS3 protein expressed in *E. coli* [14], which is reported to differentiate between horses vaccinated with a commercially produced inactivated AHSV serotype 4 (AHSV 4) vaccine [4] and those which were naturally infected.

This paper reports on the application of Western blotting using AHSV 4 grown in Vero cells as an antigen to detect antibodies to the non-structural proteins. The presence of antibodies to any of these proteins could potentially be used to discriminate between animals vaccinated with commercially prepared and purified inactivated AHSV 4 vaccine and those that were either vaccinated with live, attenuated AHSV 4 vaccine or were naturally infected with a wild type virus.

Materials and methods

Animal sera

The serological status of all test sera had been determined previously using the type-specific AHSV 4 virus neutralisation test (VNT) [6]. A total of 40 AHSV antibody positive sera were available for assay which had been collected from vaccinated animals and from animals in the field. These included sera from 10 non-vaccinated, seropositive horses that had been collected during the AHSV 4 outbreak in Morocco in 1989, sera from 20 horses that had been vaccinated on 2–3 occasions with the recommended dose of AHSV 4 live attenuated vaccine (Horse Lyopest, Biopharma, Morocco) and sera from 10 horses that had been vaccinated on 2–3 occasions with the recommended dose AHSV 4 inactivated vaccine (Equipest, Rhône Mérieux). One serum from a non-vaccinated, seronegative horse was used as a control.

AHSV antigen preparation

The AHSV 4 used to prepare antigen for the Western blotting had been isolated from spleen tissue taken from a horse that had died in Morocco during the 1989 epizootic and passaged 8 times in Vero tissue culture cells.

Five confluent monolayers of Vero cells, grown in 850 cm^3 roller bottles with Eagles minimum essential medium (Gibco-BRL) supplemented with 10% foetal bovine serum and containing 100 iu/ml of penicillin, 2.5 μg/ml of fungizone and 60 μg/ml of Tylocine, were each inoculated with 2.0 ml of AHSV 4 $Vero_8$ containing $10^{7.0}$ $TCID_{50}$/ml. Cultures were harvested when showing 80 to 100% cytopathic effect (CPE). Supernatant fluids were frozen and thawed, nuclei and cell debris were removed by centrifugation at 1200 **g** for 5 min, and the supernatant fluids containing both intra- and extracellular virus were concentrated by precipitation using an equal volume of saturated ammonium sulphate for 4 h at 4°C. The precipitate was pelleted by centrifugation at 10 000 **g** for 30 min and then resuspended in 25 ml of 2mM Tris, pH 8.0. The virus was then pelleted through a 40% (w/v) sucrose cushion by ultra centrifugation at 100000 **g** (rotor SW28) for 90 min. The product represented a relatively crude mixture of cellular and viral proteins (both structural and non-structural proteins; data not shown). Cellular proteins prepared from non-infected Vero cells were used as a negative control antigen.

SDS-PAGE

Viral proteins were separated by SDS-PAGE on 12.5% polyacrylamide gel [11]. Separated viral proteins were transferred onto a nitrocellulose membrane (NCM) by electrophoresis at 100 v for 1 h as described elsewhere [23]. The NCM was subsequently cut vertically into 0.5 cm wide strips and stored at 4°C.

Western immunoblotting assay

Sera were assayed using an immunoblotting technique described previously [1, 13]. Briefly, the NCM strips were blocked with 3% bovine serum albumin in 20 mM Tris and 500 mM NaCl, pH 7.5 (TBSS) containing 0.05% Tween 20 (diluent and washing buffer) for 2 h at room temperature. Test sera, diluted 1 in 20, were added to individual strips which were then incubated overnight at ambient temperature (18–20°C). The strips were washed 3 times (10 min per wash) and then immersed in peroxidase-conjugated rabbit anti-horse IgG diluted 1 in 1000, for 1 h at ambient temperature. After washing, the strips were developed with 4-chloro-1-naphthol developer-substrate solution.

The molecular weights (M_r) of the stained protein bands were compared to viral proteins present in purified virus from radio labelled infected cells, or synthesised by in vitro translation of the corresponding double stranded RNA segments of the virus. The following relationships were established:

1. M_r of 62K corresponded to VP5. In the present study this protein resolved as a doublet.
2. M_r of 54K, 48K and 32K/22K proteins corresponded to the non-structural proteins NS1, NS2 and NS3/3a, respectively.
3. The AHSV VP2 and VP7 proteins which correspond to M_r of 98K and 38K, respectively, were observed infrequently and at a low level.

Results

Antibodies against at least 4 distinct AHSV proteins (VP5, NS1, NS2 and NS3/3a) were detected by Western blotting in sera collected from the field and from horses vaccinated with either attenuated or inactivated AHSV 4 vaccine (Figs. 1–4). Variable levels of secondary antibody binding to other immune complexes were observed as bands which did not correspond to structural AHSV proteins, for example, M_r of 32K. These may therefore represent antibodies that were interacting with minor translation or degradation products of viral proteins in the standard AHSV 4 antigen preparation. Since the attenuated live vaccine (Horse Lyopest, Biopharma) was produced in Vero tissue culture cells and was not purified, the vaccine might have been contaminated with cellular proteins. These contaminants could have induced immune responses in some of the vaccinated animals, which could account for some of the additional protein bands seen in Fig. 2. However, no such bands were detected in antigen preparations from non-infected cells.

Figure 1 shows the presence of antibodies to the 3 non-structural proteins in all 10 convalescent field sera. Of the 20 horses vaccinated with live attenuated AHSV 4 vaccine, 19 developed antibodies to NS1 and NS2 while all 20 developed antibodies to NS3/3a (Figs. 2 and 3). Although antibodies against NS1 and NS2 were detected in sera from the 10 horses vaccinated with the inactivated AHSV vaccine, antibodies against the NS 3/3a protein were only detected in 7 of these sera (Fig. 4). The antibody response against NS3/3a that was detected in 70% of the animals vaccinated with the inactivated vaccine, were less abundant than those observed with sera from the field cases or those vaccinated with attenuated live vaccine.

No AHSV specific protein bands were detected with sera from the seronegative control horse or when the cellular proteins from non-infected Vero cells were used as an antigen.

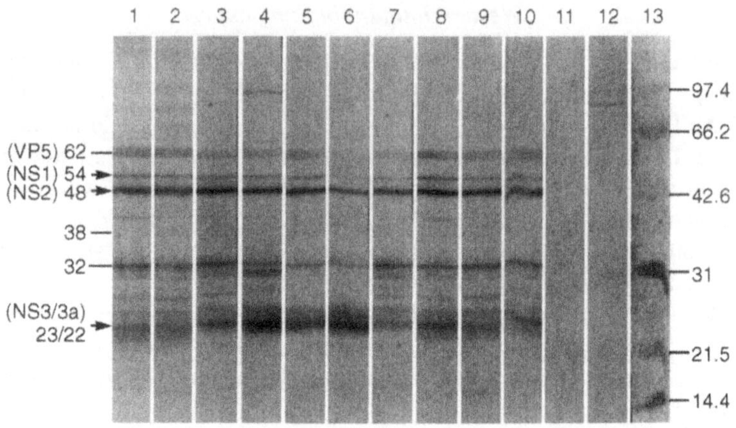

Fig. 1. Western immunoblotting of AHSV 4 proteins with sera from 10 AHSV 4 convalescent horses (*1–10*) and with serum from a non-vaccinated AHSV free horse (*11*). *12* shows Western immunoblotting of cellular proteins from non infected cells with a convalescent serum. *13* indicates molecular weight (M_r) markers. The sizes (M_r) of the markers are shown to the right of the figure (x 10^{-3}). The calculated M_r (x 10^{-3}) of the detected AHSV bands is given to the left of the gel and the equivalent AHSV protein nomenclature is in brackets

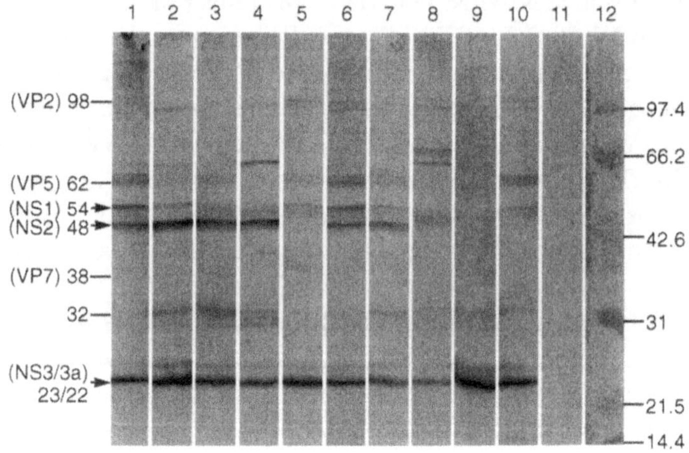

Fig. 2. Western immunoblotting of AHSV 4 proteins with sera from horses, numbered 1 to 10, vaccinated with attenuated, live AHSV 4 vaccine (*1–10*) and with serum from a non-vaccinated AHSV free horse (*11*). *12* shows the molecular weight (M_r) markers. The sizes (M_r) of the markers are shown to the right of the figure (x 10^{-3}). The calculated M_r (x 10^{-3}) of the detected AHSV bands is given to the left of the gel and the equivalent AHSV protein nomenclature is in brackets

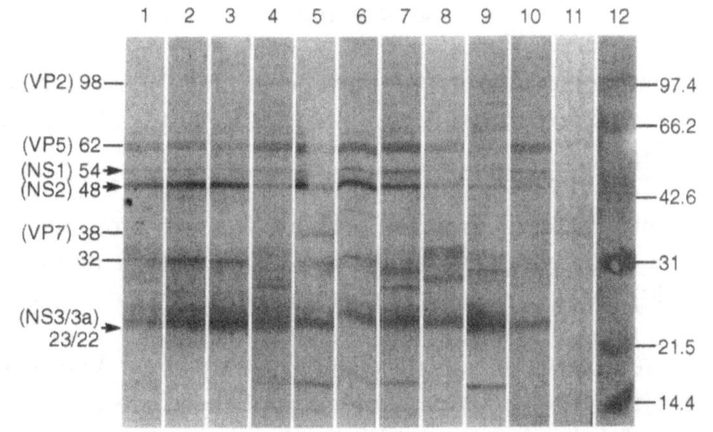

Fig. 3. Western immunoblotting of AHSV 4 proteins with sera from horses, numbered 11 to 20, vaccinated with attenuated, live AHSV 4 vaccine (*1–10*) and with serum from a non-vaccinated AHSV free horse (*11*). *12* shows the molecular weight (M_r) markers. The sizes (M_r) of the markers are shown to the right of the figure (x 10^{-3}). The calculated M_r (x 10^{-3}) of the detected AHSV bands is given to the left of the gel and the equivalent AHSV protein nomenclature is in brackets

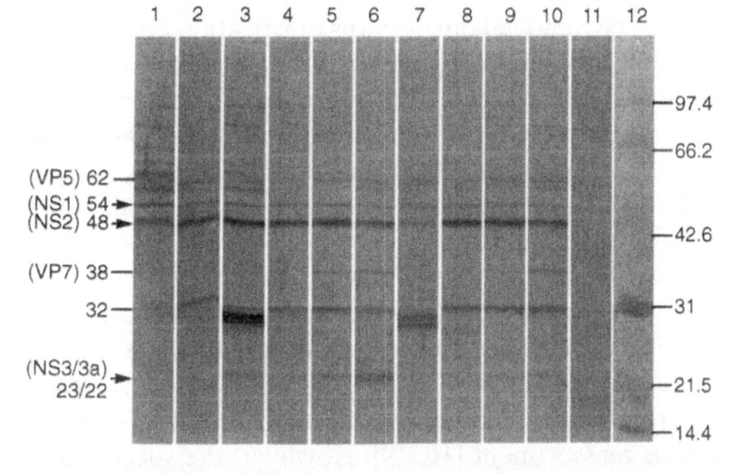

Fig. 4. Western immunoblotting of AHSV 4 proteins with sera from 10 horses vaccinated with inactivated AHSV4 vaccine (*1 10*) and with serum from a non-vaccinated AHSV free horse (*11*). *12* shows the molecular weight (M_r) markers. The sizes (M_r) of the markers are shown to the right of the figure (x 10^{-3}). The calculated M_r (x 10^{-3}) of the detected AHSV bands is given to the left of the gel and the equivalent AHSV protein nomenclature is in brackets

Discussion

The Western immunoblotting procedure reported here identified specific antibodies against up to 6 recognised AHSV proteins VP2, VP5, VP7, NS1, NS2 and NS3/NS3a [18, 24].

Although VP2 and VP7 are known to be antigenic [2, 13, 20, 21], these proteins were only detected in a few sera by Western blotting. This may be because important antigenic sites on VP2 and VP7 tend to be denatured by SDS-PAGE [12, 13, 20]. Only those sites that escape denaturation would be available for binding. This may constitute a major constraint in the use of this test when attempting to detect antibodies to these and possibly other AHSV proteins, for example, VP1, VP3, VP4 and VP6. Detection of these proteins may require a more sensitive technique such as the Biotin-Avidin method [1] where up to 10 bluetongue virus associated proteins have been identified.

Antibodies against three non-structural proteins were detected in most sera collected from the field and from horses given live attenuated vaccine. Failure to detect antibodies against NS1 or NS2 in two separate horses could be attributed to individual animal variation. This result emphasises the importance of including more than one viral protein as a 'marker' and the value of an assay system such as Western blot which can readily identify antibodies to all these 'markers'.

The non-structural proteins, NS1, NS2 and NS3, are produced during virus replication but can be removed during virus purification [3]. Hence, no antibodies should be raised against these 3 proteins if animals are vaccinated with a fully purified, then inactivated, vaccine. The presence of antibodies against the three non-structural proteins in sera from animals administered the inactivated vaccine Equipest (Rhône Mérieux, France) in this study, shows that the vaccine was not purified sufficiently to exclude these proteins. Laviada et al. [14] also showed a weak but positive response against NS3 by ELISA in 10–20% of sera taken from horses after primary vaccination with the same inactivated vaccine.

Clearly, the use of inactivated vaccines have many advantages over the more traditional attenuated vaccines since vaccinated animals are able to develop a protective immune response [3, 9] in the absence of any virus replication within the host and therefore no possibility of viraemia, vector transmission, reversion to virulence or virus reassortment [10, 19]. However, the successful discrimination between infected and vaccinated animals using AHSV non-structural proteins as 'markers' is dependent on the level of purification of the vaccine virus. Related data presented by other authors at this meeting [8, 22, 25] have demonstrated that protective immune responses may also be generated using recombinant, expressed structural proteins as a sub unit vaccine. Such antigen preparations cannot replicate and do not contain NS1, NS2, or NS3/3a. Consequently they are not only completely safe but there is no possibility of generating a response to AHSV non-structural proteins. Therefore, these sub unit vaccines would also have a significant advantage over non-purified, inactivated vaccines if used in conjunction with tests such as Western blotting in order to discriminate between infected and vaccinated animals.

Acknowledgements

The authors thank Dr. El Harrak for providing the virus and sera, and also for his useful discussion. We would also like to thank Dr. J. Berrada for his help in preparing the manuscript. This work was financially supported by Association des Universités Partiellement ou entièrement de Langue Française-Université des Réseaux d'Expression Française.

References

1. Adkison MA, Stott JL, Osburn BI (1987) Identification of bluetongue virus protein-specific antibody responses in sheep by immunoblotting. Am J Vet Res 48: 1194–1198
2. Burrage TG, Travejo R, Stone-Marschat M, Laegreid WW (1993) Neutralising epitopes of African horse sickness virus serotype 4 are located on VP2. Virology 196: 799–803
3. Burroughs JN, O'Hara RS, Smale CJ, Hamblin C, Walton A, Armstrong R, Mertens PPC (1994) Purification and properties of virus particles, infectious subviral particles, cores and VP7 crystals of African horse sickness virus serotype 9. J Gen Virol 75: 1849–1857
4. Dubourget P, Preaud JM, Detraz N, Lacoste F, Fabry AC, Erasmus BJ, Lombard M (1992) Development, production and quality control of an industrial inactivated vaccine against African horse sickness virus serotype 4. In: Walton TE, Osburn BI (eds) Bluetongue, African horse sickness and related orbiviruses. Proceedings of the Second International Symposium. Paris 1991. CRC Press, Boca Raton, pp 874–888
5. Grubman MJ, Lewis SA (1992) Identification and characterisation of the structural and non-structural proteins of African horse sickness virus and determination of the genome coding assignments. Virology 186: 444–451
6. Herniman KAJ, Gumm ID, Owen L, Taylor WP, Sellers RF (1980) Distribution of bluetongue virus and antibodies in some countries of the eastern hemisphere. Bull Off Int Epiz 92: 581–586
7. Hess WR (1988) African horse sickness. In: Monath TP (ed) The arboviruses: epidemiology and ecology, vol 2. CRC Press, Boca Raton, pp 1–18
8. van Staden V, Smit CC, Stolz M, Maree FF, Huismans H (1998) Characterisation of two African horse sickness virus nonstructural proteins, NS3 and NS 1. In: Mellor PS, Baylis M, Hamblin C, Calisher CH, Mertens PPC (eds) African horse sickness. Springer, Wien New York, pp 251–258 (Arch Virol [Suppl] 14)
9. House JA, Lombard M, Dubourget P, House C, Mebus CA (1994) Further studies on the efficacy of an inactivated African horse sickness serotype 4 vaccine. Vaccine 12: 142–144
10. House JA (1998) Future international management of African horse sickness vaccines. In: Mellor PS, Baylis M, Hamblin C, Calisher CH, Mertens PPC (eds) African horse sickness. Springer, Wien New York, pp 297–304 (Arch Virol [Suppl] 14)
11. Laemmli UK (1970) Cleavage of structural proteins during the assembly of the head of bacteriophage T4. Nature 227: 680–685
12. Laviada MD, Babin M, Dominguez J, Sanchez-Vizcaino JM (1992) Detection of African horse sickness in infected spleens by sandwich ELISA using two monoclonal antibodies specific for VP7. J Virol Methods 38: 229–242
13. Laviada MD, Arias M, Sanchez-Vizcaino JM (1993) Characterisation of African horse sickness virus serotype 4 induced polypeptides and their reactivity in Western immunoblotting. J Gen Virol 74: 81–87
14. Laviada MD, Roy P, Sanchez-Vizcaino JM, Casal JI (1995) The use of African horse sickness virus NS3 protein, expressed in bacteria, as a marker to differentiate infected from vaccinated horses. Virus Res 38: 205–221

15. Lhafi A, Tber A, Fikri A, Laghzaoui K (1992) African horse sickness in Morocco: the epizootics of 1989 and 1990. In: Walton TE, Osburn BI (eds) Bluetongue, African horse sickness and related orbiviruses. Proceedings of the Second International Symposium, Paris 1991. CRC Press, Boca Raton, pp 205–216
16. Lubroth J (1988) African horse sickness and the epizootic in Spain 1987. Equine Pract 10: 26–33
17. Mellor PS (1993) African horse sickness: transmission and epidemiology. Vet Res 24: 199–212
18. O'Hara RS (1994) Identification of the genome segments and proteins controlling the virulence of African horse sickness. PhD Thesis, Reading University
19. O'Hara RS, Meyer AJ, Pullen L, Martin LA, Mertens PPC (1998) Development of a mouse model system and identification of the individual genome segments of African horse sickness virus serotypes 3 and 8 involved in determination of virulence. In: Mellor PS, Baylis M, Hamblin C, Calisher CH, Mertens PPC (eds) African horse sickness. Springer, Wien New York, pp 259–279 (Arch Virol [Suppl] 14)
20. Ranz AI, Miguet JG, Anaya C, Venteo A, Cortes E, Vela C, Sanz A (1992) Diagnostic methods for African horse sickness virus using monoclonal antibodies to structural and non-structural proteins. Vet Microbiol 33: 143–153
21. Roy P, Chuma T, Le Bois H, Hirasawa T (1992) Molecular cloning of RNA segment 7 and synthesis of the major core antigen VP7 of African horse sickness virus by recombinant Baculovirus: its use as a group-specific diagnostic reagent. In: Walton TE, Osburn BI (eds) Bluetongue, African horse sickness and related orbiviruses. Proceedings of the Second International Symposium, Paris 1991. CRC Press, Boca Raton, pp 461–470
22. Roy P (1998) New generation of African horse sickness vaccines based on structural and molecular studies of the virus particles. In: Mellor PS, Baylis M, Hamblin C, Calisher CH, Mertens PPC (eds) African horse sickness. Springer, Wien New York, pp 177–202 (Arch Virol [Suppl] 14)
23. Towbin H, Staehelin T, Gordon J (1979) Electrophoretic transfer of proteins from polyacrylamide gels to nitrocellulose sheets: procedures and some applications. Proc Natl Acad Sci USA 76: 4350–4354
24. Van Staden V, Huismans H (1991) A comparison of the genes which encoded non-structural protein NS3 of different Orbiviruses. J Gen Virol 72: 1073–1090
25. Wade-Evans AM, Pullen L, Hamblin C, O'Hara RS, Burroughs JN, Mertens PPC (1998) VP7, from African horse sickness virus serotype 9, protects mice against a lethal heterologous serotype challenge. In: Mellor PS, Baylis M, Hamblin C, Calisher CH, Mertens PPC (eds) African horse sickness. Springer, Wien New York, pp 211–219 (Arch Virol [Suppl] 14)

Authors' address: Dr. S. Idrissi Bougrine, Département de Microbiologie Immunologie Maladies Contagieuses, Institut Agronomique et Vétérinaire Hassan II, B.P. 6202-Instituts, 10101 Rabat, Morocco.

Conclusions and recommendations of the meeting

Conclusions

1. All equine species are likely to be susceptible to infection with AHSV although donkeys and zebra rarely exhibit clinical signs. Nevertheless both of these species are able to mount a viraemia which may be of longer duration than that in horses and can be of sufficiently high titre to infect vector *Culicoides*. Donkeys and zebra are therefore able to play a role in the epidemiology of AHS. The importance of that role, in comparison to horses, is likely to be dependent upon the local situation.
2. Although donkeys can be infected and develop a viraemia, they do not seem to be long-term reservoirs for AHSV.
3. AHSV isolated from infected donkeys in the absence of clinical signs can cause a lethal infection in susceptible horses
4. The precise role of mules in the epidemiology of AHS has yet to be established.
5. Under certain climatic conditions and in regions where efficient adult, vector *Culicoides* are active throughout the year AHSV is able to persist beyond sub-Saharan Africa for several years. However, persistence should not be interpreted as meaning "static", the virus is peripatetic and is able to move between adjacent areas in response to the changing immunological status of the local vertebrate host population, as brought about through vaccination or infection.
6. In those regions specified in 5 above, the absence of clinical cases of AHS in horses for much of the year (approximately 9 months) has been highlighted and has not been fully explained. It may be that during these "silent periods" virus circulates, subclinically in non-horse equines only, though this has not been confirmed. Such a situation is likely to be encouraged by a number of factors including: better vaccination cover in the horse population and a greater accessibility of non-horse equines (which are usually stabled outside) to vector *Culicoides*, at times of low vector abundance (winter and spring).
7. *Culicoides imicola ss*[1] is probably the only important AHSV vector in southern Europe, North Africa and most of sub-Saharan Africa. However in certain areas of Africa *C. imicola* sibling species might also be important in AHSV transmission, although the role (if any) of these species remains to be clarified.

---

[1] *Sensu stricto* (in the strict sense).

8. The differentiation of the sibling species of *C. imicola* is difficult when based solely upon morphological characters and therefore other methods of differentiation (e.g. RAPD-PCR) are required to facilitate identification by non-specialist staff.
9. High abundances of *C. imicola* in Morocco and Iberia (and presumably elsewhere) generally occur where climatic and geographic conditions favour both the establishment of larval breeding sites and low adult mortality rates.
10. Larval breeding sites are favoured by high soil moisture, especially during dry periods of the year. Soils with an ability to retain high levels of moisture for periods of time sufficient to allow completion of the developmental stages are usually clay soils. Particularly favourable positions are on level ground (often at the bottom of slopes from which they receive run-off) and in areas of higher rainfall.
11. Such conditions also encourage vegetation growth, and it appears that the remotely-sensed variable, NDVI (normalised difference vegetation index), which provides a measure of green vegetation, can be used to map the distribution of *C. imicola.* In countries surrounding the Mediterranean, the distribution of *C. imicola* is largely described by higher levels of the annual minimum NDVI (ie. those areas with higher soil moisture and vegetation levels during the driest, most barren period of the year – late summer).
12. Other factors may also be important for the establishment of productive *C. imicola* breeding sites. These may include the presence of specific soil flora and fauna, to provide food for the *Culicoides larvae*, and the mineral content of the soil.
13. The most important climatic variable examined so far, which affects the mortality rate of adult *C. imicola*, is the wind speed during the night time activity period. Higher night-time wind speeds decrease the vectorial capacity of adult *C. imicola* in two ways; first, the abundance of the vector is decreased; second, longevity is reduced, leading to a smaller proportion of the population living long enough to transmit AHSV.
14. There is evidence for a northern limit to the distribution of *C. imicola* that is independent of both NDVI and wind speed. This is probably a "cold" temperature limit, although the precise level/duration of the limiting temperatures remain to be determined.
15. The effects of temperature upon the transmission of AHSV by *C. imicola* are likely to be complex. Higher temperatures increase the replication rate of the virus in the vector which facilitates earlier transmission but at higher temperatures the vector lives for a shorter time. At lower temperatures AHS virogenesis slows and ceases altogether at about 15°C but at lower temperatures the lifespan of the vector is extended. The likelihood of AHSV transmission by *C. imicola* is therefore a function of the interaction of these two opposing sets of trends.
16. Furthermore, AHSV may persist in infected vectors kept at low temperatures (< 15°C), for extended periods of time and is then capable of replication and transmission when the temperature rises above the threshold level. This set of

circumstances could provide an overwintering mechanism for AHSV in the absence of vertebrate involvement.

17. The successful overwintering of AHSV in Morocco 1989–1991 occurred only in areas where adult *C. imicola* are present throughout the year. These areas are those where this species is generally most abundant. Therefore during epizootics these are the areas where AHSV surveillance and control measures should be concentrated.
18. Structural analysis of orbivirus particles and proteins by X-ray crystallography is advancing dramatically. The major structural proteins in the native icosahedral core particle of bluetongue virus (BTV) serotype 1 (South Africa) have recently been resolved at the atomic level for the first time. The core particle of BTV is approximately 10 times larger than any other structure so far analysed to comparable resolution. This work will provide a model for the structure of other closely related orbiviruses.
19. RNA sequence analysis of the AHSV genome has progressed to the stage where at least one representative copy of each of the 10 segments has been fully analysed. Therefore the full genome of an AHSV has now been analysed.
20. Representative examples of all of the different AHSV proteins have now been synthesised in at least one expression system. Recombinant baculoviruses expressing individual or multiple proteins have proved to be particularly effective for this purpose.
21. A mouse model has been developed for the study of both the virulence characteristics of AHSV and immune responses to the virus. In this model, genome segments 2, 5 and 10 encoding the outer capsid proteins VP2, VP5 and nonstructural protein (NS)3, respectively, were shown to be involved in virulence. Reassortment between virulent and non-virulent strains of AHSV generated progeny viruses with novel virulence characteristics. The outer core protein VP7, when used as a subunit vaccine in this model, was shown to induce a significant level of cross-serotype protection. This protection was not due to the presence of humeral antibody.
22. Further studies involving the mouse model and *Culicoides* cell lines confirmed the involvement of NS3 in a virus release mechanism from cells, and suggested that segment 10 may influence virulence by affecting the timing of release and hence spread to other cells. The outer capsid proteins VP2 and VP5 are thought to influence virulence by affecting cell attachment and entry.
23. Good AHSV vaccines should be safe, elicit a rapid and enduring protective immunity, preferably in the absence of viraemia and should prevent the development of a secondary viraemia following infection in order to avoid the possibility of infecting vector insects.
24. Vaccines should be produced following high international standards.
25. Live, attenuated vaccines should not be capable of reversion to virulence as determined by backpassage through the target vertebrates and by passage through vector *Culicoides*.
26. Additional data on efficacy, safety and duration of immunity in all domestic equine species are required for all currently available live AHSV vaccines.

27. Virtually all AHSV vaccines that are available at present contain live attenuated virus. Although good control has been achieved with these vaccines in the past, many uninfected countries are reluctant to accept the importation of animals vaccinated with them. There is therefore an urgent requirement for the development of effective inactivated or recombinant vaccines.
28. The use of effective inactivated or recombinant vaccines removes any possibility of reversion to virulence or of infecting vector insects and also affords the opportunity to differentiate between vaccinated and naturally infected animals.
29. The presence of an international AHSV vaccine bank would be of value in controlling and eradicating the virus from epizootic areas by ensuring the rapid and ready availability of vaccine in emergencies.
30. The development of an effective vaccine against *Culicoides spp.*, similar to the ones already developed against ticks, would have significant advantages including protection against all AHSV serotypes and against all other *Culicoides*-borne pathogens.
31. New group-specific ELISA's for the detection of AHSV antibodies and antigens have been developed and validated, and are now accepted by the OIE.
32. A second generation ELISA based upon the expressed NS3 of AHSV-4 has been developed. This assay has been used to successfully differentiate between animals naturally infected with AHSV-4 and those vaccinated with an inactivated vaccine against the same serotype. Further work is necessary to confirm that this assay is able to detect reliably animals infected with heterologous strains and serotypes of AHSV.
33. There is a requirement for the development of diagnostic tests suitable for use in field laboratories that will allow the early diagnosis of AHSV infection.
34. The development of a PCR technique which is capable of distinguishing between all serotypes of AHSV would be a welcome addition to the battery of diagnostic tests available for this virus. The results of existing studies in this area should be confirmed and the work extended.
35. Harmonised sampling methods, levels of surveillance and statistical analyses necessary to confirm the absence of AHSV from an area should be agreed internationally.
36. Mathematical modelling studies on aspects of AHSV epidemiology and in the design of vaccination strategies have provided valuable insights into how this virus spreads in the field and how it may best be eliminated from areas or controlled.

Recommendations

1. The role of the mule in the epidemiology of AHS should be defined.
2. The vector capacity of the sibling species of *C. imicola* for AHSV should be established.
3. Non-morphological (e.g. RAPD-PCR or other molecular) methods for the differentiation of the sibling species of *C. imicola sl*[2] should be developed.

[2] *Sensu lato* (in the broad sense).

4. Areas of high *C. imicola* abundance in the Mediterranean basin should be identified as a matter of priority, to allow more focussed surveillance and control during future epizootics of AHS (and bluetongue). Areas of high annual minimum NDVI should be investigated first.
5. The northern limit to the distribution of *C. imicola* in Iberia (and Greece), and its putative temperature dependence, should be characterised in detail.
6. Research should be directed towards elucidating those conditions of temperature and wind speed that optimise the transmission of AHSV via *C. imicola*, taking into account vector longevity and virus replication rates.
7. The ability of AHSV to survive for "long periods" in *Culicoides* vectors maintained at low temperatures should be investigated in detail.
8. The molecular biological basis and the genetics of heredity of the infection, dissemination and transmission barriers which have been identified in a variable proportion of individuals within vector species of *Culicoides* should be studied in detail.
9. The epidemiological mechanisms underlying the ability of AHSV to overwinter in epizootic areas, particularly in regard to the location of virus during the AHS "silent periods", should be investigated in more detail.
10. Further study is required of the AHSV structure and function particularly with regard to the outer capsid proteins, interactions between proteins and RNA and the minor structural and non-structural proteins. X-ray crystallography and cryo-electron microscopy have considerable potential in these areas.
11. Further comparative RNA sequence analysis, particularly of the genome segments encoding the variable proteins of AHSV (segments 2, 5 and 10) is required to fully evaluate the significance and extent of variation.
12. The basis of protection afforded by VP7 of AHSV and the immunodominant character of VP7 should be studied in more detail.
13. As VP2 of AHSV can generate neutralising antibodies which may be important in protection and in the development of a serotype specific ELISA, genome segment 2 from all 9 AHSV serotypes should be used in sequence analysis and expression studies.
14. Efforts to develop and market safe and effective inactivated, recombinant or sub-unit vaccines as an economic alternative to attenuated vaccines should be accelerated.
15. The potency of AHSV vaccines should be assessed in all intended equine target species (including donkeys) to determine levels of efficacy.
16. Consideration should be given to establishing an AHSV emergency vaccine bank.
17. The further development of improved assays capable of differentiating between naturally infected and vaccinated animals (using inactivated or sub-unit vaccines) should continue. This work should explore the potential for assays based upon NS 1 and NS2 in addition to those based upon NS3.
18. Work towards the development of diagnostic tests that will allow the early diagnosis of AHS should be continued. These tests should be sufficiently

robust as to permit their use under field situations or in unsophisticated laboratories.

19. Work on the development and validation of AHSV type-specific PCR variants should continue.
20. All new AHSV diagnostic tests and assays should be confirmed as being specific for AHSV and without significant cross reactions with the related equine encephalosis serogroup of orbiviruses.
21. The chapters on AHS in the OIE *International Health Code* and the OIE *Manual of Standards for Diagnostic Tests and Vaccines* should be revised to take into account recent epidemiological findings and the new developments in diagnostic tests, making clear the range of assays that are now prescribed tests.
22. Work into developing accurate and robust mathematical models describing the epidemiology of AHS and on designing the most effective control strategies should be encouraged as a matter of priority. Such work should be continually updated in the light of new research findings.
23. Communication between AHS research workers in different institutions and between research workers and other concerned parties should continue to be encouraged.

P. S. Mellor
M. Baylis
C. Hamblin
C. H. Calisher
P. P. C. Mertens

GPSR Compliance

The European Union's (EU) General Product Safety Regulation (GPSR) is a set of rules that requires consumer products to be safe and our obligations to ensure this.

If you have any concerns about our products, you can contact us on ProductSafety@springernature.com

In case Publisher is established outside the EU, the EU authorized representative is:

Springer Nature Customer Service Center GmbH
Europaplatz 3
69115 Heidelberg, Germany

Batch number: 10732720

Printed by Printforce, the Netherlands